RACE and CRIME

SECOND EDITION

For my loving parents, Daphne and Patrick Gabbidon,
who continue to be supportive of all my endeavors.
SLG

To my ancestors for paving the way.
HTG

RACE and CRIME

SECOND EDITION

Shaun L. Gabbidon

Pennsylvania State University, Harrisburg

Helen Taylor Greene

Texas Southern University

Los Angeles • London • New Delhi • Singapore • Washington DC

For information:

SAGE Publications, Inc.
2455 Teller Road
Thousand Oaks, California 91320
E-mail: order@sagepub.com

SAGE Publications Ltd.
1 Oliver's Yard
55 City Road
London EC1Y 1SP
United Kingdom

SAGE Publications India Pvt. Ltd.
B 1/I 1 Mohan Cooperative Industrial Area
Mathura Road, New Delhi 110 044
India

SAGE Publications Asia-Pacific Pte. Ltd.
33 Pekin Street #02-01
Far East Square
Singapore 048763

Printed in the United States of America.

Library of Congress Cataloging-in-Publication Data

Gabbidon, Shaun L., 1967-
   Race and crime / Shaun L. Gabbidon, Helen Taylor Greene. —2nd ed.
      p. cm.
   Includes bibliographical references and index.
   ISBN 978-1-4129-6778-5 (pbk.)
      1. Crime and race—United States. 2. Criminal justice, Administration of—United States.
   3. Discrimination in criminal justice administration—United States. 4. Minorities—United States.
   I. Greene, Helen Taylor, 1949- II. Title.

HV6789.G32 2009
364.973089—dc22   2008030696

This book is printed on acid-free paper.

09   10   11   12   10   9   8   7   6   5   4   3   2

Acquisitions Editor:	Jerry Westby
Editorial Assistant:	Eve Oettinger
Production Editor:	Karen Wiley
Copy Editor:	Heather Jefferson
Typesetter:	C&M Digitals (P) Ltd.
Proofreader:	Andrea Martin
Indexer:	Molly Hall
Cover Designer:	Edgar Abarca
Marketing Manager:	Christy Guilbalt

Contents

Preface   xi

Acknowledgments   xv

1. **Overview of Race and Crime**   1
   Prejudice and Discrimination   5
   Historical Antecedents of Race and Crime in America   7
       Native Americans   7
       African Americans   9
       White Ethnics   19
       Latino Americans   25
       Asian Americans   29
   Conclusion   35
   Discussion Questions   36
   Internet Exercise   36
   Internet Sites   36

2. **Extent of Crime and Victimization**   37
   History of Crime and Victimization Statistics in the United States   40
   The Uniform Crime Reporting Program   41
   Victimization Surveys   45
   Limitations of Arrest and Victimization Data   46
       Definitions of Racial Categories   46
       Variations in Reporting and Recording   48
       Utilization of Population, Crime, Arrest, and Victimization Estimates   49
   Arrest Trends   49
   Victimization Trends   52
   Homicide Victimizations   53
   Hate Crime Trends   55
   Conclusion   57
   Discussion Questions   59
   Internet Exercises   60
   Internet Sites   60

3. **Theoretical Perspectives on Race and Crime**   61
   What Is Theory?   62
   Biology, Race, and Crime   63

Crime and Human Nature 64
Intelligence, Race, and Crime 65
r/K Life History Theory 66
Sociological Explanations 69
Social Disorganization 70
    Contemporary Social Disorganization Theory 72
    Mass Incarceration and Social Disorganization 75
Collective Efficacy 75
Culture Conflict Theory 76
Strain/Anomie Theory 78
    Limitations of Strain/Anomie Theory 79
General Strain Theory 80
Subcultural Theory 81
The Subculture of Violence Theory 81
    The Code of the Streets 85
Conflict Theory 88
    Conflict Theory, Race, and Crime 89
The Colonial Model 92
Integrated and Nontraditional Theories on Race and Crime 94
    Structural-Cultural Theory 94
    Abortion, Race, and Crime 95
    Critical Race Theory 96
Conclusion 96
Discussion Questions 98
Internet Exercise 98

4. **Policing** **99**
Overview of Policing in America 100
Historical Overview of Race and Policing 102
    Native Americans 104
    African Americans 107
    Asian Americans 108
    Latinos 109
    White Immigrants 110
Contemporary Issues in Race and Policing 111
    Citizen Satisfaction With Police 112
    Police Deviance 115
    Racial Profiling 118
    Police Accountability 126
    Police Innovations 127
Conclusion 129
Discussion Questions 129
Internet Exercise 129
Internet Sites 130

5. **Courts** **131**
Overview of American Courts: Actors and Processes 134
    A Note on the Philosophy, Operation, and Structure of
      Native American Courts 136

Historical Overview of Race and the Courts in America 138
   Native Americans 138
   African Americans 139
   Latinos 141
   Asian Americans 142
Contemporary Issues in Race and the Courts 143
   Bail and the Pretrial Process 143
   Race and Pretrial Release 144
   Scholarship on Bail and Pretrial Release 145
   Legal Counsel 147
   Defense Counsel 147
   Statewide Study of Pennsylvania's Public Defense System 148
   Plea Bargaining 149
   Jury Selection 150
   Voir Dire 152
   Jury Nullification 155
Drug Courts 157
   Structure and Philosophy of Drug Courts 158
   Effectiveness of Drug Courts 159
Conclusion 160
Discussion Questions 161
Internet Exercise 161
Internet Sites 161

6. Sentencing **163**
Sentencing Philosophies 164
   Sentencing and Politics 164
Historical Overview of Race and Sentencing 165
   Punishment in Colonial America 165
   Early Colonial Cases 166
   Crime and Justice in Colonial New York 169
   Early National Sentencing Statistics 172
   The 1980s to 2000s: The Changing Nature
      of Sentencing Practices 173
   The "War on Drugs" 176
Contemporary Issues in Race and Sentencing 177
   Scholarship on Race and Sentencing 177
   Race and Misdemeanor Sentencing 182
   Sentencing Disparities and the "War on Drugs" 183
   Minority Judges 189
Conclusion 197
Discussion Questions 198
Internet Exercise 198
Internet Sites 199

7. The Death Penalty **201**
Significant Death Penalty Cases 201
Historical Overview of Race and the Death Penalty 204
Current Statistics on the Death Penalty 208

State Death Penalty Statistics   208
Federal Death Penalty Statistics   208
Scholarship on Race and the Death Penalty   214
    Substitution Thesis/Zimring Analysis   215
Public Opinion and the Death Penalty   217
    Race and Support for the Death Penalty   218
    The Marshall Hypotheses   220
Contemporary Issues in Race and the Death Penalty   221
    Capital Jury Project   221
    Wrongful Convictions   222
    Death Penalty Moratorium Movement   228
Conclusion   229
Discussion Questions   229
Internet Exercise   230
Internet Sites   230

8. Corrections   **231**
Overview of American Corrections   232
    Public Opinion and Corrections   233
Historical Overview of Race and Corrections   234
Early National Prison Statistics   238
Prison Gangs   242
Contemporary State of Corrections   245
    Jails   247
    Probation and Parole   249
Contemporary Issues in Race and Corrections   250
    Explaining Racial Disparities in Corrections   250
    Prisoner Reentry Concerns   253
    Felon Disenfranchisement   258
    Political Prisoners   260
Conclusion   261
Discussion Questions   262
Internet Exercise   262
Internet Sites   262

9. Juvenile Justice   **263**
Overview of Juvenile Justice   264
Historical Overview of Race and Juvenile Justice   267
    The Child Savers   268
    Juvenile Courts   269
Juvenile Crime and Victimization   270
Data on Youth in the Juvenile Justice System   280
Contemporary Issues in Race and Juvenile Justice   281
    Disproportionate Minority Confinement   281
    The Future of DMC   285
    Minority Female Delinquency   286
    Juveniles, Race, and the Death Penalty   288
    Delinquency Prevention   290

Conclusion   292
Discussion Questions   293
Internet Exercises   293
Internet Sites   293

Conclusion   **295**

Appendix   **299**

References   **307**

Index   **345**

About the Authors   **367**

Photo Credits and Permissions   **369**

Preface

"Justice is Blind" represents the basic motto and principle of our criminal justice system. It symbolizes equity in the administration of justice and represents our basic rights in a free society.

For many in the minority community, however, society is not that free and justice is far from blind. Justice in many cases has perfect 20/20 vision that distinguishes people on the basis of race, ethnicity, gender, religious beliefs and social and economic status.

—National Organization of Black Law Enforcement Executives (2001, p. 4)

Race and crime continues to be a contemporary issues in many societies where there is a diverse population and racial minorities. Since the colonial era, race and crime in America have been inextricably linked; there has been a belief that minorities, especially Blacks, are more criminal. At first, support for this belief was the result of racist ideologies that labeled minorities as both "criminal" and "inferior." More recently, support for this erroneous belief was based on the number of racial and ethnic minorities who are disproportionately arrested and imprisoned. After the 1960s, the relationship between race and crime became more ambiguous as we learned about the role of justice practitioners and their use of discretion.

The opening quotation captures the beliefs of many racial and ethnic minorities about justice in the United States in the past and present. Whites, who form the majority of the U.S. population, are less likely to believe there is discrimination in the administration of justice. Because the news media usually focus on persons who commit crimes, especially serious crimes like murder and rape, it is easy to lose sight of the fact that the majority of Americans, regardless of their race or ethnicity, are law-abiding citizens. It seems that we have just as easily lost sight of the historical context of race and crime in the United States. Why do racial minorities, most of whom are law-abiding citizens, continue to be labeled as criminals? The study of race and crime has a long history in the discipline of criminology and the study of criminal justice. In the 19th century, positivist scholars (those who explained crime using biological, sociological, or psychological factors) deemed the physical characteristics of racial minorities as being associated with crime. Thus, when a prisoner had features like African Americans and Asians, their criminality was often linked to these characteristics (Gabbidon, 2007a; Gabbidon & Taylor Greene, 2005).

Early criminology texts devoted whole chapters to race and crime that not only presented crime figures, but also sought to explain the trends related to race and crime (Gabbidon & Taylor Greene, 2001). Interestingly, contemporary criminology textbooks do not devote as much attention to race and crime as did earlier texts (Gabbidon & Taylor Greene, 2001). Even many of the early textbooks omitted many important topics like slave patrols, lynching, race riots, and legal segregation, which often resulted in socially disorganized communities. More recently, despite a strong argument for studying race and crime put forth by LaFree and Russell (1993), only a few comprehensive books on this topic are available (Barak, Flavin, & Leighton, 2006; Gabbidon, 2007a; Mann, 1993; Tarver, Walker, & Wallace, 2002; Tonry, 1995; Walker, Spohn, & Delone, 2007).

Most of the early scholarly research that is available on race refers primarily to Blacks. This is due, at least in part, to the fact that until recently, Blacks were the largest minority group in the United States and therefore the most visible. Another important factor in the focus on Blacks probably has to do with their foray into higher education, especially into the discipline of criminology. Most majority scholars were uninterested in studying race and crime. Blacks, in contrast, were interested. Even before the emergence of Black criminologists, Black scholars at historically Black colleges and universities were studying Black issues, including crime (Taylor Greene & Gabbidon, 2000). It is only recently that other minorities have received increased attention. At the same time, Latinos are now the largest minority group and also have more scholars interested in race, ethnicity, and crime; as a result, more research is being published on this group. This does not mean that other racial and ethnic groups have not been subjected to differential treatment in society and the administration of justice. It means that the historical record of their experiences is less complete. Notably, although interest in Latino and Native American crime has increased, the research on Asian Americans and crime is still limited.

Despite more research, books, and government documents about race and crime, we are still unable to explain and adequately address the continuous pattern of over-representation of some minorities in arrest and victimization statistics, corrections, persons under sentence of death, and juvenile delinquency.

We believe that prior attempts to make sense of the disproportionate number of minorities in the administration of justice are incomplete because they fail to consider relevant historical information.

One of our goals in writing this book is to put the study of race and crime in a more complete historical context. Another goal is to examine several contemporary issues relevant to understanding race and crime. To achieve these goals, we utilize a limited-systems approach to examine policing, courts, sentencing, the death penalty, and corrections in the past and the present. An additional chapter examines the juvenile justice system. We include an issues approach to focus on several contemporary challenges in the study of race and crime, including hate/bias crimes, racial profiling, sentencing disparities, wrongful convictions, felon disenfranchisement, political prisoners, disproportionate minority confinement, minority female delinquency, juveniles and the death penalty, and delinquency prevention. We include the major racial and ethnic groups in the United States—Asians, Blacks,

Latino/as, Native Americans, and Whites—although not as much information is available on all groups.

Various terms are used to refer to these groups. Some are the terms preferred in present-day usage, whereas others also are utilized to preserve their temporal context, especially in direct quotations. For example, you will see Blacks referred to as *Negroes*, *African Americans*, and *colored*; Native Americans referred to as *American Indians*; and Latinos referred to as *Hispanics*.

The book is divided into nine chapters that present historical details and contemporary information on both the administration of justice and related issues. Chapter 1 provides an overview of race and crime. It begins with a discussion of what many have referred to as the "invention of race." The remainder of the chapter highlights the historical experiences of Native Americans, African Americans, White ethnics, Latino Americans, and Asian Americans. The chapter pays particular attention to how crime has intersected with each of their experiences. Chapter 2 examines the extent of crime and victimization. It includes an overview of the history of the collection of crime data in the United States, a discussion of the limitations of crime statistics, the reported extent of crime and victimization for various racial groups, and an analysis of hate crime statistics. Chapter 3 presents theoretical perspectives on race and crime and provides a discussion of biological, sociological, subcultural, and nontraditional theoretical perspectives, including the colonial model and countercolonial criminology.

Chapters 4 through 9 examine race and several key components of the administration of justice: police, courts, sentencing, the death penalty, and corrections. An overview of policing in the United States is presented in Chapter 4. Minority employment data and an analysis of the history of race and policing are also presented. Contemporary issues presented in this chapter include citizen satisfaction with the police, police deviance (brutality and use of excessive force), racial profiling, police accountability, and recent innovations in policing. Chapter 5 examines the history of race and the courts in America. The chapter also examines how race impacts various facets of the American court system (i.e., bail, legal counsel, plea bargaining, etc.). A portion of the chapter also looks at the promise of drug courts.

Chapter 6 includes historical information and a comprehensive discussion of sentencing disparities. The chapter provides an overview of the sentencing process, along with a discussion of sentencing philosophies and contemporary issues related to race and sentencing. Chapter 7 examines race and the death penalty. Following an examination of the key Supreme Court death penalty cases the chapter examines the history of the death penalty in America, and also public opinion on the death penalty. Other contemporary issues discussed include the Capital Jury Project, wrongful convictions, and the death penalty moratorium movement. Chapter 8 provides a review of the history of corrections and the overrepresentation of racial minorities in jails and prisons. The chapter also examines prisoner reentry concerns, felon disenfranchisement, and political prisoners.

The issue of race and juvenile justice is presented in Chapter 9. The chapter presents an overview of juvenile justice in the United States and the historical context of race effects in juvenile justice, an explanation of the extent of juvenile delinquency and

victimization, and a discussion of several contemporary issues, including disproportionate minority confinement, minority female delinquency, the death penalty, and
delinquency prevention. The book closes with a conclusion that provides a brief reflection on the findings from the various chapters. The chapter also discusses prospects
for study and the future of race and crime.

Overall, we envision this book as an addition to the body of knowledge in the area
of race and crime. With our expanded historical coverage, we hope those who read this
work leave with an appreciation for the similar historical experiences of most
American racial and ethnic groups. We also hope that readers will see how race and
ethnicity have mattered and continue to matter in the administration of justice.

Acknowledgments

There are numerous individuals who have assisted us in the completion of this project. First, we would like to express our appreciation to our editor, Jerry Westby, for his continued support and encouragement. His assistant, Eve Oettinger, also provided assistance in the completion of this manuscript. We thank the following original reviewers for their constructive comments and suggestions that produced a well-received first edition:

Dr. Mary Atwell, Radford University

Dr. Stephanie Bush-Baskette, Rugers-Newark University

Dr. Charles Crawford, Western Michigan University

Dr. Alex del Carmen, University of Texas at Arlington

Dr. Roland Chilton, University of Massachusetts

Dr. Martha L. Henderson, Southern Illinois University

Dr. D. Kall Loper, University of North Texas

Dr. Mike Males, University of California, Santa Cruz

Dr. Michael A. McMorris

Dr. Karen Parker, University of Delaware

Dr. Charles Reasons, Central Washington University

Dr. Katheryn Russell-Brown, University of Florida

Dr. Adina Schwartz, John Jay College of Criminal Justice

Dr. Susan F. Sharp, University of Oklahoma

Dr. Shirley Williams, New Jersey City University

Dr. Bill Wells, Southern Indiana University

Dr. Ernest Uwazie, California State University, Sacramento

For the second edition, we thank the following reviewers who provided great suggestions to improve the text:

Dr. Tony Barringer, Florida Gulf Coast University

Dr. Dawn Beicher, Illinois State University

Dr. Lorenzo Boyd, Fayetteville State University

Dr. Roland Chilton, University of Massachusetts, Amherst

Dr. Ben Fluery-Steiner, University of Delaware

Dr. Kareem Jordan, University of North Florida

Dr. Peter C. Kratcoski, Kent State University

Dr. Everette B. Penn, University of Houston, Clear Lake

Dr. Carolyn Petrosino, Bridgewater State College

Dr. Robert Sigler, University of Alabama

Dr. Ernest Uwazie, California State University, Sacramento

I would like to thank my family for their continued encouragement and support. At Penn State, I would like to thank Dr. Steve Peterson, Director of the School of Public Affairs, who has provided a supportive environment for conducting research, in which the possibilities continue to seem endless. During the completion of both editions of this work, the assistance of several research assistants proved invaluable. Specifically, Nora Carerras, Nancy McGee, Patricia Patrick, and Leslie Kowal are acknowledged for their contributions. I acknowledge my son, Jini Gabbidon, for his efforts organizing the References into one file. Finally, I would like to acknowledge my mentor and intellectual partner, Dr. Helen Taylor Greene, for her continued guidance and support.

—Shaun L. Gabbidon

I thank my family for their support and appreciation of my scholarly endeavors. I thank Dr. Gabbidon, my coauthor, for his vision and commitment to the study of race and crime, and for inspiring me to persevere no matter what. I also thank my colleagues at Texas Southern University and elsewhere for their support. A special thanks to Ms. Pinkie Cotton and Mr. Brian Durham for assistance with constructing many of the tables in Chapters 2 and 4.

—Helen Taylor Greene

Overview of Race and Crime

1

> [B]ecause skin color is socially constructed, it can also be reconstructed. Thus, when the descendants of the European immigrants began to move up economically and socially, their skins apparently began to look lighter to the whites who had come to America before them. When enough of these descendants became visibly middle class, their skin was seen as fully white. The biological skin color of the second and third generations had not changed, but it was socially blanched or whitened.
>
> —Herbert J. Gans (2005)

At a time when the United States is more diverse than ever, with the minority population topping 100 million (one in every three U.S. residents) in 2006 (U.S. Census Bureau, 2007), the notion of race seems to permeate almost every facet of American life. Certainly, one of the more highly charged aspects of the race dialogue relates to crime. Before embarking on an overview of race and crime, we must first set the parameters of the discussion, which include relevant definitions and the scope of our review. It is always important when speaking of race to remind readers of the history of the concept and some current definitions.

Gossett (1963) has asserted that the idea of race originated 5,000 years ago in India, but was also prevalent among the Chinese, Egyptians, and Jews. Other scholars continue to believe that Francois Beniean was the first to categorize humans, but that the concept of race was invented by Carolus Linnaeus (1707–1778), a Swedish botanist who invented a system of classifying plants and animals. It was, however, Johan Frederick Blumenbach (1752–1840), a German specialist in anatomy and an anthropologist, who laid out the first categorization of the five races in his 1795 work, "On the Natural Variety of Mankind." In the work, he separated the inhabitants of the earth into five races: Ethiopian (African or Negroid), Mongolian (Asian), American (Native American), and Malaysian (Pacific Islander). When categorizing the fifth group, Whites, Blumenbach coined the term *Caucasian* (Feagin & Booher Feagin, 2008). Relying on Blumenbach's work, European scholars created a categorization that led to the belief that the differences among the groups were biological—and from the beginning Europeans placed themselves at the apex of the racial hierarchy (Feagin & Booher Feagin, 2008). It is widely believed, however, that the biological differences among racial groups are attributable to the migration patterns of people out of Africa (Dulaney, 1879; Shane, 1999) (see Figure 1.1).

Today, social scientists refer to race as a social construct. Gallagher (1997) expanded on this notion, writing, "Race and ethnicity are social constructions because their meanings are derived by focusing on arbitrary characteristics that a given society deems socially important. Race and ethnicity are social products based on cultural values; they are not scientific facts" (p. 2). Another relevant definition has been provided by Flowers (1988): "Race . . . refers to a group of persons characterized by common physical and/or biological traits that are transmitted in descent" (p. xiv). Finally, the U.S. Census Bureau (2000) has added the following:

> The concept of race . . . reflects self-identification by people according to the race or races with which they most closely identify. These categories are socio-political constructs and should not be interpreted as being scientific or anthropological in nature. (www.census.gov/main/www/cen2000.html)

Recent criminal justice investigations involving the use of DNA have challenged the notion that there are no biological differences between races (see Highlight Box 1.1).

Figure 1.1 Migration Patterns Out of Africa

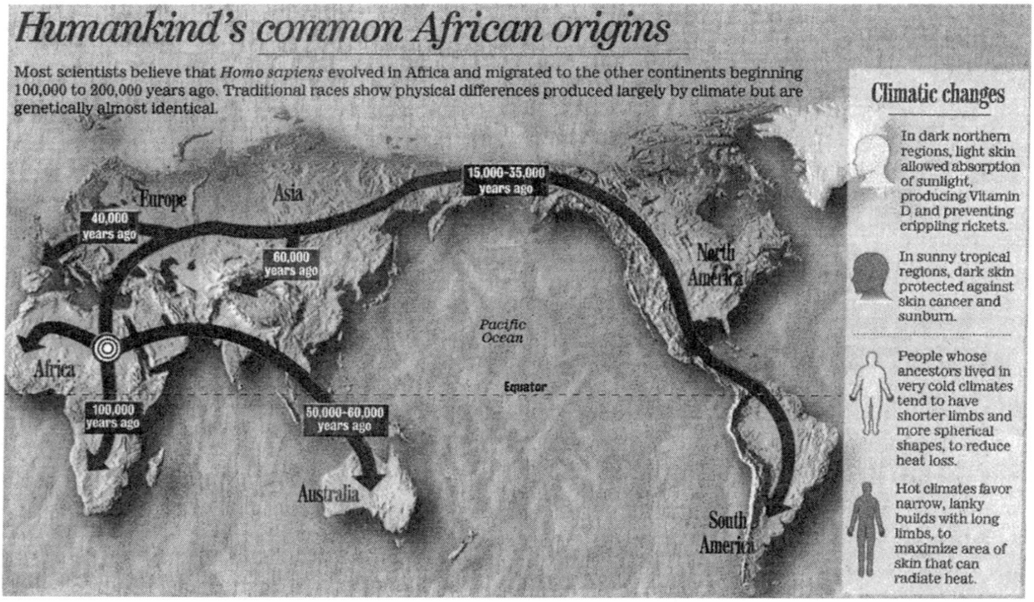
Humankind's common African origins
Most scientists believe that Homo sapiens evolved in Africa and migrated to the other continents beginning 100,000 to 200,000 years ago. Traditional races show physical differences produced largely by climate but are genetically almost identical.
Europe
Asia
40,000 years ago
60,000 years ago
15,000-35,000 years ago
North America
Pacific Ocean
Equator
Africa
100,000 years ago
50,000-60,000 years ago
Australia
South America
Climatic changes
In dark northern regions, light skin allowed absorption of sunlight, producing Vitamin D and preventing crippling rickets.
In sunny tropical regions, dark skin protected against skin cancer and sunburn.
People whose ancestors lived in very cold climates tend to have shorter limbs and more spherical shapes, to reduce heat loss.
Hot climates favor narrow, lanky builds with long limbs, to maximize area of skin that can radiate heat.

SOURCE: Shane, 1999, p. 6A.

Racial categories in American federal statistics are guided by Directive No. 15, issued by the Office of Federal Statistical Policy and Standards in 1977. Directive No. 15 replaced the Office of Management and Budget (OMB) Circular No. A-46, entitled "Race and Color Designations in Federal Statistics," issued in 1952 and revised on May 3, 1974 (Knepper, 1996).

Highlight Box 1.1

Getting DNA to Bear Witness

Police were desperately hunting a serial killer who had murdered five women in Louisiana. Leads were turning cold. DNA analysis of tissue shed at a crime scene did not match profiles in the FBI's database of DNA from known felons. Then investigators sent the tissue to a private lab in Sarasota, Fla., for further analysis. In a conference call in March with the Louisiana investigators, Tony Frudakis, the founder and chief scientific officer of DNAPrint Genomics, reported his lab's results: The suspect was a black male. The phone line fell silent. After all, eyewitnesses had described the suspect as white, and, historically, few serial killers are black.

When Frudakis was asked if he was sure, he replied: "I'm positive. You're wasting your time dragneting Caucasians; your killer is African-American." Investigators refocused their search and last month arrested the alleged killer, Derrick Todd Lee. The first reported successful use of DNA in the United States to provide clues about a suspect's ancestry has impressed some experts and unsettled others. "It's logical to use these technologies to identify leads in criminal cases—I have no qualms with that," says Barry Scheck, an expert on the use of DNA in criminal investigations. "But there are potential dangers as well," he says, citing risks to a suspect's privacy. Others are concerned that ancestry is often a poor guide to appearance.

Roots. Behind the controversy are recent studies showing it's possible to identify people's likely ancestral roots by looking for tiny variants in the sequence of DNA "letters" found most often in specific groups. "Races do exist, and they have some biological meaning," says Mark Batzer, a human geneticist at Louisiana State University and an author of one such study.

DNAPrint, which offers its service to both police and genealogy buffs, extracts and analyzes DNA from tissue left at crime scenes or cells swabbed from inside the cheek. It estimates a person's "biogeographical ancestry admixture"—an ethnic recipe giving the person's fraction of African, East Asian, Indo-European, and American Indian ancestry. No one has independently studied the accuracy of DNAPrint tests, but Frudakis says that in over 3,000 trials, "we've yet to confirm an error." Louisiana police aren't the first to use what's known as "DNA photofitting" technology to narrow a list of suspects. Last year, Britain's Forensic Science Service started making rough estimates of suspects' likely ethnicity. It compares markers in crime-scene DNA with its own database of DNA sequences that are more common in one ethnic group than another. The FSS also tests the DNA of suspects for a gene associated with red hair—a useful clue for police in a country where red hair is more common than in the United States.

(Continued)

(Continued)

But DNA analysis of any kind raises concerns about genetic privacy—for example, the possibility that a suspect's DNA will reveal not just ancestral markers but clues to the person's health. And the test may not always be as helpful as it seemed in the Louisiana investigation, say some experts. "Ancestry doesn't necessarily correlate with appearance, so this isn't going to be very useful," says Pilar Ossorio, an expert on bioethics. The University of Wisconsin–Madison professor cites herself as an example. Her mother is of Russian and German ancestry, her father is Mexican, and she says she's often mistaken as being from the Indian subcontinent. "Race is not something you can define in genetic terms," she says. The very idea of measuring race—an effort long associated with bad science and bigotry—makes many people uncomfortable. DNAPrint, however, argues that the biogeographical ancestry its test measures is different from "race" and that the results confirm how little racial categories actually mean. "By showing the continuum of genetic variation among people, our test dispels race as a scientific way of categorizing people," says Mark Shriver, an expert on human population genetics at Pennsylvania State University and the developer of the DNA print test. And in response to worries about medical privacy, the company says the sequences it analyzes for ancestry are not known to be linked to disease-related genes.

The controversy is about to heat up as more criminal investigators turn to the technology and it becomes more powerful. "We'd like to push the limits," says Frudakis. His lab and others are closing in on genes that affect traits including skin pigmentation and iris and hair color. Tests for those genes might give better clues to suspects' appearance, and Frudakis projects that they could be available later this year. On the horizon are screens for genetic variations that affect height and the shape of facial features. Frudakis says these are all fair game for DNA testing. He points out that they are, after all, the same traits police ask about when they have a human eyewitness.

SOURCE: Simons, D. H. (2003, June 23).

Directive No. 15 defined five racial/ethnic categories for federal statistics: American Indians/Native Americans, Asian/Pacific Islanders, Blacks, Hispanics, and Whites. Starting with the 2000 census, the OMB requires federal agencies to use no less than five racial categories: White; Black, African American, and Negro; American Indian and Alaska native; and Asian, Native Hawaiian, and other Pacific Islanders. A "some other race" category is also included. Respondents also could choose more than one category to indicate their racial identities. Although there are possibly 63 combinations of racial categories, the majority of the respondents reported only one racial group. "Hispanic origin" is viewed as an ethnic rather than a racial category in federal statistics. In the 2000 census, for the first time, the ordering of inquiries placed the question about Hispanic origin before the question about race (U.S. Bureau of the Census, 2000, 2001).

The term *ethnicity* comes from the Greek word *ethnos*, which means "nation." Generally, ethnic groups are defined by their similar genetic inheritances or some identifiable traits visible among most members of a particular group. Ethnic groups are also generally held together by a common language, culture, group spirit (nationalism or group solidarity), or geography (most typically originate from the same region) (Marger, 1997). Therefore, in the case of race and ethnicity, most anthropologists generally see these terms as more culturally relevant as opposed to biologically relevant.

We separate the American population into five groups: Native Americans, Whites, African Americans, Hispanic/Latino(a) Americans, and Asian Americans. We acknowledge that there are limitations to these categories. First, these categories do not take into account the ethnic variation within each race. For example, Table 1.1 provides a breakdown of the U.S. population by race. As you can see in Table 1.2, when we refer to "Latino Americans," there are a number of ethnic groups within this racial classification. This is true of other races as well. Another example is the category of African American/Black. There is also ethnic diversity within this category, with it often encompassing people from the Caribbean (e.g., Jamaica, Haiti, etc.), African countries, and other parts of the world. Because each of these groups has had a unique experience in America, it is, at times, presumptive for researchers to presume that the experience of one African/Black American is representative of so many diverse groups. Nevertheless, although we are aware of the problems with these classifications, the research and data we review follow this classification approach. Second, and relatedly, with the use of the multiracial category in the 2000 census, the lines between racial groups have become rather blurred, so population and crime data increasingly have considerable limits (this topic is discussed further in Chap. 2).

Since the 2000 census, as has been their tradition, the U.S. Census Bureau has continued to provide population estimates. In 2006, they released figures that revealed that the minority population had topped 100 million. By this time, Hispanics/Latinos had become the largest minority group (14.8%, 44 million), with Blacks now being the second largest (13.4%, 40 million). Given the changing demographics of the United States, some have called for the discontinuance of the term *minority* (Texeira, 2005). In place of *minority*, which some believe is a "term of oppression" or a term that seeks to minimize the collective aspirations of a group, the term *people of color* has been suggested (Texaira, 2005). Whatever the term to be used, if current estimates are correct, it is clear that one day racial and ethnic groups now considered to be minorities will become nearly half the U.S. population (U.S. Census Bureau, 2004). In fact, current estimates are that Whites will represent only 50% of the population in 2050, with Hispanics representing nearly a quarter of the population and other racial and ethnic minorities comprising the remainder of the populace (U.S. Census Bureau, 2004).

Prejudice and Discrimination

Even with the growth in the minority population, a central concern has been prejudice and discrimination. Prejudice is when someone fosters a negative attitude toward a

Table 1.1   Estimates of Population by Race/Ethnicity, 2006

Race/Ethnicity	Population
White	198,744,494
Hispanic	44,321,038
Black/African American	40,240,898
Asian	14,907,198
American Indian/Alaskan Native	4,497,895
Native Hawaiian/Pacific Islander	1,007,644

SOURCE: U.S. Bureau of the Census (2007).

Table 1.2   Hispanic or Latino and Race, 2000

Classification	Number	Percent
Total population	281,421,906	100.0
Hispanic or Latino (of any race)	35,305,818	12.5
Mexican	20,640,711	7.3
Puerto Rican	3,406,178	1.2
Cuban	1,241,685	0.4
Other Hispanic or Latino	10,017,244	3.6
Not Hispanic or Latino	246,116,088	87.5
White alone	194,552,774	69.1

SOURCE: U.S. Bureau of the Census (2000).

particular group. This is usually in the form of stereotypes that often result in people making negative generalizations about an entire group (Healey, 2007). Discrimination is considered the "unequal treatment of a person or persons based on group membership" (Healey, 2007, p. 20). As you can imagine, having prejudicial attitudes toward a particular group, in many instances, can lead to discriminatory actions in areas such as employment, housing, and the criminal justice system. Thus, determining whether prejudice and discrimination permeates the criminal justice system is a critical part of this work.

The remainder of this chapter provides a brief historical overview of each major racial/ethnic group, highlighting the complex history of race in America and how this history is intertwined with race, crime, and the criminal justice system. Readers should keep in mind that our historical review is not meant to be comprehensive. Rather, we see our review as illustrative of how concerns regarding race and crime are not new and have been the norm since the European "discovery" of America.

Historical Antecedents of Race and Crime in America

NATIVE AMERICANS

Prior to the arrival of Europeans in the Americas, the native people had existed on the continent for thousands of years. It is believed that they originated from eastern Asia. More specifically, it is believed that they have been in North America for the last 30,000 years, having crossed over from Asia into America on glaciers that, due to warming trends, later melted (Polk, 2006, pp. 3–4). Over time, they built complex societies throughout the Americas. Even so, on arrival in the Americas (South America and the West Indies), it is clear from their actions that Christopher Columbus and his followers viewed the native people (then referred to as "Indians," now referred to as "Native Americans") as inferior (H. Clarke, 1992). The brutality that followed has been painstakingly documented by firsthand observers of the massacres (De Las Casas, 1552/1993). Sale (1990) has suggested that, prior to the arrival of Europeans, there were about 15 million Native Americans in North America. According to Healey (2003), nearly four centuries later, in 1890, only 250,000 remained. Today, there are nearly 4.5 million American Indians/Alaskan Natives in the United States. Nonetheless, considering the historical decimation of the Native American population, some criminologists have viewed their massacre as genocide (Barak, Flavin, & Leighton, 2006).

Although some have categorized all Native Americans into one group, they represent a diverse set of societies separated by "language, economy, polity, and customs" (Feagin & Booher Feagin, 1996, p. 197). It has been noted that their societies were more advanced than those of the Europeans who colonized them. As a result, Europeans borrowed much from Native American agriculture and pharmacology. Furthermore, some have noted that "Benjamin Franklin and Thomas Jefferson both admired and learned from the democratic political institutions of major North American tribes" (Feagin & Booher Feagin, 1996, p. 197).

During their initial contact with Europeans, Native Americans assisted the newcomers with advice on how to survive in their new environment. However, once colonists became comfortable with the surroundings, they began to displace, enslave, and destroy Native American societies. Over time, massacres of Native Americans became commonplace throughout the colonies, but once the Constitution was ratified (with little mention of Native Americans), treaties were enacted with the aim of ending massacres and also protecting Native American lands from further pillage. However, the government did not honor the treaties. According to Feagin and Booher

Feagin (1996), "By action or inaction, [the government] supported the recurrent theft of Native American lands. In practice, they approved of ignoring boundary rights wherever necessary" (p. 200). Such actions were sanctioned at the highest levels, with presidents such as Andrew Jackson encouraging the defiance of Supreme Court rulings related to Native Americans. From 1790 to the mid-1800s, there were more than 300 treaties signed between Whites and Native Americans, most of which were not honored. As a result, conflicts persisted, which led to concerns regarding "criminal aggression" and the subsequent enactment of another approach: removal. Healey (2003) wrote,

> East of the Mississippi, the period of open conflict was brought to a close by the Indian Removal of 1830, which dictated a policy of forced emigration to the tribes. The law required all eastern tribes to move to new lands west of the Mississippi. Some of the affected tribes went without resistance, others fought, and still others fled to Canada rather than move to a new territory. (p. 190)

This infamous "Trail of Tears," as it became known, resulted in the death of thousands of Native Americans. Nearly 40 years later, in 1867, the Dolittle Committee, which was investigating several recent massacres of Native Americans, found that much of the aggression by Native Americans around that time had occurred in response to White aggression (Harjo, 2002).

The same year of this massive removal of Native Americans, the Bureau of Indian Affairs (BIA) was established to handle matters related to this population. Following the creation of the BIA, the agency had to deal with the competing aims of the federal government. On the one hand, the government created the agency to help Native Americans; on the other hand, the military had a policy of "genocidal extermination." Nearly 60 years after the creation of the BIA, the 1887 Dawes Act provided that individual families be provided with reservation lands. While well meaning, as Feagin and Booher Feagin (2008) observed, "This new policy soon resulted in a large-scale land sale to Whites. Through means fair and foul, the remaining 140 million acres of Indian lands were further reduced to 50 million acres by the mid-1930s" (p. 144). In the early part of the 20th century, the government tried to assimilate Native Americans by sending them to Indian boarding schools that were Christian-based and used to indoctrinate Native Americans with American culture. In the process, Native Americans were forced to abandon their native language and customs. The attempt to assimilate Native Americans culminated during the 1920s with the passage of the Indian Citizenship Act of 1924, which granted all Native Americans citizenship. The end of this period saw Native Americans calling for new policies, one of which came in the form of the 1934 Indian Reorganization Act. The act, which essentially ended the Dawes Act, allowed Native Americans "to establish Indian civil and cultural rights, allow for semiautonomous tribal governments similar in legal status to counties and municipalities, and foster economic development on reservations" (Feagin & Booher Feagin, 2008, p. 144). As with all legislation, there were problems. Most notably, Native Americans saw this act as giving too much power to the secretary of the interior. In addition, many Native Americans believed the act violated the sovereignty or their right to govern themselves provided by previously enacted treaties.

The second half of the 20th century spurred more attempts by Native Americans to rid themselves of governmental control. In the early 1950s, Congress enacted legislation called *termination*, which "call[ed] for an end to the reservation system and to the special relationships between the tribes and the federal government" (Healey, 2004, p. 134). This process also negated previous treaties, a policy that was vigorously opposed by Native Americans. In addition, based on the specifics of the policy, "Tribes would no longer exist as legally recognized entities, and tribal lands and other resources would be placed in private hands" (Healey, 2004, p. 134). Because of this policy, many Native Americans moved to urban areas.

The decades following the enactment of the termination policy saw increasing opposition from Native Americans. After about 25 years, the policy was repealed. In 1975, the Indian Self-Determination and Education Assistance Act "increased aid to reservation schools and Native American students and increased the tribes' control over the administration of the reservations, from police forces to schools and road maintenance" (Healey, 2004, p. 136). This act provides much of the basis under which many tribes now operate. Recent federal legislation has allowed some tribes to open gambling facilities on reservations, which in 2000 generated more than $10 billion in revenues (Spilde, 2001). Other tribes have invested in other ways to generate revenue (e.g., tax-free cigarette sales). Native Americans' move to self-determination also has resulted in suits against the federal government seeking reparations for past broken treaties. With 561 recognized tribes and a population of more than 4 million, Native Americans remain a notable presence in the United States.

AFRICAN AMERICANS

African Americans are another group that has had a long and arduous relationship with the United States. With the Native American population nearly completely decimated because of the brutality, enslavement, and diseases that were brought to the Americas by the Spanish, Bartolome De Las Casas, the priest who accompanied Columbus to America, sought a way to stem their extermination.

De Las Casas' idea centered on not ending the slave system, but instead replacing the Native Americans with another labor force: Africans. Of De Las Casas' thinking, Finger (1959) wrote,

> Having heard that the Negroes of the Portuguese colonies in Africa were more robust than the natives of the West Indies Islands, he [De Las Casas] recommended that Black slaves be imported to take the place of Indians in server tasks of the plantations and mines. (p. 716)

Finger (1959) also described the results of De Las Casas' suggestion:

> A terrible traffic in human flesh ensued. Portuguese raiders carried the Africans
> from their homes, and English sailors conveyed them across the Atlantic. Spanish,
> Portuguese, and later English slave-owners worked the poor Black men as though
> they possessed no natural rights as human beings. (pp. 716–717)

As with the decimation of the Native American population, the slave trade involving
Africans has been viewed as genocidal and referred to as the "African holocaust"
(Clarke, 1992).

Africans initially arrived in the colonies as a result of piracy (Higginbotham, 1996).
When a slave ship carrying Africans headed to the West Indies was taken over by
pirates and ran out of supplies, the pirates landed in Jamestown, Virginia, where they
sold the Africans for food and supplies. It is important to note that, prior to their movement into perpetual slavery, Africans had existed much like the other citizens in the
colony. Thus, from their arrival in 1619 to the 1660s, Africans were not considered
slaves in colonial America; they were able to fulfill indentures and were fairly integrated into the life of the colony. After 1660, however, colonial legislation made it clear
that Africans were to be considered slaves.

McIntyre (1992) believes the leaders of the colony came to a juncture where they
needed to decide the best way to further the economic fortunes of its citizens, and
they came up with several potential options. The first involved the continued use of
the indentured servant system for Blacks and Whites. Second, the colonists, like the
Spaniards earlier, thought about enslaving the Native Americans. Third, both Native
Americans and Blacks could be enslaved. Fourth, the colonists could create a free
labor system for Blacks, Whites, Indians, and immigrants. Eventually, they chose the
fifth option: the enslavement of Blacks. McIntyre (1992) has suggested that this was
the case because Whites had the option to appeal for protection from the British
monarchy; in addition, they could appeal to general White public opinion. Enslaving
Native Americans did not appeal to the colonists because besides feeling that they
would not hold up under slave conditions, they were aware that the natives were
familiar with the terrain, which would have permitted easy escape. For the next two
centuries, African Americans would serve as the primary labor force keeping the
southern economy afloat.

Although much of the slave system was kept intact by "plantation justice," there was
little interference in these matters from outside developing criminal justice institutions except when slaves escaped or there was a slave revolt. In times of escapes, slave
owners cooperated by enlisting "slave patrollers" to ensure slaves were quickly captured and returned to their owners. Similarly, when slave revolts occurred, slave owners worked together to expeditiously bring a close to the uprisings that threatened the
stability of the slave system (Aptheker, 1943/1993). Slave owners were so committed
to quelling escapes and revolts that they enacted widespread slave codes to reduce
their likelihood. Describing the slave codes, Russell (1998) wrote,

Slave codes embodied the criminal law and procedure applied against enslaved Africans. The codes, which regulated slave life from cradle to grave, were virtually uniform across states—each with the overriding goal of upholding chattel slavery. The codes not only enumerated the applicable law but also prescribed the social boundaries for slaves—where they could go, what types of activity they could engage in, and what type of contracts they could enter into. Under the codes, the harshest criminal penalties were reserved for those acts that threatened the institution of slavery (e.g., the murder of someone White or a slave insurrection). The slave codes also penalized Whites who opposed slavery. (pp. 14–15)

In addition to the slave codes, Whites used psychology to keep the slave system intact. Claude Anderson (1994) wrote that "this process was designed to instill in Blacks strict discipline, a sense of inferiority, belief in the slave owners' superior power, acceptance of the owners' standards and a deep sense of a slave's helplessness and dependence" (p. 165). Moreover, Anderson noted that "the slave owners strove to cut Blacks off from their own history, culture, language and community, and to inculcate White society's value system" (p. 165).

Another telling dynamic during the slave era was the way in which punishment was exacted for crimes committed by African Americans in comparison with Whites. After reviewing nearly every appellate case on antebellum slavery and race relations from 1630 to 1865, A. Leon Higginbotham, the late jurist and scholar, formulated his "Ten Precepts of American Slavery Jurisprudence" (Higginbotham, 1996; see Highlight Box 1.2). These precepts describe the foundations on which justice was distributed during this era. Most notably, to maintain the slave system, White supremacy called for little justice to be distributed to African Americans, whereas Whites were indifferent to their own criminal activity. This was most pronounced in the crime of rape. Whites might rape Black women with impunity; however, if Blacks so much as looked at White women in an unacceptable way, they were subjected to severe beatings. Table 1.3 highlights the differential punishments for African American and White crimes during the slave era.

The 1700s brought similar race and crime concerns. Some Whites, however, continued to show indifference toward their own criminal activity. Although the slave system began to expand under the encouragement of the colonial aristocracy, the slave trade began to be shunned in the international community. Subsequently, there was a movement to stop the trade, although slavery continued for those slaves already in America. Du Bois (1891) wrote about the movement to stop the slave trade as having four periods, and these were tied to large-scale efforts by Whites to circumvent the law. In his assessment of the four periods, Du Bois wrote that there were varying levels of commitment to this initiative. The compromise of the Constitutional convention allowed the slave trade to continue until 1808; however, Du Bois' research showed that Whites never took the prohibition seriously, considering the large numbers of persons who were actively involved in trading slaves even with the threat of imprisonment.

Table 1.3   Criminal Punishments by Race in Slave Era Virginia

Crime	White Offender	Black Slave Offender
Murder (White victim) Petit treason (murder of slave owner)	Maximum penalty: death	Death
Murder (Black victim)	Rarely prosecuted	If prosecuted, whipping, hard labor, or death
Rape (White victim)	10–20 years, whipping, or death if minor victim	Death or castration (same penalty for attempted rape)
Rape (Black victim)	No crime	No crime, exile, or death (If rape of free Black women, penalty could be death)
Assault (White victim)	1–10 years (if done with intent to kill)	Whipping, exile, mutilation, or death

SOURCE: Higginbotham and Jacobs (1992), as presented in Russell (1998, p. 16).

Highlight Box 1.2

The 10 Precepts of American Slavery Jurisprudence

1. Inferiority: Presume, preserve, protect, and defend the ideal of the superiority of Whites and the inferiority of Blacks.

2. Property: Define the slave as the master's property, maximize the master's economic interest, disregard the humanity of the slave except when it serves the master's interest, and deny slaves the fruits of their labor.

3. Powerlessness: Keep Blacks—whether slave or free—as powerless as possible so they will be submissive and dependent in every respect, not only to the master, but to Whites in general. Limit Blacks' accessibility to the courts and subject Blacks to an inferior system of justice with lesser rights and protections and greater punishments. Utilize violence and the powers of government to ensure the submissiveness of Blacks.

4. Racial "purity": Always preserve White male sexual dominance. Draw an arbitrary racial line and preserve White racial purity as thus defined. Tolerate sexual relations between White men and Black women; punish severely relations between White women and non-White men. As to children who are products of interracial sexual relations, the freedom or enslavement of the Black child is determined by the status of the mother.

5. Manumission and free Blacks: Limit and discourage manumission; minimize the number of free Blacks in the state. Confine free Blacks to a status as close to slavery as possible.

6. Family: Recognize no rights of the Black family; destroy the unity of the Black family; deny slaves the right of marriage; demean and degrade Black women, Black men, Black parents, and Black children; and then condemn them for their conduct and state of mind.

7. Education and culture: Deny Blacks any education, deny them knowledge of their culture, and make it a crime to teach those who are slaves how to read and write.

8. Religion: Recognize no rights of slaves to define or practice their own religions, choose their own religious leaders, or worship with other Blacks. Encourage them to adopt the religion of the White master, and teach them that God, who is White, will reward the slave who obeys the commands of his master here on earth. Use religion to justify the slave's status on earth.

9. Liberty—resistance: Limit Blacks' opportunity to resist, bear arms, rebel, or flee; curtail their freedom of movement, freedom of association, and freedom of expression. Deny Blacks the right to vote and to participate in government.

10. By any means possible: Support all measures, including the use of violence, that maximize the profitability of slavery and that legitimize racism. Oppose, by the use of violence if necessary, all measures that advocate the abolition of slavery of the diminution of White supremacy.

SOURCE: Higginbotham (1996).

Du Bois found that when the U.S. government signed the Treaty of Ghent in 1814, it further committed to ending the international slave trade. As a part of this commitment, participating nations were asked to engage in searches of vessels abroad; however, America was unwilling to agree to this stipulation. Hence, many ships that flew the American flag were not American; they were slave traders who sought refuge by using the American flag. Du Bois also noted that even after the death penalty was instituted for slave trading, during the four periods, he found few instances where Whites had been convicted, much less executed, for being connected to the slave trade. In the end, this early form of White crime in America, which was particularly tied to the ruling class of slaveholders in the South, was allowed to persist because Whites were unwilling to give up the financial benefit derived from the slave trade and system (C. Anderson, 1994; E. Williams, 1944).

During the mid-1850s, there was a crisis brewing regarding slavery. Although a civil war seemed imminent, the North and South tried to delay the inevitable. Of particular concern during this period was the acquisition of territories in the southwest portion of the United States. The debate centered on which states should be slave states—if any at all. Predictably, northerners argued to keep such states free, whereas southerners wanted to preserve the institution of slavery, so they argued the reverse. Vigorous debate led to the well-known Compromise of 1850, which essentially gave each side a portion of what they wanted. For example, California entered the Union as a Free State, while other territories would enter the Union without mention of slavery (Franklin & Moss, 2000). One of the provisions of the Compromise led to the enactment of the Fugitive Slave Law of 1850.

A revision of the 1793 Fugitive Slave Act, the Fugitive Slave Law (or Act) was structured to ensure the return of runaway slaves. The act called for the appointment of numerous commissioners who were authorized to hire deputies who all could "enlist the aid of bystanders or posses to enforce the act" (Kennedy, 1997, p. 83). Furthermore, monetary incentives were tied to this process. For example, "commissioners would be paid a fee of $5 in each case in which he determined that a slave master was *not* entitled to an alleged fugitive slave, and would be paid a fee of $10 in each case in which he determined that a master was entitled to the accused person" (Kennedy, 1997, pp. 83–84; italics added). Finally, to illustrate the seriousness in which the enforcement of the act was to be taken, there was a stipulation that if a U.S. Marshall refused or neglected to execute warrants issued by commissioners, they would be fined $1,000 (Kennedy, 1997). The enactment of this Act and other provisions of the Compromise still could not stop the move toward civil war. Thus, not long after the notorious 1857 Dred Scott decision that continued to increase the tensions between North and South, the country headed into the Civil War in 1861.

Following the Emancipation Proclamation in 1863, which freed the slaves in the Confederate states, and the enactment of the Thirteenth Amendment in 1865, which ended slavery throughout the United States, many African Americans chose to remain in the South. Others had dreams of going north and starting anew. Unfortunately, southern landowners were unwilling to part so easily with their former free labor force. Therefore, following emancipation, they enacted the "Black codes." These codes were an assortment of laws that targeted poor Whites and African Americans. Some scholars have argued that the laws were specifically created so that a significant number of African Americans could be returned to plantation owners through the convict-lease system (Du Bois, 1901/2002; Myers, 1998; Oshinsky, 1996). The convict-lease system allowed states to lease convict labor to private landowners. Although some poor Whites also were caught up in this legal system, many of those being leased out to southern landowners were African Americans. Before long, whereas previously they had engaged only in trivial offenses, African Americans began to engage in more bold and brutal offenses, which shocked southern Whites who had created the system (Du Bois, 1901/2002).

Prior to the Civil War, primarily Whites had been incarcerated in southern penal institutions, and one product of the massive changes in the South was the increasing number of African Americans found in prisons. Following this period, along with the convict-lease system, states such as Mississippi ran notorious state prisons that put the prisoners to work. Parchman Farm was one of the most infamous (Oshinsky, 1996).

The Reconstruction Era also brought the formal advent of hate groups. Groups such as the Knights of White Camellia, the Constitutional Union Guards, the Pale Faces, the White Brotherhood, the Council of Safety, the '76 Association, and the infamous Knights of the Ku Klux Klan were all formed to ensure White supremacy ruled in the South following Emancipation and the passage of the Thirteenth Amendment in 1865, which officially abolished slavery. These groups wreaked havoc on African American and other citizens, who were targets of their hatred. Lynching became the means used to intimidate and handle those who challenged the racist White power structure (see Figure 1.2). It is generally accepted that, between 1882 and 1930, "At least three thousand Black men, women, and children were murdered by White gangs during this era of the lynch mob, and this toll does not count other racially motivated murders or Black deaths from race riots" (Beck & Tolnay, 1995, p. 121). These indiscriminate killings of African Americans (and some Native Americans and Spanish-speaking minorities), usually by hanging, were typically carried out to avenge some exaggerated crime committed by an African American or other "undesirable" minority against a White person (Zangrando, 1980). In most instances, the alleged crime of rape was used to justify these horrific actions.

The Ku Klux Klan emerged as the leading hate organization. In an effort to suppress African American economic equality and pride, the Klan beat African Americans for minor things, such as "Black women . . . dressing in brightly-colored clothes, and men for being impolite, talking back to Whites or failing to say 'Yes Sir'" (Katz, 1986, p. 39). In many jurisdictions, Klan activities were condoned by local law enforcement. As a result, many African Americans lost faith in the justice system and stopped reporting crimes altogether (Katz, 1986).

On the eve of the 20th century, the *Plessy v. Ferguson* (1896) "separate but equal" decision was hailed by southern bigots. This decision was significant in that now Whites had legal support to enforce some of their ideas concerning White supremacy and the separation of the races. Furthermore, this decision allowed law enforcement officials to take action against African Americans who sought basic services now reserved for Whites. Of this separation of the color line, Du Bois (1899) clearly saw the dangerousness of state-sanctioned segregation, writing,

> [Another] cause of Negro crime is the exaggerated and unnatural separation in the South of the best classes of Whites and Blacks. A drawing of the color line, that extends to street-cars, elevators, and cemeteries, which leaves no common ground of meeting, no medium for communication, no ties of sympathy between two races who live together, and whose interests are at bottom one—such a discrimination is more than silly, it is dangerous. (p. 59)

Ten years after the turn of the 20th century, African Americans were primarily southern. Meier and Rudwick (1970) observed that "approximately three out of four lived in rural areas and nine out of ten lived in the South" (p. 213). The "Great Migration," however, changed the landscape of the North and South. By the 1950s,

Figure 1.2 Oliver Cox's Lynching Cycle

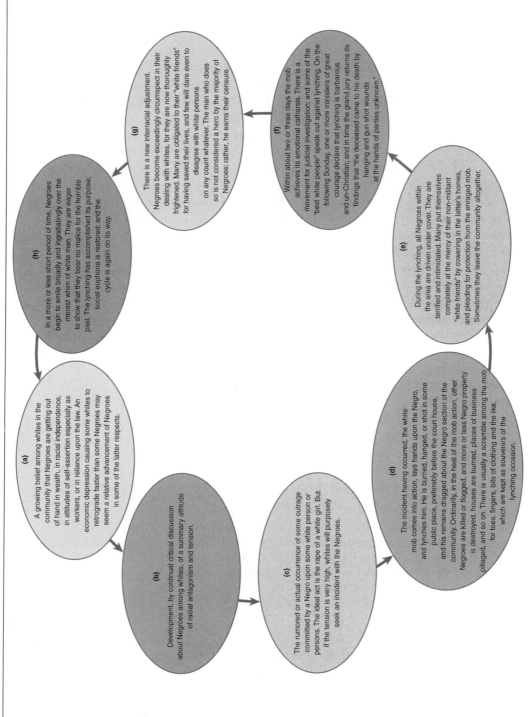

(a) A growing belief among whites in the community that Negroes are getting out of hand in wealth, in racial independence, in attitudes of self-assertion especially as workers, or in reliance upon the law. An economic depression causing some whites to retrograde faster than some Negroes may seem a relative advancement of Negroes in some of the latter respects.

(b) Development, by continual critical discussion about Negroes among whites, of a summary attitude of racial antagonism and tension.

(c) The rumored or actual occurrence of some outrage committed by a Negro upon some white person or persons. The ideal act is the rape of a white girl. But if the tension is very high, whites will purposely seek an incident with the Negroes.

(d) The incident having occurred, the white mob comes into action, lays hands upon the Negro, and lynches him. He is burned, hanged, or shot in some public place, preferably before the court house, and his remains dragged about the Negro section of the community. Ordinarily, in the heat of the mob action, other Negroes are killed or flogged, and more or less Negro property is destroyed, houses are burned, places of business pillaged, and so on. There is usually a scramble among the mob for toes, fingers, bits of clothing and the like, which are kept as souvenirs of the lynching occasion.

(e) During the lynching, all Negroes within the area are driven under cover. They are terrified and intimidated. Many put themselves completely at the mercy of their non-militant "white friends" by cowering in the latter's homes, and pleading for protection from the enraged mob. Sometimes they leave the community altogether.

(f) Within about two or three days the mob achieves its emotional catharsis. There is a movement for judicial investigation; and some of the "best white people" speak out against lynching. On the following Sunday, one or more ministers of great courage declare that lynching is barbarous and un-Christian; and in time the grand jury returns its findings that "the deceased came to his death by hanging and gun shot wounds at the hands of parties unknown."

(g) There is a new interracial adjustment. Negroes become exceedingly circumspect in their dealing with whites, for they are now thoroughly frightened. Many are obligated to their "white friends" for having saved their lives, and few will dare even to disagree with white persons on any count whatever. The man who does so is not considered a hero by the majority of Negroes; rather, he earns their censure.

(h) In a more or less short period of time, Negroes begin to smile broadly and ingratialingly over the merest whim of white men. They are eager to show that they bear no malice for the horrible past. The lynching has accomplished its purpose; and the social euphoria is restored; and the cycle is again on its way.

SOURCE: Adatped from Cox (1945).

"Negroes were mainly an urban population, almost three fourths of them being city-dwellers" (Meier & Rudwick, 1970, p. 213). During this era, African Americans crowded into northern cities in search of job opportunities; what they found, however, were overcrowded urban areas with assorted European immigrants either seeking similar opportunities or already established in the low-skill, low-wage jobs that African Americans had hoped to receive. African American women were able to secure employment in domestic service, where, unfortunately, White men often sexually assaulted them. Writing of the dilemma this posed, scholar activist Angela Davis (1981) noted,

> From Reconstruction to the present, Black women household workers have considered sexual abuse perpetrated by the "man of the house" as one of their major occupational hazards. Time after time they have been victims of extortion on the job, compelled to choose between sexual submission and absolute poverty for themselves and their families. (p. 91)

African American men who did find work were also relegated to menial jobs and, from 1890 to 1930, were often used as strikebreakers (Massey & Denton, 1993). Their role as strikebreakers often led to racial violence in the North, which repeatedly culminated in race riots. From 1900 to 1919, there was a steady stream of race riots throughout the North. The riots continued into the 1920s, with Whites resisting integration "by any means necessary." As Massey and Denton (1993) documented,

> A wave of bombings followed the expansion of Black residential areas in the cities throughout the north. In Chicago, fifty eight homes were bombed between 1917 and 1921, one every twenty days; and one Black real estate agent, Jesse Binga, had his home and office bombed seven times in one year. (p. 35)

Devastating riots followed in Tulsa, Oklahoma, in 1921 (Hirsch, 2002) and Rosewood, Florida, in 1923 (Russell, 1998). Because of the continuing racial tensions related to labor competition and integration attempts, race riots persisted well into the 1960s (Grimshaw, 1969).

In the 1930s, the "Scottsboro Boys" shed international attention on the plight of African Americans. The case involved several African American boys who were traveling in a freight train with several White boys and two White girls. After a fight ensued, the White boys were ejected from the train.

At the next stop in Scottsboro, Alabama, the girls got off the train and claimed they had been gang-raped by the nine African American boys. Playing on the worst fears of southern White men, the girls' accusations resulted in a mob being quickly formed in anticipation of the lynching of the boys (Carter, 1969). With the protection of law enforcement, however, the boys made it to trial. Following several trials, the boys were found guilty and sentenced to the death penalty. Although it was later revealed that the claims were a hoax, the boys served a combined 104 years in prison (Walker, Spohn, & DeLone, 2004).

During the 1930s and 1940s, there was continued interest in the subject of crime among African Americans. In the last edition of his landmark text, *Principles of Criminology* (1947), pioneering criminologist Edwin Sutherland devoted a chapter to "crime in relation to race and nativity." He first noted that, much like today, African Americans were "arrested, convicted, and committed to prisons approximately three times as frequently as White persons" (Sutherland, 1947, p. 121). Sutherland also cautioned that some of these statistics "probably reflect a bias against all of the minority races but especially against the Negro" (p. 121).

By the early 1950s, African Americans and other ethnic groups were still struggling to survive in an increasingly segregated and hostile America. Some turned to crime, whereas others turned to the United Nations for assistance. In 1951, African Americans petitioned the United Nations and charged the U.S. government with genocide against African Americans (Patterson, 1951/1970). Although the United Nations did not respond to the petition, African Americans had made the commitment to try and change their position within American society. This movement was given a further push by the 1955 kidnapping and slaying of Emmett Till in Mississippi.

The shocking and brutal killing of the 14-year-old boy for "disrespecting" a White woman spurred a movement that picked up steam with the Montgomery boycott, which started on December 5, 1955. The civil rights movement showed the national and international communities the depth of racial hatred and interracial strife in America. The demonstrations that defined the movement were seen by millions on TV, and the brutality of the police toward nonviolent demonstrators spoke to the oppressive role the police played in the African American and other minority communities.

By the 1960s, according to figures from Tuskegee Institute (Zangrando, 1980), lynchings were rare events; however, Whites had successfully used the practice to discourage any serious level of integration. Therefore, although Thurgood Marshall and his colleagues were successful in the landmark *Brown v. Board of Education* (1954) case, minority communities did not substantially change for decades to come. Because of "the White strategy of ghetto containment and tactical retreat before an advancing color line" (Massey & Denton, 1993, p. 45), substantial underclass communities were in existence by the 1970s. This bred a level of poverty and despair that fostered the continuation of the African American criminal classes and organized crime. The riots of the 1960s were a response to the long-standing troublesome conditions in some of these cities (National Advisory Commission on Civil Disorders, 1968).

When African Americans were finally able to take advantage of the opportunities forged by the civil rights movement and desegregation, many of them left inner-city areas (Black flight). Unfortunately, they were among the ones who had brought an important level of stability to these communities. As a result of this exodus, these communities are now heavily comprised with what Wilson (1987) describes as "the truly disadvantaged." They are heavily dependent on the underground economy for survival (see Venkatesh, 2006), which has likely contributed to the overrepresentation of African Americans throughout the U.S. criminal justice system.

Just recently, the plight of "the truly disadvantaged" was brought to the forefront of American consciousness with the 2005 Hurricane Katrina fiasco, in which the government—at all levels—failed to provide an adequate response to the needs of poor and mostly black New Orleans residents (Dyson, 2006; Potter, 2007). Moreover, in the absence of the government response, citizens who took matters into their own hands have been portrayed as criminals (Russell-Brown, 2006). Although there has been some indication that crime has increased in cities where evacuees were relocated, this "truly disadvantaged" population continues to be stereotyped and faced with few options to survive in the aftermath of the devastation left by Hurricane Katrina.

Even with the many struggles encountered by African Americans and other Black ethnic groups, and the historical fixation on their criminality, they have contributed to every aspect of American life, from the toiling of the soil in the south and factory work in the north, to produce the wealth that made America what it is, to the innumerable scientific, musical, and artistic contributions that are now considered staples of American culture.

WHITE ETHNICS

During the early 1600s, while the slave trade in South America and the West Indies was carrying on, the British colonized parts of what would later become the American colonies. This led to many of the same kinds of conflicts with Native Americans that the Spanish had quelled with unimaginable brutality. Although the British saw the colonies as somewhere they could send criminals and other undesirables, they also saw opportunity for monetary gain, so they encouraged immigration to the colonies. Some came freely, whereas others used indentures to get them to the New World. These arrangements allowed them to work for a period of time to pay for their travel expenses to the colonies. Once their indentures were completed, immigrants were free to pursue whatever opportunities they desired. In addition to British immigrants, Germans and Italians were among the first to immigrate to America. Many began to arrive in the early 1600s, settling first in New Amsterdam (New York) and later in Pennsylvania (Sowell, 1981). Given this rich history of European immigration to the United States, we briefly review the history of several White ethnic groups. Although our review does not cover every White ethnic group that immigrated to America, we provide discussions of several of the major White ethnic groups. We begin with an overview of the experience of German Americans. This is followed by a review of the experiences of Italian Americans, Irish Americans, and Jewish Americans. As you will see, many of these groups have similar stories regarding their reason for making the long journey to America. In addition, many have nearly identical experiences on their arrival in America.

German Americans

Faust (1927) places the first German in America at the time of Leif Eriksson's pioneering journey that landed him in North America 500 years prior to Columbus' arrival. Among Eriksson's crew was a German named Tyrker, who "is credited with discovering grapes in North America and therefore also naming the new land Vineland" (Rippley, 1976, p. 22). Not until the 1500s was there a settlement of Germans in America. Located in Port Royal, South Carolina, the settlement was comprised of Huguenots (French Protestants) and Alsatian and Hessian Protestants (both of German origin). The settlement, however, was destroyed by the Spaniards, and, thus, only lasted 4 years from 1562 to 1566. The next wave of German immigrants arrived with the first settlers in Jamestown in 1607. Often referred to as the "Dutch," which is likely "a linguistic slip that occurred because the word 'Dutch' so closely resembles a German's designation for himself, *Deutsch*" (Rippley, 1976, p. 24), they were often mistreated in the early colonial period. Consequently, they sympathized with the plight of Native Americans and "chose to remain with the Indians, preferring their friendship to that of the 'gentlemen' of Jamestown" (Faust, 1927, p. 8).

In the late 1600s, thirteen German families arrived in Philadelphia and represented the beginning of mass German immigration to the United States (Coppa & Curran, 1976). Many of these immigrants came at the urging of William Penn, who told them of the religious freedoms in his colony of Pennsylvania (Sowell, 1981). Others came as a result of the disarray in their homeland. Of this, Coppa and Curran (1976) wrote: "The havoc wrought by the Thirty Years' War (1618–1648) devastated Germany for many decades: commerce declined; industry was crippled; and intellectual life sustained a deep if not mortal blow" (p. 45). The German population also increased because of the use of indentures to get them to America. Hence, those who wanted to immigrate to America signed contracts that paid their way to America. As one might imagine, this was shady business. Sowell (1981) writes that: "the indentured servants were preyed upon by the dishonest. Some ship captains provided inadequate food or sold them into longer periods of bondage than actually required to work off the cost off their transportation. Germans who could not understand English were particularly vulnerable" (p. 49). As a consequence of all these events, by the time of the Revolutionary War, there were about 225,000 German Americans in the colonies (Rippley, 1976, p. 29).

Immigration from Germany in the 1800s began slowly; however, because of continuing issues in the homeland, Germans continued to hear from other groups of the promise of America. As such, around the 1830s, the number of German immigrants rose again and continued to increase throughout the 19th century. By the 1900 census, there were more than 2.6 million Germans in America (Faust, 1927). These formidable numbers made them a significant force in American culture and politics. They were outstanding farmers and glassmakers and have been credited with setting up the first paper mill. Culturally, they incorporated German chocolate cake, coleslaw, sauerkraut, hotdogs, and hamburgers into American life. Well-known Germans such as Albert Einstein, Babe Ruth, Lou Gehrig, and Presidents Hoover and Eisenhower, among others, helped shape sports, science, and political life in America.

Given their large numbers in the American colonies following the Revolutionary War, Germans, unlike some other ethnic groups, were accepted early in the development of the country. Consequently, throughout the 1800s and 1900s, there were few bumps along the path toward full assimilation. An exception to this was during World War I, when America went to war with Germany. The anti-German sentiment was strong, but as Sowell (1981) notes, the animus was not restricted to Germans in Germany:

> Anti-German feeling among Americans was not confined to Germany, but extended quickly to the whole German culture and to German Americans, many of whom were sympathetic to their former homeland. German books were removed from the shelves of American libraries, German-language courses were canceled from the public schools, readers and advertisers boycotted German-American newspapers. (p. 65)

Anti-German sentiment returned with World War II; however, it never approached the level of World War I. Also, it was Japanese Americans who caught the ire of patriotic Americans. After World War II, German Americans further assimilated by inter-marriage and their increasing advancement within key institutions in American society. Today, Germans are no longer a distinct segment of the population. In fact, looking back at their history, they have long been considered a significant segment of the White American population.

Italian Americans

Centuries after Christopher Columbus "discovered" the New World, other Italians would take advantage of his discovery by immigrating to the American colonies. Although low in number, Italians were among the earliest immigrants to colonial America. The small numbers were not simply because of the disinterest in immigrating to America. Some jurisdictions, such as Maryland, only allowed the settlement of immigrants from Britain (Iorizzo & Mondello, 2006). But as a result of labor shortages, such laws started to disappear in the colonies. By 1648, Maryland had also changed its practice and passed legislation that "encouraged French, Dutch and Italians to come to its shores" (Iorizzo & Mondello, 2006, p. 26). To further encourage immigration to the colonies, in 1649, the Toleration Act was passed, which ensured religious freedom for Catholics. From the 1600s through the mid-1800s, the immigration from Italy was steady, but, mirroring the trend of other White ethnic groups, really picked up in the late 1800s. Those Italians who immigrated were trying to escape the turmoil in their homeland or simply looking for better economic opportunities. Among them were not only poor people, but various artists and political dissidents who were middle class and others who were revolutionaries. Settling in mostly northern cities, they contributed to the diversity of cities such as Boston, New York, and Philadelphia (Iorizzo & Mondello, 2006).

By 1920, more than 4 million Italians had arrived in the United States. This was not necessarily a welcome development. Leading up to this period, during the late 1800s and early 1900s, heavy anti-Italian sentiment had led to numerous killings and hangings (Marger, 1997). Therefore, to stem Italian immigration to the United States, the Immigration Act of 1924 placed a stringent quota on the number of Italians who could immigrate to the country. In 1929, that number "was only 5,802, compared with 65,721 for British Immigrants" (Feagin & Booher Feagin, 2008, p. 96). Similar to the experience of other ethnic immigrant groups, their religion, Catholicism, also became a point of contention, along with stinging stereotypes, which, as noted in the experience of other ethnic groups, have often been created to demonize new immigrants. Italians were perceived by many to be "dangerous" and "inferior" to other European immigrants. The perception was enhanced by the image of the Italian Mafia (also referred to as the "Black Hand") (Marger, 1997).

The belief that Italians were heavily involved in organized crime likely originated from the fact that many of the immigrants had originated from Sicily, where the mafia was a social institution. However, in America, Italian organized crime became an obsession. The terms *organized crime* and *mafia* became synonymous with Italians. They were considered a lawless race. One congressional report described them as morally deficient, excitable, superstitious, and revengeful (Iorizzo & Mondello, 2006). These negative and racist characterizations were clearly unfair considering that the Irish, German, Jewish, and Polish immigrants had preceded them in organized criminal activity (Iorizzo & Mondello, 2006). In fact, as Sowell (1981) has aptly noted: "Organized crime was an existing American Institution, and the Italian Americans had to literally fight their way into it" (p. 125). Even so, in the early part of the 20th century, Italians had "*lower* crime rates than other Americans" (Sowell, 1981, p. 125; italics added). Although Italians eventually assimilated into American society and are presently subsumed under the White racial category, some of the early stereotypes remain.

Irish Americans

According to Meagher (2005), "The first Irishman came to America in 1584 as part of Sir Walter Raleigh's ill-fated expedition to the Outer Banks of North Carolina" (p. 1). Later, they came in great numbers to America looking for opportunities and escaping extreme poverty in Ireland. Meagher has noted that 60% of those who came in the 17th century did so by way of indentures. Others were given the option of leaving Ireland in place of a prison sentence for a criminal conviction. Those who came in the mid-1800s as a result of the potato famine in Ireland, which killed (through starvation and disease) an estimated 1 million people, contributed to the exponential increase of Irish Americans. For example, during the 100-year period from 1820 to 1920, about 5 million Irish arrived in America (Meagher, 2005). They settled in areas throughout the country; however, many landed in northern states such as New York, Massachusetts, Pennsylvania, and Illinois. In addition, by the early 1860s, one third of the Irish population could be found in the western and midwestern parts of the United States. Wherever the Irish settled, because of the prevailing nativist views and their predominantly Catholic backgrounds (some were Protestant), they often were ostracized and relinquished to the worst areas of cities.

Historians have generally agreed that few immigrant groups have encountered such harsh conditions as did the Irish in 19th-century America. Many of them brought alcohol and fighting habits to American shores. As a result, they often caught the attention of police officials, who called police vans "paddy wagons" because so many Irish were occupants. In some cities, such as New York, the areas where the Irish dominated were some of the toughest (see Highlight Box 1.3). Not until the second- and third-generation families did the Irish truly start to become a part of the American fabric. In fact, the Irish become major contributors to American culture. During the early and mid-20th century, they became major contributors to the arts and were prominently featured in major motion pictures. Nevertheless, they were still faced with challenges.

Restrictive immigration quotas in the 1920s also hit them hard, and there were still barriers in place that restricted them from reaching their full potential occupationally. For example, Irish women, unlike other White ethnic females, had to take jobs as domestic servants to make ends meet. As noted previously with the experience of Black female domestics, these were dangerous jobs that often resulted in sexual harassment, rape, or, out of desperation, a descent into prostitution (Meagher, 2005). Nevertheless, large numbers of them headed to college, and research shows that in the 1920s and 1940s, they were as successful as the native-born European immigrants. By 1960, "Irish occupational status exceeded national averages and was higher than every other white ethnic group except Jews" (Meagher, 2005, p. 132). In short, after initial resistance to their presence in America, the Irish had fulfilled the promise of the "American Dream." It is noteworthy, however, that the swiftness in which the Irish were able to rise out of the doldrums of their early American experience has much to do with the fact that, as time went on, the Irish became integrated into the fabric of American society and assimilated into the status of White Americans (T. Allen, 1994; Ignatiev, 1996).

Highlight Box 1.3

Leonardo DiCaprio at the Tokyo Premiere of *Gangs of New York*

The movie *Gangs of New York* (2002) depicts the immigration of the Irish to New York during a period when there was a strong sense of resentment and hate directed toward immigrants. Largely based on actual events, the movie shows how ethnic antagonism between the native population (English) and newest immigrant group (Irish) resulted in brutal gang wars. The Irish are portrayed as a criminogenic ethnic group who bring their bad habits to an already overcrowded and notorious district of New York. The movie culminates with the "Draft Riots," which were provoked by ethnic tensions and Whites who objected to being drafted into the Union army to fight for the liberation of African American slaves, while they themselves were struggling to survive. Prior to the September 11, 2001, terrorist attack on the World Trade Center buildings, the Draft Riot was considered the single event to have caused the largest loss of life in New York City history (more than 1,000 deaths).

Jewish Americans

Interestingly, the first Jews who arrived in America were of Hispanic origin. In 1654, twenty-three Sephardic Jews from Spain and Portugal arrived in New Amsterdam (Finkelstein, 2007). Their arrival in the New World would begin with controversy. On arrival, they were sued by the captain of the ship that brought them to America because their fares had not been paid. To pay their fares, "The court ordered two of the new arrivals imprisoned and the belongings of all 23 passengers sold at auction" (Finkelstein, 2007, p. 31). Moreover, the governor of New Amsterdam, Peter Stuyvesant, wanted them to leave. In short, he viewed Jews as repugnant and originating from a "deceitful race" (Finkelstein, 2007, p. 31). Stuyvesant was so anti-Semitic that he banned Jews from building a synagogue and restricted their enlistment in the military. Thus, the first American synagogue was not built until the 1720s. Henceforth, Jews began to branch out and started to become somewhat more accepted within American society. This was fostered by the advent of American Freemasonry, in which Christians and Jews interacted. Although discrimination remained a part of the Jewish American landscape, Article VI of the U.S. Constitution, which banned religious discrimination, provided some respite for Jews who aspired for public office.

The 19th century saw a considerable increase in the Jewish presence in America. Whereas there were only 3,000 Jews in America in 1820, forty years later there were 200,000 (Finkelstein, 2007). Tied by religious and cultural traditions, many arrived from Russia, Poland, and other Eastern European countries, where they had long been persecuted for their religious beliefs and customs. To preserve their culture, in 1843, twelve German Jews gathered in a New York café and founded B'nai B'rith, which means "Sons of the Covenant." The mission of the organization was ambitious, but laid the grounds for an organization that, by 1861, was "operating in every major Jewish community in America" (Sachar, 1993, p. 71). The mission of the organization was as follows:

> uniting Israelites in the work of promoting their highest interests and those of humanity; of developing and elevating the mental and moral character of the people of our faith; of inculcating the purest principles of philanthropy, honor, and patriotism; of supporting science and art, of alleviating the wants of the victims of persecution; providing for, protecting and assisting the widow and orphan on the broadest principles of humanity. (Finkelstein, 2007, p. 64)

Recounting Jewish history, Feagin and Booher Feagin (2008) write that, "From the Egyptian and Roman persecutions in ancient times to massacres in Spain in the 1400s, to brutal pogroms in Russia in the 1880s, to the infamous German massacres, Jews might be regarded as the most widely oppressed racial or ethnic group in world history" (p. 113). Seeking relief from persecution in European countries, Jews continued to arrive in America en mass. In the 40 years from 1880 to 1920, 2 million Jews arrived in America. As the persecution continued, many more arrived and eventually assimilated into the American way of life while maintaining their Jewish traditions. However, coinciding with this significant wave of immigration was an increase in anti-Semitism. Describing this turbulent period for Jews, Finkelstein (2007) writes: "Much of this was fueled by the stereotypes brought over from Europe by the large numbers of newly arrived Christian immigrants. Jews faced growing restrictions in housing, employment, and education" (p. 79).

In the first quarter of the 20th century, the mass immigration and squalid living conditions of Jews resulted in abundant numbers of Jewish youth hanging out on the streets. This produced rising juvenile delinquency rates, which became the target of a number of Jewish organizations. In a similar vein, whereas the 1920s and 1930s were periods of considerable Jewish progress, Brodkin Sacks (1997) noted that Jewish success in organized crime was also critical in their upward mobility. She specifically mentioned that "Arnold Rothstein transformed crime from a haphazard, small-scale activity into a well-organized and well-financed business operation. Consider also Detroit's Purple Gang, Murder Incorporated in New York, and a host of other big-city Jewish gangs in organized crime" (p. 399). Such activities were also found among other ethnic groups striving to move up the social ladder albeit through criminality within urban areas.

The period also saw quotas established restricting the number of Jews who could attend prestigious universities such as Harvard. Thus, although they were progressing in terms of their status in American society, there remained barriers to full assimilation. Jews, however, continued to be successful in educational pursuits and small businesses. In 1921, Albert Einstein won the Nobel Prize in Physics, and Jews were among the most successful immigrants. Because of their success in education, Finkelstein (2007) notes that, "By the end of World War II . . . most Jews had established themselves firmly into the middle class, with large numbers employed in 'economically secure' jobs as civil servants. Teachers, accountants, lawyers, and medical professionals" (pp. 129–130). As a result, many moved out of the ghettos and into the suburbs, where they were largely unwelcome. In time, however, Jews assimilated and were also categorized as White Americans (Brodkin, 1999; Brodkin Sacks, 1997).

Each of the aforementioned White ethnic groups came to America seeking prosperity, but was immediately thrust into dire socioeconomic conditions. In many instances, crime provided the means to rise above their condition (Bell, 1960; Light, 1977). Initially, each group was labeled as *criminal*, but after a period of decades, most were able to rise out of their situations and assimilate into America—as White Americans (Gans, 2005).

LATINO AMERICANS

Prior to the 2000 census, the term *Hispanic* was used to refer to persons from Mexico, Puerto Rico, Cuba, and Central and South America. Feagin and Booher Feagin (2008) noted that the term *Latino* emerged because it "recognizes the complex Latin American origins of these groups. It is a Spanish language word and is preferred by many Spanish speaking Americans" (p. 206). Our review of their history focuses on the two largest ethnic groups under the Latino category: Mexicans and Puerto Ricans.

Mexicans

From 1500 to 1853, the Spanish conquered and ruled Mexico. For much of this period, Mexicans were exploited for their labor by the Spanish. Many Mexicans became Americans with the annexation of Texas. Following the Mexican-American War (1846–1848) and the Treaty of Guadalupe Hidalgo (1848), Mexicans had the option to stay in the United States or return to Mexico. According to Feagin and Booher Feagin (2008), although many stayed, others immigrated to America.

Sowell (1981) wrote that Mexicans immigrated to America in three great waves. The first wave of Mexicans came to America by railroad—and ironically, over the years, railroads become one of the largest employers of Mexicans. Specifically, they were employed "as construction workers, as watchmen, or as laborers maintaining the tracks. Many lived in boxcars or in shacks near the railroads—primitive settlements that were the beginning of many Mexican-American communities today" (Sowell, 1981, p. 249). Before World War I, other industries employing Mexicans were agriculture and mining. Mexican workers in America were paid considerably more than they were in Mexico. As a result, there was a steady flow of seasonal workers crossing the Mexican border to earn money to take back home to Mexico. Labor shortages caused by World War I resulted in formalized programs to encourage such practices. About 500,000 Mexicans came to America to work during this period (Tarver, Walker, & Wallace, 2002). Beginning in this period, Mexicans also were subject to negative stereotypes, such as being considered "dirty," "ignorant," and lacking standards of appropriate behavior (Sowell, 1981). Even so, they were tolerated because of the dire need for their labor. With the arrival of the Depression, "Fears of the unemployed created an anti-immigrant movement, and immigration laws were modified to deport the 'undesirables' and restrict the numbers of foreign-contract laborers" (Tarver et al., 2002, p. 54).

About the same time of the notorious Scottsboro cases, the federal government, under the direction of President Herbert Hoover, commissioned the first national crime commission. Commonly referred to as the "Wickersham Report," for its director, George Wickersham, the report, published in 1931, covered almost every aspect of American criminal justice. The report included a review of the state of Mexicans and crime. These reports found that there were varying levels of crime among Mexicans in California and Texas. In general, however, the report noted that, like African Americans, Mexicans were treated with considerable prejudice by the justice system (Abbott, 1931). The report suggested that the criminality of the Mexicans was overstated. There was also brief mention of Filipinos, who were overrepresented in offenses related to gambling, and Japanese, who were "among the most law abiding of all population groups" (p. 415).

The second wave of Mexican immigrants came to the United States during World War II. Another war had resulted in another labor shortage, which produced

> the Bracero Program, which brought in thousands of agricultural workers to help with the labor shortage. *Bracero* is a Spanish term that was used to describe guest workers coming from Mexico to the United States. By the end of the Bracero Program in 1964, 5 million Mexican workers were imported into this country. (Tarver et al., 2002, p. 54)

The third wave of Mexican immigration is tied to the various immigration laws from the 1970s to the present, which have sought to protect, defend, or curtail Mexican immigration to the United States. One such law, the Immigration and Reform and Control Act of 1986, provided temporary residency for some illegal aliens. Furthermore, those who came to America before 1982 were provided with permanent resident status. According to Tarver et al. (2002),

> This act had an enormous impact on Mexican immigration, with 1,655,842 people entering the United States during the decade of the 1980s. Since the first decade of the twentieth century, this was the largest number of immigrants from a single country. (p. 55)

Another law aimed at Mexican illegal immigration is the Illegal Immigration Reform and Immigrant Responsibility Act of 1996. In addition to shoring up the borders in California and Texas, the act "increased the number of investigators monitoring workplace employment of aliens, passport fraud, and alien smuggling" (Tarver et al., 2002, p. 55). Today, the fears concerning illegal immigration continue. Americans believe that the heavy influx of Mexicans is changing the fabric of the country. Besides concerns about job competition and the strain on social services caused by the considerable illegal immigration, Americans have continued their fascination with the perceived connection between immigration and crime (Hickman & Suttorp, 2008; Martinez & Valenzuela, 2006; Stowell, 2007). As you should know by now, this fear-based fascination is not new—it is the American way (Martinez, 2006).

Puerto Ricans and Other Latino Groups

Puerto Rico has a history dating to the early 1500s. The island was colonized by the Spanish in the late 1400s. Not until 1897, however, did Puerto Ricans gain their independence. However, the Spanish-American War resulted in America taking over the island in 1898. In the 1950s, Puerto Rico became a commonwealth of the United States, granting Puerto Ricans more independence in their governance. From 1945 to the 1970s, the high unemployment rate on the island resulted in one in three Puerto Ricans leaving the island (Feagin & Booher Feagin, 2008). Significant numbers of Puerto Ricans headed to New York and other states, such as New Jersey and Delaware. Thus, after having only 2,000 Puerto Ricans in New York in 1900, there was significant Puerto Rican immigration to the United States, which resulted in an increase to 70,000 in 1940 and 887,000 by 1960 (Feagin & Booher Feagin, 2008). Upon their arrival, as with other immigrants who headed to the "promised land," they were faced with high levels of unemployment and poverty. In fact, these dire circumstances result in what has been referred to as *circular migration*. That is, after the opportunities they were seeking did not materialize, Puerto Ricans would head home, but then return because of the lack of opportunities in Puerto Rico. Mirroring the experience of other racial and ethnic groups, over time, Puerto Ricans were also saddled with negative stereotypes, such as "lazy," "submissive," "violent," and "criminal." Moreover, because they cannot always "pass" as White, they have been unable to assimilate like some other ethnic groups. As a result of their varying skin tones and backgrounds, they can be categorized as either White or Black.

Cubans have also been a force among the Latino population. With much of their immigration coming after Fidel Castro's takeover of the government in 1959, they currently number about 1.5 million. Combined, South Americans from the Dominican Republic, El Salvador, and Colombia also represent another 2 million Latinos (Healey, 2006). Given these figures, it is no wonder that Latinos have become the largest minority group in the United States. In the process, they have surpassed African Americans, who have long held that title. They have also, however, suffered from some of the same crime-related concerns as other ethnic groups before them. Notably, however, they have not experienced the same levels of crime and violence as African Americans (Martinez, 2002). This may reflect the fact that many Latinos have come to the United States specifically seeking opportunities for employment, with a willingness to take the most undesirable jobs in the labor market. For many, these jobs provide much more financial compensation than the available employment in the various Latin American countries from which a substantial portion of Latino immigrants originate. Nevertheless, some Latinos have drifted into gangs and other criminal activities as a way to survive in America. Unfortunately, their criminal activities have been exaggerated by the news media and Hollywood, which has resulted in continuing stereotypes (Martinez, Lee, & Nielsen, 2001; see Highlight Box 1.4).

Highlight Box 1.4

Minorities Suspicious of Ethnic Groups: Poll Finds Stereotypes, Resentment Are Mutual

By Lesley Clark
McClatchy Newspapers
Thursday, Dec. 13, 2007

WASHINGTON: The nation's three largest minority groups—blacks, Hispanics and Asians—view one another with deep suspicion, though there's evidence that the divide could be bridged, a new poll finds.

Billed as the first of its kind, the nationwide poll of 1,105 blacks, Hispanics and Asians found that all three groups held negative stereotypes of one another—though in some cases, a majority or nearly as many respondents rejected such beliefs.

Pollster Sergio Bendixen said the mixed results "reflect the extent to which the poll is capturing not a static picture, but a racial landscape in flux."

The poll, conducted in August and September, was sponsored by New America Media, an association of more than 700 ethnic media outlets. It follows an uproar over a San Francisco Asian weekly's decision in February to publish a column titled, "Why I Hate Blacks."

The ensuing furor sparked a community forum and a decision to try to better understand the tension among the largest ethnic groups in the United States, said Sandy Close, executive editor and director of NAM.

"It's a small opening in an otherwise convoluted, complicated landscape of 'Can we all get along?' Close said of the poll. "We see this information as better to know—and discuss—than not to know."

The poll found that friction among ethnic and racial groups is "rooted in the mistrust that the groups harbor towards each other," as well as a belief that the other groups are "mistreating them or are detrimental to their own future."

More than seven in 10 respondents consider "racial tension" to be an important problem, and many subscribe to racial or ethnic stereotypes that Bendixen said he discovered in earlier focus groups.

For example, the poll found that 44 percent of Hispanics and 47 percent of Asians said they're "generally afraid of African-Americans because they are responsible for most of the crime."

Bendixen noted that half of Hispanics rejected the statement, as did 45 percent of Asians.

The poll also found that just more than half of blacks feel threatened by Hispanic immigrants, agreeing with the statement that "they are taking jobs, housing and political power away from the black community." But 45 percent disagreed. Only 34 percent of Asians believe Hispanics are displacing blacks.

But majorities of Hispanics and blacks believe that "most Asian business owners do not treat them with respect."

Moreover, the three groups appear more trusting of whites than of one another, with majorities in each group saying they felt "more comfortable doing business" with whites than with any of the other minority groups.

The poll showed high levels of "ethnic isolation" among the groups, with a majority of each group reporting that most of their friends, neighbors and people they associate with are of the same ethnic background.

SOURCE: Clark (2007).

ASIAN AMERICANS

Asian Americans provide another interesting case study of ethnic group acculturation in America. Like Latinos, they belong to a number of ethnic groups, such as Filipino, Korean, Japanese, and Vietnamese. Table 1.4 provides an overview of the population for the various Asian American groups. We begin our review with a brief discussion of the Chinese American experience.

Chinese Americans

According to Daniels (1988), there were Chinese in America as early as the late 1700s. However, not until the California gold rush of the mid-1850s was there any significant Chinese immigration to America: Between 1849 and 1882, nearly 300,000 Chinese came to America (Daniels, 1988). However, the Chinese Exclusion Act of 1882 limited immigration until the 1940s (Tarver et al., 2002).

Most of the early Chinese immigrants were male (90%) and came to work in America temporarily. However, they came in significant enough numbers to represent

Table 1.4   Asian American Population

Classification	Number	Percent*
Asian	10,242,998	3.6
Asian Indian	1,678,765	.6
Chinese	2,432,585	.9
Filipino	1,850,314	.7
Japanese	796,700	.3
Korean	1,076,872	.4
Vietnamese	1,122,528	.4
Other Asian	1,285,234	.5

SOURCE: U.S. Bureau of the Census (2000).

*This percentage is out of the total U.S. population.

nearly 10% of California's population between 1860 and 1880 (Daniels, 1988). Those who did stay were subjected to considerable violence due to anti-Chinese sentiment. Chinatowns had existed since the arrival of the Chinese in America; they embraced these areas because there they were free to maintain their own culture without fear of hostility—although some areas occupied exclusively by Chinese inhabitants were "shabby looking, vice-infested, and violence prone" (Sowell, 1981, p. 141).

The Chinese were quite successful as laborers as well as in independent businesses such as restaurants and laundries (Daniels, 1988; Sowell, 1981). Yet as with other immigrant groups, the Chinese were not immune from engaging in illegal activities. Daniels (1988) wrote that prostitution and gambling flourished in the "bachelor society" created by the dearth of Asian women in America. In 1870, "More than 75% of the nearly 3,000 Chinese women workers in the United States identified themselves as prostitutes" (Perry, 2000, p. 104). Brothels and opium-smoking establishments became popular among both Asians and Whites. Regarding opium use among early Chinese immigrants, Mann (1993) suggested that 35% of the Chinese immigrants smoked opium regularly, which "led to the first national campaign against narcotics" (p. 59), and the subsequent legislation was aimed at "excluding Chinese participation in American society" (p. 59). On the participation of the Chinese in these illegal activities, Daniels (1988) noted, "Since all of these activities were both lucrative and illegal, it seems clear that police and politicians in the White community were involved in sanctioning and profiting from them" (p. 22).

Eventually, following the pattern of other immigrants, Asian organized crime emerged, and secret societies such as "tongs" were formed. Describing these organizations, Perry (2000) indicated that such societies were originally created to assist Asian men in adjusting to America. But, as Perry notes, over time, many evolved into criminal organizations or developed links with Chinese triads. Consequently, the tongs came to dominate prostitution, along with gambling, drugs, and other vice crimes. So, in addition to providing sexual outlets, they also created other opportunities for recreation and escapist behavior. Despite the profits reaped by Whites from the legal and illicit activities of the Chinese, heavy anti-Chinese sentiment persisted in California, which led to numerous negative campaigns against the population. Pointing to the roots of this negative sentiment, Sowell (1981) wrote, "The Chinese were both non-White and non-Christian, at a time when either trait alone was a serious handicap. They looked different, dressed differently, ate differently, and followed customs wholly unfamiliar to Americans" (pp. 136–137). Once they began to receive jobs in competition with Whites, they became targets of increasing violence and, in several instances, were massacred. By and large, the Chinese were generally relegated to the most menial and "dirty" occupations, such as mining, laying railroad tracks, and agricultural work. As a result of the Chinese Exclusion Act of 1882, unlike other ethnic minorities, the Chinese population decreased from the late 1880s through the mid-1940s. Since then, their numbers have increased, and they have remained the largest segment of the Asian American population. Until the last 30 years of the 20th century, Japanese Americans represented the second-largest group among Asians in the United States. Several other Asian groups have now surpassed them in population (most notably, Filipinos). We review the Japanese American experience below.

Japanese Americans

Before arriving on the shores of North America in the last quarter of the 19th century, a considerable contingent of Japanese workers (30,000) arrived in Hawaii. They were contract workers who came to the island to provide much-needed labor for sugar plantations and "to serve as a counterweight to the relatively large number of Chinese in the islands" (Daniels, 1988, pp. 100–101). Like the Chinese before them, the Japanese also arrived on North American shores as a result of labor needs, and the relatively small number of Japanese men who made it to America (about 2,200 by 1890) filled the continuing need for laborers on California farms (Daniels, 1988). Like the Chinese and other groups, some Japanese immigrants turned to illicit activities, such as prostitution and other petty crimes, to survive.

Over time, the number of Japanese in America began to increase, with 24,326 in 1900, 72,157 in 1910, and nearly 127,000 by 1940. Mirroring the experience of the Chinese, anti-Japanese sentiment arose in the United States, culminating with the arrival of World War II. During World War II, negative sentiment toward the Japanese reached new heights; they were hated and mistrusted by many Americans. Once the attack on Pearl Harbor commenced, in December 1941, life for Japanese Americans would never be the same. In February 1942, President Roosevelt issued Executive Order 9066 (Dinnerstein & Reimers, 1982). The order, which was upheld by the Supreme Court, required that all Japanese from the West Coast be rounded up and placed in camps called *relocation centers*. In all, about 110,000 were rounded up on 5 days' notice and were told they could take only what they could carry. The camps were nothing more than prison facilities with armed military police on patrol watching for escapes.

Following the war, the Japanese population remained low in the United States due to immigration restrictions that were not lifted until the 1960s. At that time, Japanese Americans represented 52% of the Asian American population. However, over the next 20 years, the number of Japanese who immigrated to America declined. This trend was largely a result of the increased need for labor in Japan, which stunted the immigration of the Japanese to America (Takaki, 1989). For those Japanese who were already here or among those who came after stringent quotas were lifted in the 1960s, they would go on to become some of the most successful immigrants. Today, economic indicators related to income and unemployment levels all reveal a positive trend for Japanese Americans. Nonetheless, Japanese Americans "still face exclusion from certain positions in many business, entertainment, political, and civil service areas, regardless of abilities" (Feagin & Booher Feagin, 2008, p. 293). Two other Asian groups whose numbers have increased over the last few decades are Filipinos and Koreans. We provide brief overviews of their American experiences in the next section.

Filipinos and Koreans

Filipinos have been in the United States since the 1700s. But, as you might expect, much of their most significant immigration to the United States occurred in the 19th and 20th centuries. Many headed to plantations in Hawaii due to labor shortages. Unfortunately, when they arrived in America, they encountered violent attacks from Whites. In California, they competed with White farm workers; besides receiving lower wages than their counterparts, they were the target of continuing violence. In fact, in 1929 and 1930, there were brutal riots that were brought on by anti-Filipino sentiment (Feagin & Booher Feagin, 2008). During this same time, the 1924 immigration law restricted the number of Filipinos that could enter the country to 50 (Kim, 2001). Since this early period, although their population has increased precipitously (an estimated 4 million in 2007), they remain the targets of violence, and in post-9/11 America, some have been targeted as potential terrorists.

Like Filipino Americans, Korean Americans headed to Hawaii in the early part of the 20th century to fill labor shortages. Koreans also followed other Asian groups to California. In the case of Koreans, the place of choice was San Francisco. Limited by immigration restrictions, much of Korean immigration followed World War II. Not until 1965 were the stringent immigration restrictions lifted. This policy change coincided with more Koreans (mostly from South Korea) arriving in America. Looking for opportunities, Koreans headed to inner-city communities, where they set up dry cleaners and convenience stores. Unfortunately, the relations between Koreans and urban residents are tenuous at best, but are better described as resentful for entering largely African American communities and "setting up shop," as some have noted. Therefore, besides feeling that they were mistreated by clerks in Korean establishments, some have felt that such businesses should be owned by community members. This sentiment spilled over in the Los Angeles riot of 1992 (Kim, 1999). There, tensions remain between the two communities, but the dialogue continues. In 2007, Korean Americans again received negative attention because the perpetrator of the Virginia Tech massacre was an immigrant from South Korea (see Highlight Box 1.5).

Highlight Box 1.5

Madman, not Koreans, to Blame for Shootings

By DeWayne Wickham

It seems this has to be said. What happened at Virginia Tech University had nothing to do with Korean-Americans. The carnage loosed by Seung Hui Cho had nothing to do with his national origin and everything to do with his dementedness.

Cho, who was born in South Korea but lived 15 of his 23 years in the USA, was a madman who took the lives of 32 people in a senseless spasm of violence before committing suicide. His victims were randomly chosen. They were black and white, Asian and Hispanic, native- and foreign-born.

The guilt for his actions is his alone to bear. But sadly, there are some in this country who think otherwise—people incapable of seeing Cho as simply a deranged young man.

After the Virginia Tech gunman's identity became known, some Korean schoolchildren in Los Angeles were taunted with chants of "go back to Korea," said Eun Sook Lee, executive director of the National Korean-American Services & Education Consortium.

That was one of many acts of misplaced rage that Korean-Americans have been subjected to in the wake of the Virginia Tech massacre.

Incidences abound

Lee told me that a Korean-American woman who expressed concern about poor service at a Los Angeles post office said she heard someone respond contemptuously: "Don't go shooting up this place."

In upstate New York, the Korean owner of a dry cleaner in a white neighborhood said not a single customer came into his business the day after Cho was identified as the man responsible for the carnage at Virginia Tech. Korean-American residents of a San Francisco senior citizen building reported being told by some other residents to "go back home." A maintenance worker in that building said to Korean residents that when he returned to work on Monday, "I hope you are all not here." And Lee said she has also gotten reports from Korean-Americans in Michigan of the windows being smashed on Hyundai cars, which are made in Korea.

While these bad acts should pain us, they shouldn't surprise us. Many Muslim Americans were the targets of hateful acts after the 9/11 attacks. In both cases, the people behind this intolerance are far from rational thinkers.

Not responsible

Anyone who takes out his anger over the shootings at Virginia Tech on people who had nothing to do with that senseless act makes little sense himself. Fearing just such a reaction, South Korean President Roh Moo Hyun has held an emergency Cabinet meeting and several times has offered his condolences to the shooting victims, their families and the American people.

Cho's family was just as contrite. His uncle told the Associated Press "I sincerely apologize" for his nephew's killing spree. Cho's sister, Sun Kyung Cho, also said in a statement that her family "is so very sorry for my brother's unspeakable actions." He "has made the world weep," she said, and her family prays for the "loved ones who are experiencing so much excruciating grief."

It is this grief, and not the misguided efforts to blame other Korean-Americans for Cho's crimes, that should move us to action. We need to do more to keep guns out of the hands of those who are mentally deranged—and to see that those who are mentally ill get the treatment they need to keep them and others safe.

(Continued)

(Continued)

"We are as shocked and as grieving as anyone else by his actions," Lee said of Cho's shooting rampage. But as troubled as she is by the scattered reports of people blaming other Koreans for what Cho did, Lee doesn't want her concerns to deflect attention away from those who were killed or wounded. "We should not be considered a victim," she said. "The victims were at Virginia Tech. But at the same time, we should not be seen as the indirect perpetrators of that attack either."

Of course not. Seung Hui Cho was a mass murderer. He—and he alone—should be reviled for what happened at Virginia Tech.

SOURCE: USA TODAY, April 24, 2007, p. 11A.

In closing, the difference between Asians and ethnic groups who came to be classified as White is that, although they have attained high levels of achievement, Asians have never fully assimilated. This leaves them, as one author put it, "as perpetual outsiders" (Perry, 2000). Like African Americans, Native Americans, and some Latinos, Asian Americans have maintained a distinct racial categorization. Gould (2000) has suggested that physical characteristics unique to their race (e.g., skin color, facial characteristics, size) have barred them from full assimilation and acceptance in America.

In recent years, Asian Americans have been labeled the *model minority* because of their success in education. Some see their success as proof that all groups can succeed if they "put their best foot forward." Others see this label as problematic (Wu, 2002), noting that all Asians are not equally successful. For example, as Perry (2000) noted, "Koreans and Vietnamese consistently lag behind Chinese, Japanese, and Asian Indians on most indicators of socioeconomic status" (p. 100). Furthermore, the continuing discrimination in employment, income, and education are masked by such a label (Perry, 2000). Nevertheless, over the last century, Asian Americans have been a productive force in the United States.

Photo 1.1   Japanese American internees await processing in 1942.

Conclusion

Since the categorization of races in the late 1700s, societies have, unfortunately, used the social construct to divide populations. In America, the notion of race was not of considerable use until the 1660s, when color was one of the deciding factors in the creation of the slave system. It was at this time in history that the category "White" began to take on increased importance.

Along with "Whiteness" came racism, which justified the system from the point of view of the dominant population. For the next two and a half centuries, as more White ethnic immigrants came to the United States looking for opportunities, they were looked down on as well. However, at some point, each group was allowed to fully assimilate and truly "become White," and over time, the stereotypes with which they had been identified eventually dissipated (see Table 1.5). In the case of Native Americans, African Americans, Asians, and Latinos, however, this process has been more difficult because of distinct physical traits that have limited their ability to fully assimilate.

Our review of the historical antecedents of race and crime in America has revealed that, over the past few centuries, although the level of crime in each group has varied over time, most racial/ethnic groups have committed the same kinds of offenses and have had similar offenses perpetrated against them by the dominant culture. Initially, Whites criminally brutalized Native Americans and African Americans. As time went on, ethnic immigrants such as the Germans, Italians, Irish, Jews, Asians, and Latinos also were subjected to harsh treatment and sometimes violence. As these "White ethnic" groups assimilated into the populace, they, in turn, became part of the oppressive White population, continuing at times to engage in racial violence against other minority groups.

In short, the history of race and crime in America is a story of exploitation, violence, and, in the case of most racial/ethnic groups, the common use of crime as a way to ascend from the lower rungs of American society. The next chapter examines official crime and victimization data for the various races.

Table 1.5   Early Stereotypes of Racial and Ethnic Minorities* (1600s–1900s)

Native American	Irish	Jewish	African American	Mexican American	Puerto Rican	Chinese/ Japanese
childlike	temperamental	too intelligent	bad odor	lazy	emotional	devious
cruel	dangerous	crafty	lazy	backwards	lazy	corrupt
thieves	quarrelsome	clumsy	criminal	lawless	criminal	dirty
wild beasts	idle		apelike	violent		crafty
exotic	apelike			shiftless		docile
powerful				improvident		dangerous

SOURCE: Adapted from Feagin and Booher Feagin (2008) and Sowell (1981).

*All groups were thought to be "biologically inferior" to the native White population.

Discussion Questions

1. Explain the origin of race and its implications for race and crime.

2. What is the significance of the increasing racial and ethnic population for race and crime in America?

3. What role has the law played in the experiences of the groups portrayed in the chapter? Provide some examples using specific laws.

4. How does slavery intersect with the study of race and crime?

5. What role does "Whiteness" play in understanding race and crime?

Internet Exercise

Visit the U.S. Bureau of the Census Web site (http://www.census.gov) and review the many definitions of race. Provide another logical way of categorizing the various races.

Internet Site

U.S. Bureau of the Census: http://www.census.gov

Extent of Crime and Victimization 2

Present statistics which pretend to report the criminal behavior of minority ethnic and racial groups both reflect and perpetuate a large number of errors and myths, which can be, in their most innocent form, misleading, and, in their least innocent, both vicious and malevolent.

—Geis (1972, p. 61)

Race and crime have been inextricably linked throughout American history. As noted in Chapter 1, early stereotypes of some Americans implied notions of their criminality. Over time, beliefs about the inferiority and criminality of certain groups, including African Americans, Native Americans, White immigrants, and others, fostered the eugenics movement of the early 20th century and the "law and order" campaigns that came later. Although not created to do so, crime data often are used to support (erroneous) beliefs about minorities and crime. Between 1850 and the early 1900s, census data about convicted persons were the primary source of criminal statistics. At that time, unlike today, distinctions were made between foreign and native-born convicts. Most foreign convicts were immigrants from European countries and were classified based on their place of origin (French, German, etc.). Despite opposition, in 1930, the U.S. Congress mandated that the Federal Bureau of Investigation (FBI) collect and report crime data. By the 1960s, the increase in crime recorded in the FBI's Uniform Crime Reports (UCR)—especially in urban areas—was used to justify the implementation of more punitive crime control policies for more than two decades. Today, there are several sources of statistics on crime and victimization (see Table 2.1). Most are either funded and/or collected by federal agencies. The Bureau of Justice Statistics (BJS) is the primary statistical agency in the U.S. Department of Justice (DOJ). Other DOJ agencies, including the FBI and the Office of Juvenile Justice and Delinquency Prevention (OJJDP), provide statistical reports as well. Unlike most nations, the United States continues to include racial categories in crime and victimization data.

Table 2.1   Sources of Crime and Victimization Statistics

Source	Sponsor	Inception	Methodology	Scope	Race Included
Crime in the United States (UCRs)	FBI	1930	Crimes reported to the police	Crimes reported Crimes cleared Persons arrested Law enforcement personnel	Yes
National Crime Victimization Survey (NCVS)	Bureau of Justice Statistics (BJS)	1972	Interviews	Victims Offenders Offenses	Yes
Sourcebook of Criminal Justice Statistics	BJS	1973	Compendium of statistics	Criminal Justice System Public Attitudes Arrests Judicial Processing Prisoners	Yes
Law Enforcement Management and Administrative Statistics (LEMAS)	BJS	1987	Surveys of police agencies	Personnel Expenditures Operations Equipment	Yes
National Incident-Based Reporting System (NIBRS)	FBI	1988	Crime incidents	Victims Offenders Offenses Circumstances	Yes
Hate Crime Statistics	FBI	1991	Crimes reported to the police	Victims Offender Motivation Location	Yes

Although the majority of persons arrested in the United States are White, crime sta-tistics indicate disproportionality for some racial and ethnic minorities. This means that some minorities comprise a larger proportion of persons arrested and victimized in comparison to their representation in the population. For example, although African Americans make up about 13% of the population in the United States, in 2006, they comprised 30.5% of all persons arrested, 56.2% of persons arrested for murder, 53.3%

of persons arrested for robbery, and 36.2% of persons arrested for motor vehicle theft (Federal Bureau of Investigation, 2006). Some of these arrest patterns persisted throughout the 20th century and into the 21st century.

Victimization data show that American Indians and African Americans have the highest violent victimization rates (Harrell, 2007). Most troubling is the fact that, according to the National Crime Victimization Survey (NCVS), only about half of violent crime victimizations that occur are reported to the police (Hart & Rennison, 2003; Rand & Catalano, 2007). Although the prospect of there being more crime than we can ever measure or know is daunting, it must be remembered that the majority of Americans are neither arrested nor victimized regardless of race. In 2006, the population of the United States estimated in the UCR was 299,398,484, and the estimated number of persons arrested was 13.7 million (Federal Bureau of Investigation, 2006). Of persons who are arrested, regardless of their racial categories, fewer are arrested for murder, rape, robbery, or aggravated assault, which are considered the most serious crimes against the person, than for property offenses, most notably larceny/theft. In that same year, there were an estimated 25 million victims of crimes of violence and theft (Rand & Catalano, 2007). Victims are more likely to report property victimizations (burglaries and thefts) than violent crime victimizations (aggravated and simple assaults, rape/sexual assaults, robberies).

As the quote at the beginning of this chapter suggests, crime statistics can be misleading. Despite the fact that most Americans are not involved in crime, over time, both crime and the administration of justice have become racialized (Covington, 1995; Keith, 1996). Covington (1995) used the concept of racialized crime to describe the process of generalizing the traits, motives, or experiences of individual Black criminals to the whole race or communities of noncriminal Blacks. The concept also can be applied to other minorities. For example, according to the *2005 National Gang Threat Assessment* (Bureau of Justice Assistance, 2007), 49% of gang members are Latinos. This might lead some to think that most Latinos are gang members, especially in communities with large Latino populations. Although crime statistics were not initially designed to label certain groups of people as criminals, this is exactly what has occurred. The overrepresentation of some racial and ethnic minorities in arrest data, especially for violent crimes, has led to misperceptions about race and crime.

In this chapter, we present a brief overview of the history of crime and victimization statistics, some limitations of crime and victimization data, and analyses of data that provide information about race and crime. Here, we focus on two major sources of statistics: the UCR, compiled by the FBI and entitled *Crime in the United States*, and the NCVS, an annual publication funded by the DOJ (BJS) and compiled by the U.S. Bureau of the Census entitled *Criminal Victimization*. The goal of the chapter is to assess arrest and victimization trends in the United States to better understand what they do and do not tell us about race and crime. We also seek to clarify some arrest and victimization trends during the 20th and early 21st centuries. Finally, we summarize data from the FBI National Hate Crime Data Collection Program. At the outset, we acknowledge that there are variations by race, class, and gender that are often overlooked and difficult to assess with currently available crime and victimization data sources. First, an overview of the history of collecting crime and victimization data in our country is presented.

History of Crime and Victimization Statistics in the United States

The history of crime statistics in the United States dates back to the 19th century. Robinson (1911) was one of the first to analyze crime statistics and noted,

> The purpose of criminal statistics is two-fold: (1) that one may judge of the nature and extent of criminality in a given geographical area, and (2) that one may determine the transformation, if any, which is occurring in these two phases. The results, when known, may give direction to many movements of one kind or another, but the purpose of the statistics is to furnish these two sets of data. Their application is another question. (pp. 27–28)

Some states started collecting crime data in the early 19th century. Several state legislatures mandated the collection of statistics on crime and criminals in two categories: judicial and prison statistics. Judicial statistics included information on persons appearing before the courts and their offenses. New York (1829) and Pennsylvania (1847) required clerks of the courts to submit transcripts or statements of convictions and/or criminal business. In 1832, Massachusetts mandated that the attorney general report his work and the work of the district attorneys to the legislature. In Maine (1839), county attorneys were required to report the number of persons prosecuted and their offenses to the attorney general, who also was required to submit a report to the governor. Twenty-five states legislated the collection of judicial criminal statistics between 1829 and 1905 (Robinson, 1911). State prison statistics of criminals were collected in two ways: Sheriff and prison officials sent information to either the secretary of state or a state board of charities and corrections. Massachusetts required reports as early as 1834; most other states mandated prison statistics later in the 1800s or early in the 20th century. Robinson described these statistics as less comprehensive than judicial statistics.

At the federal level, the collection of crime statistics was the responsibility of the U.S. Bureau of the Census during the 1800s. Beginning in 1850, the sixth census of the population required U.S. Marshals to collect population statistics for free inhabitants of jails and penitentiaries. In addition to counting inmates, data were collected on the sex, age, nativity, and color (of the native-born) of convicted persons and prisoners (Robinson, 1911). For the native-born, the categories of Native, White, and Colored were included. Knepper and Potter (1998) attributed distinctions between foreign-born and native-born criminals to immigration in the last two decades of the 1800s. A shift from nativity to biological conceptions of race in crime statistics paralleled developments in eugenics and criminal anthropology (Knepper, 1996; Knepper & Potter, 1998). According to Robinson (1911), federal crime statistics really began with the 1880 census, when prisoner, judicial, and police statistics were collected and published for the first time. Noted penologist Frederick H. Wines was given the responsibility of revising the crime statistics for the 1880 census. Since 1880, there have been several changes in the collection of census crime statistics, but, for the most part, they continue to be prison criminal statistics.

In 1870, Congress passed a law requiring the attorney general to report statistics on crime annually, including crimes under federal and state laws. A year later, in 1871, the organizing conference that created the National Police Association (now known as the International Association of Chiefs of Police) called for "crime statistics for police use" (Maltz, 1977, p. 33). The 1880 and 1890 censuses collected police statistics, although they were not reported.

By the 1880s, lynching, a form of social control, was a crude, cruel, and often ritualized form of murder, especially in the South. Due to prevailing attitudes about both race and crime, lynching was not always viewed as a crime, and lynchers were not always viewed as criminals. There is no mention of lynching in the early historical analyses of crime statistics (see e.g., Maltz, 1977; Robinson, 1911; Wolfgang, 1963). Data on lynchings were collected by the *Chicago Tribune*, and in 1882, the newspaper began publishing the number of persons lynched and the reasons why (Perloff, 2000). During the early 20th century, Tuskegee Institute and the National Association for the Advancement of Colored People (NAACP) also collected lynching data.

Between 1880 and 1891, it is estimated that 100 Negroes (the classification at the time) were lynched each year (Perloff, 2000). Reported lynchings appeared to be most frequent between 1884 and 1901 (Raper, 1933; Zangrando, 1980). In 1892, the largest number of lynchings (255) was reported, and in 1932, the lowest number (8) was reported. Between 1889 and 1932, there were 3,745 lynchings reported; 2,954 were Negroes, and 791 were Whites (Raper, 1933) (see Table 2.2). Between 1882 and 1968, there were 4,742 lynchings recorded, and 3,445 (73%) victims were Black (Perloff, 2000). It remains unclear how lynchings were recorded in the major source of crime data described next.

The Uniform Crime Reporting Program

The history of the FBI's Uniform Crime Reporting program began in 1927, when a subcommittee of the International Association of Chiefs of Police (IACP) was charged with the task of studying uniform crime reporting. In 1930, the FBI began collecting data from police departments. At the time, there was considerable debate about what data should be collected, the responsible federal agency, and the reliability of the data (Maltz, 1977). Wolfgang (1963) noted that Warner (1929, 1931) was opposed to the federal government collecting statistics on crimes known to the police. Warner (1931), in a report on crime statistics for the National Commission on Law Observance and Enforcement (the Wickersham Commission), recommended that police crime statistics not be collected by the federal government. He believed that the UCRs would do more harm than good because they were both inaccurate and incomplete. By publishing crime statistics, the federal government would give credence to the UCRs, which, in turn, would influence public opinion and legislation.

Despite support for the U.S. Bureau of the Census to collect crime data, in 1930, the FBI was mandated by Congress to compile crime data collected by the police. Reports were issued monthly (1930–1931), quarterly (1932–1941), semi-annually with annual

Table 2.2   Reported Lynchings, Whites and Negroes, 1889–1932*

Year	Whites	Negroes	Total
1889	81	95	176
1890	37	90	127
1891	71	121	192
1892	100	155	255
1893	46	155	201
1894	56	134	190
1895	59	112	171
1896	51	80	131
1897	44	122	166
1898	25	108	127
1899	23	84	107
1900	8	107	115
1901	28	107	135
1902	11	86	97
1903	17	86	103
1904	4	83	87
1905	5	61	66
1906	8	65	73
1907	3	60	63
1908	7	93	100
1909	14	73	87
1910	9	65	74
1911	8	63	71
1912	4	61	65
1913	1	50	51
1914	3	49	52
1915	13	54	67
1916	4	50	54
1917	2	36	38
1918	4	60	64

Year	Whites	Negroes	Total
1919	7	76	83
1920	8	53	61
1921	5	59	64
1922	6	51	57
1923	4	29	33
1924	0	16	16
1925	0	17	17
1926	7	23	30
1927	0	16	16
1928	1	10	11
1929	3	7	10
1930	1	20	21
1931	1	12	13
1932	2	6	8
TOTAL	791	2,954	3,745

SOURCE: Raper (1933), see Appendix C (in original source), pp. 448–480.

NOTE: The deaths from the race riots in Atlanta, Chicago, Tulsa, and other places are not classified as lynchings; neither are gang murders in southern or northern cities. As popularly defined, a chief distinction between a lynching and a gang murder is that whereas the latter is premeditated and carried out by a few people in conspired secrecy from constituted authorities, the former is usually spontaneous and carried out in public fashion with scores, hundreds, and not uncommonly thousands of eyewitnesses. Gang murderers—like other murderers—operate in secrecy to evade the law; lynchers operate in the open and publicly defy the law.

*Data secured from *The Negro Year Book* (1931–1932, p. 293), and from materials subsequently secured from the Department of Records and Research, Tuskegee Institute.

accumulations (1942–1957), and annually beginning in 1958. UCR contributors compile information on crimes reported, cleared, persons arrested, and law enforcement personnel, and they forward it on to either a state UCR program or directly to the FBI. Seven offenses comprised the crime index—murder/nonnegligent manslaughter, forcible rape, robbery, aggravated assault, burglary, larceny/theft, and motor vehicle theft—until 1979, when arson was added. The UCR program's Supplementary Homicide Report (SHR) provides information about murder victims, offenders, and incidents.

During its history, several changes have occurred in the UCR program. In 1957, a Consultant Committee on Uniform Crime Reporting made 22 recommendations for changes, although only 2 were implemented including changes in statistical presentation and revision in classification of the crime index (Wolfgang, 1963). In 1980, the UCR renamed arrest categories to include American Indian or Alaskan Native and Asian or Pacific Islander. During the 1980s, efforts began to modernize the UCR program, which culminated in implementation of the National Incident-Based Reporting System (NIBRS) in 1988. According to the Federal Bureau of Investigation (1998), the NIBRS was created to enhance the quantity, quality, and timeliness of crime statistical data collected by the law enforcement community and to improve the methodology used for compiling, analyzing, auditing, and publishing the collected crime data.

Data are collected on eight index crimes and 38 other offenses, victims, offenders, and circumstances of crime incidents. Unlike the UCR, the NIBRS records each offense occurring, attempted and completed crimes, and weapons information for all violent offenses (Rantala & Edwards, 2000). It also includes information on multiple victims, offenders, and crimes that are part of the same incident. Today, only 19 states and all local agencies in only 3 states (Idaho, Iowa, and South Carolina) participate in the NIBRS. Thus, they are not representative of the nation as a whole (Finkelhor & Ormrod, 2004). Although informative, NIBRS data are weakened by the low participation of police agencies and most states.

Another addition to data collected and reported by the FBI is hate crime statistics. A year after President George H. W. Bush signed the Hate Crime Statistics Act into law in April 1990, the FBI developed the National Hate Crime Data Collection Program (NHCDCP) and began reporting hate crime statistics in 1992 (Nolan, Akiyama, & Berhanu, 2002).

Photo 2.1 At its headquarters, 69 Fifth Avenue, New York City, the NAACP flew a flag to report lynchings until, in 1938, the threat of losing its lease forced the association to discontinue the practice.

Hate/bias crime statistics include information on victims, offenders, and incidents for eight index crimes, as well as simple assault, intimidation, and destruction/damage/vandalism. An incident is considered a hate crime whenever the facts indicate actions by an offender were motivated by bias against certain individuals or groups. The number of police agencies participating in the NHCDCP has increased since 1991. In 2006, 12,620 agencies were included and represented about 85.2% of the total U.S. population (Federal Bureau of Investigation, 2006). Hate/bias crime information is available in both the FBI UCR program and the NCVS discussed next.

Victimization Surveys

One of the earliest victimization surveys was conducted by the National Opinion Research Center for the President's Commission on Law Enforcement and the Administration of Justice in the 1960s. Letting citizens self-report their victimizations was viewed as another way to determine the extent of crime, especially crime that never came to the attention of the police. Today, the NCVS is the primary source of information on victims of crime in the United States. In 1972, the Law Enforcement Assistance Administration implemented the NCVS, originally known as the National Crime Survey. Since its inception, the U.S. Bureau of the Census has administered the NCVS. Initially, there were four surveys, a complex survey design, sampling, and estimating schemes. Criminal victimization reports have been issued annually since 1973 and are now prepared by BJS statisticians.

The early surveys included about 100,000 persons and 50,000 households in 26 cities. Information about victims, offenders, and reporting was collected from a representative sample of households. Interviews of persons 12 years old and over identifies rape, robbery, assault, personal and household larceny, burglary, and auto theft victimizations. The NCVS includes both attempted and completed crimes that "are of major concern to the general public and law enforcement authorities" (Bureau of Justice Statistics, 1992, p. iii). Initially, businesses were included, but were dropped, along with city surveys of individuals, in the 1970s (Mosher, Miethe, & Phillips, 2002). The NCVS does not collect information on homicides, although it often includes data on homicide victims. Victimization reports are more detailed than UCR arrest data and until recently included several tables that allowed for comparisons by race, gender, and age. Information about age, gender, and race in recent annual reports is limited and less useful.

The NCVS also has undergone changes since its inception. In 1992, it was renamed the NCVS after a redesign of the survey was implemented (Kindermann, Lynch, & Cantor, 1997). The Crime Survey Redesign Consortium recommended numerous changes in the crime survey to improve its accuracy (B. Taylor, 1989). The redesign specifically focused on improving the collection of data by expanding the capability of the survey to prompt recall by respondents. In 2000, questions were added to the NCVS to identify victims of hate crimes (Harlow, 2005).

Enhanced screening questions are believed to have improved recall of respondents about domestic violence, rape, and sexual attacks, which has led to higher estimates of some victimization rates (Kindermann et al., 1997). Approximately 87,000 households and 159,000 persons were interviewed in 2000 (Mosher et al., 2002). In 2006, 135,300 persons and 76,000 households were included, and the NCVS sample was redrawn to include households in the 2000 census, resulting in a sample of first-time interviews. Unlike in the past, first-time interviews that are not comparable to interviews completed in previous years are included in the 2006 report due to insufficient funding (Rand & Catalano, 2007). In this chapter, victimization trend analyses include data from 1993 to 2005 and the 2006 reported victimizations.

Limitations of Arrest and Victimization Data

Since their inception, the problems of various sources of crime statistics have been acknowledged. One criticism of crime statistics is that they are unreliable because they cannot tell us how much crime takes place, how many persons were arrested, or how many crime victims there are. At most, arrest statistics are no more than "descriptions of the persons who, for a veritably endless array of reasons (many of which are beyond our knowledge) are subjected to arrest" (Geis, 1972, p. 65). As previously noted, there was opposition to the collection of police crime statistics early in the 20th century due in part to their limitations. Victimization surveys have several deficiencies as well (Mann, 1993; Mosher et al., 2002). Table 2.3 provides a summary of the limitations of arrest (police) statistics and victimization data. Despite improvements, each collection still has problems, including definitions of racial categories, variations in reporting and recording, and utilization of population, crime, arrest, and victimization estimates. These limitations are briefly described next (for a comprehensive overview of these limitations, see Mosher et al., 2002).

DEFINITIONS OF RACIAL CATEGORIES

As discussed in Chapter 1, racial categories in federal statistics are guided by Directive No. 15, issued by the Office of Federal Statistical Policy and Standards in 1977. The 2000 census expanded racial categories to reflect the diversity of multiracial Americans. Despite improvements in federal statistical policies, racial categories in crime and victimization statistics are still problematic. One problem is the classification of persons of Hispanic origin. Hispanics are classified as an ethnic group in federal statistics, but the category is completely omitted from arrest statistics.

Knepper (1996) noted that there is no scientific definition of race, that social categories of race are both simplistic and wrong, and that official racial categories are the result of legal definitions that date back to slavery. He stated, "Race is a political concept. . . . [It] represents a powerful means of reinforcing an ideology of distinct races that began during the colonial period and was cemented during Jim Crow" (p. 86).

Table 2.3   Limitations of Arrest and Victimization Data

Issue	Arrest (Police) Statistics	Victimization Data
Arrest estimates	x	
Citizen reporting	x	x
Counting only the most serious crime	x	
Crime estimates	x	
Definition of crime categories	x	x
Definition of racial categories	x	x
Omission of offenses		x
Police recording	x	
Population estimates	x	x
Political manipulation	x	
Social construction of crime/criminals	x	x
Victimization estimates		x

Knepper argued that contemporary race-coded statistics descended from an ancestry of scientific racism. He questioned how analyses using official statistics can provide objective findings if race cannot be objectively defined. As the quotation at the beginning of the chapter indicates, crime statistics can be misleading and perpetuate errors and myths about racial minorities (Geis, 1972).

Racial data first appeared in the UCR in 1933 and included three categories: Whites, Blacks, and Others. In 1934, the UCR included a Mexican category that was dropped in 1941. Initially, age, sex, and race of persons arrested were compiled from fingerprint cards submitted by police departments to the FBI. Since 1953, race has been taken from the reported arrest information. Arrests for Whites (foreign- and native-born), Negroes, Indians, Chinese, Japanese, Mexicans, and others were reported separately. Chinese and Japanese were counted separately until 1970, when the Asian American category was included (LaFree, 1995). Today, there are four racial categories: White, Black, American Indian and Alaskan Native, and Asian and Pacific Islander. A Hispanic ethnic category was added in 1980, although it did not distinguish between Hispanic/Latino groups (Puerto Rican, Mexican, Cuban, etc.) and was discontinued after 1985. Counting American Indians is also problematic because they are located in a variety of jurisdictions, with numerous police agencies (Greenfield & Smith, 1999). The NCVS has included racial categories since its inception in 1973. Initially, victimizations for Blacks, Whites, and Others were reported. The "Other" category was used for American Indians and Asians; Hispanics were omitted until 1977, and they are not reported in the 2006 NCVS publication.

Despite efforts to develop and improve racial categories, they are fatally flawed for two reasons. First, they are unable to capture intraracial and intraethnic heterogeneity. As Georges-Abeyie (1989) correctly noted, there is no Black ethnic monolith. All Blacks are not the same; they have different cultural and ethnic backgrounds representing numerous countries and different social classes. Other racial categories also suffer from this limitation. Asian Americans, Latinos/Latinas, Native Americans, and Whites have varying backgrounds, experiences, and cultures that cannot be captured by counting them as if they were all the same. If the racial categories are in actuality meaningless, perhaps they are unnecessary (see Highlight Box 2.1). Second, how racial categories and ethnicity are determined is questionable and often inaccurate. In the UCR, police officers often make discretionary determinations about both arrestees and victims. Race in the NCVS was initially determined by observation by the interviewer, and respondents were asked about the racial/ethnic identity of the offender.

Highlight Box 2.1

Controversial Issues: Should Race Be Excluded From Crime Statistics?

The United States is one of very few countries that collects race crime statistics. In 1992, a moratorium on the release of national race crime statistics was issued by Canada's Justice Information Council (Knepper, 1996). Should the United States follow Canada's lead and declare such a moratorium on the collection and presentation of race-coded statistics, as Knepper (1996) suggests? What are the advantages and disadvantages of collecting race crime statistics?

VARIATIONS IN REPORTING AND RECORDING

More than 40 years ago, Wolfgang (1963) noted the problems associated with efforts to obtain uniform reporting when police agencies participate and report voluntarily to the UCR program. Throughout the history of the UCR, some agencies have submitted incomplete reports or no reports. For various reasons, citizens do not report and police do not record all crimes. Likewise, and perhaps more germane, police selectively enforce the law, which contributes to variations by race. Variations in crime categories and counting only the most serious crimes are also especially problematic for the UCR. Underreporting is also common. For example, although Native American arrest rates appear to be low, an analysis of victimization trends showed much higher levels of Native American victimizations than expected (Greenfield & Smith, 1999). Recording and reporting discrepancies also affect the NCVS because the accuracy in reporting victimizations remains unknown. Recently, concern has been expressed about how the NIBRS will affect UCR crime statistics if and when it is fully implemented. Rantala and Edwards (2000) examined this issue and concluded (based on an analysis of data for 1996) that, although prone to errors, NIBRS did not report significantly higher crime rates in most cases.

UTILIZATION OF POPULATION, CRIME, ARREST, AND VICTIMIZATION ESTIMATES

Estimations are an important part of the methodology and findings reported in both the UCR and NCVS. The UCR uses population estimates to calculate the crime and arrest rates. Inaccurate population estimates affect crime and arrest statistics. According to Mosher et al. (2002), these estimates are misleading because the census counts the population only every 10 years.

The NCVS relies on a sample of the population to estimate crime victimizations experienced by citizens. Their estimation procedures are important to understanding their utility. Although efforts are made to minimize differences between the sample population and the total population, there are biases related to recall and estimates of multiple victimizations (Bureau of Justice Statistics, 1992). The impact (if any) of redrawing the 2006 survey sample on reported victimizations is unknown. It is also unknown whether the transition to automated data collection in 2006 had an impact on the findings.

Several problems have been identified that are particularly relevant to understanding racial victimizations. First, it is unclear how representative the samples are because they are based on census data that are known to undercount minorities. Second, survey estimates are based on sampling units that may not adequately capture all racial groups in the population. Third, for Asians and American Indians, the sample size is so small that it affects the reliability of the estimate (Rennison, 2001b).

Despite their limitations, scholars continue to rely on the UCR and NCVS to gauge the extent of crime and victimization. Usually, comparisons are made across racial categories, although they are misleading (Covington, 1995) and overemphasize disproportionality (Young, 1994). For decades, comparisons of White and Black arrestees and victims have been a major focus in the study of race and crime. These comparisons have resulted in misperceptions about race and crime, which is why it is important to understand not only the sources of crime data and their limitations, but also what arrest and victimization trends do and do not tell us about race/ethnicity and crime. To better understand the extent of crime and victimization, we examine arrest patterns over time (1933–2006), arrests for murder for 56 years (1950–2006), and victimizations.

Arrest Trends

Part I of the UCR provide data on four violent crimes against the person: murder (and non-negligent manslaughter), forcible rape, robbery, and aggravated assault. Historically, most persons arrested for violent person offenses are arrested for aggravated assault. For decades, Blacks have been overrepresented in arrests for violent crimes and at various times outnumbered Whites arrested for murder, robbery, and aggravated assaults. More recent trends show Whites arrested more often than Blacks for aggravated assaults and rapes. Although there are fewer homicides than other violent personal crimes, they receive considerable attention. As previously mentioned, arrest data have been reported annually since 1933. Analyzing both historical and contemporary arrest data and trends provides a better temporal perspective than just looking at the past year, the past 2 years, the past 5 years, or the past 10 years.

A summary of arrests by race between 1933 and 2006 is presented in Table A.1 (see Appendix). In the early years (1933–1941), total arrests increased steadily, as did White and Black arrests, declined for Whites between 1942 and 1944, and declined for Blacks between 1942 and 1943. There was fluctuation in arrests of Mexicans, Indians, Chinese, Japanese, and "Other" races during this period. The downward trend in total arrests and arrests of Whites ended in 1945 and ended a year earlier for Blacks. Arrests continued to increase between 1945 and 1953, and there was a substantial increase in arrests in 1952, when total arrests were more than 1 million for the first time. By 1953, there were more than 1 million arrests of Whites, and by 1960, more than 1 million arrests of Blacks. Arrest data between 1950 and 1972 provide support for the concern about crime at that time. By 1970, the number of total arrests surpassed 6 million; by 1978, the number was 9 million; and it finally peaked in 1992 at almost 12 million arrests. Between 1994 and 1999, total arrests steadily decreased, decreased for Whites and Blacks, and fluctuated for the remaining racial categories. Most recently, since 2001, total arrests have steadily increased, and since 2003, arrests have steadily increased for Whites and Blacks. During this same period, there was fluctuation in arrests of American Indian/Alaskan Natives and Asian/Pacific Islanders.

Table A.2 presents the total number of arrests and arrests for murder and non-negligent manslaughter by race between 1950 and 2006. Since 1952, with a few exceptions (1979, 1980–1981, 1984–1986, and 2004–2005), the number of Blacks arrested for murder/non-negligent manslaughter has outnumbered Whites. Between 1952 and 1962, the number of Blacks and Whites arrested for murder tripled. Between 1963 and 1972, Black arrests for murder steadily increased and tripled from 2,948 to 8,347. Arrests for Whites doubled during this same period. There were fewer arrests for American Indians, Chinese, Japanese, and Others. Between 1976 and 1978, murder arrests increased and continued to fluctuate until 1994. Between 1994 and 2000, homicide arrests steadily decreased, although there was some fluctuation within racial categories. Between 2000 and 2006, the number of persons arrested for murder has continued to fluctuate and is much lower than reported arrests since the late 1970s. In 2006, there were 9,801 reported murder arrests.

Arrest trends for murder/non-negligent manslaughter also can be analyzed by type of jurisdiction where the arrest occurred. Between 2002 and 2006, Blacks arrested for murder outnumbered the number of Whites arrested in cities. In metropolitan and nonmetropolitan counties and suburban areas, the number of Whites arrested outnumbers Blacks (see Table 2.4).

Although arrests for violent crimes are significant due to their seriousness, continuing to focus on them minimizes the importance of other arrest trends. Again, most

Table 2.4   Reported Arrests for Murder by Race, 2002–2006, in Cities, Metropolitan Counties, Nonmetropolitan Counties, and Suburban Areas

	Total	White	Black	American Indian or Alaskan Native	Asian or Pacific Islander	%W	%B	%AI	%API
2002 City	7,463	3,073	4,225	56	109	41.2	56.6	0.8	1.5
Metropolitan County	1,751	1,138	589	15	9	65	33.6	0.9	0.5
Nonmetropolitan County	885	603	233	44	5	68.1	26.3	5.0	0.6
Suburban Areas	2,665	1,634	979	18	24	61.5	36.9	0.7	0.9
2003 City	6,619	2,815	3,664	49	91	42.5	55.4	0.7	1.4
Metropolitan County	1,718	1,137	542	18	21	66.2	31.5	1.0	1.2
Nonmetropolitan County	726	502	189	34	1	69.1	26.0	4.7	0.1
Suburban Areas	2,624	1,668	899	26	31	63.6	34.3	1.0	1.2
2004 City	7,034	3,146	3,739	48	101	44.7	53.2	0.7	1.4
Metropolitan County	1,809	1,204	577	14	14	66.6	31.9	0.8	0.8
Nonmetropolitan County	696	466	188	40	2	67.0	27.0	5.7	0.3
Suburban Areas	2,813	1,755	1,010	17	31	76.5	21.9	0.8	0.7
2005 City	7,340	3,244	3,958	49	89	44.2	53.9	0.7	1.2
Metropolitan County	1,995	1,215	737	14	29	60.9	36.9	0.7	1.5
Nonmetropolitan County	748	496	203	46	3	66.3	27.1	6.1	0.4
Suburban Areas	3,060	1,827	1,168	19	46	59.7	38.2	0.6	1.5
2006 City	7,210	2,999	4,054	65	92	41.6	56.2	0.9	1.3
Metropolitan County	1,969	1,195	747	16	11	60.7	37.9	0.8	0.6
Nonmetropolitan County	622	401	189	29	3	64.5	30.4	4.7	0.5
Suburban Areas	3,073	1,775	1,254	26	18	57.8	40.8	0.8	0.6

SOURCE: FBI, *Crime in the United States* (2006, Tables 49, 55, 61, 67)

Americans are not arrested for serious violent crimes, although they are arrested for many others. Table 2.5 shows the five offenses for which individuals were most likely to be arrested by racial categories in 2006. For all racial categories, these five offenses included only one Part I offense, larceny/theft; the other four offenses were Part II offenses, including drug abuse violations, driving under the influence, and other assaults.

Table 2.5   Five Most Frequent Arrest Offenses Within Racial Categories, 2006

White		Black		American Indian/ Alaskan Native		Asian/ Pacific Islander	
Larceny/ theft	548,057	Drug abuse violations	483,886	Other assaults	13,097	Driving under the influence	13,484
Other assaults	619,825	Other assaults	306,078	Driving under the influence	13,484	Liquor law violations	12,831
Drug abuse violations	875,101	Larceny/ theft	230,980	Liquor law violations	12,831	Other assaults	
Driving under the influence	914,226	Disorderly conduct	179,733	Larceny/ theft	9,377	Larceny/ theft	9,377
Liquor law violations	398,068	Driving under the influence	95,260	Drunkeness	7,884	Drug abuse violations	8,198

SOURCE: FBI, *Crime in the United States*, 2006, Table 43.

Victimization Trends

In addition to its annual reports, the NCVS periodically publishes other reports that present victimization trends and analyses of victimization within racial groups. These special reports include *American Indians and Crime* (Greenfield & Smith, 1999; Perry, 2004); *Black Victims of Violent Crime* (Harrell, 2007; Whitaker, 1990); *Hispanic Victims* (Bastian, 1990); *Hispanic Victims of Violent Crime, 1993–2000* (Rennison, 2002b); and *Violent Victimization and Race, 1992–1998* (Rennison, 2001b). *Criminal Victimization in the United States: 1973–1990 Trends* (Bastian, 1992) provided detailed information on early victimization trends for personal crimes of violence, theft, and household crimes. The report provided victimization levels and rates for each offense broken down by gender, age, race, and combinations of these variables.

In 1973, there were an estimated 35,661,030 victimizations. Violent victimization rates (VVRs) fluctuated between 32.6 and 34.3 per 1,000 persons over the age of 12, until 1983, when it decreased to 31.0. The lowest VVR was 28.1 (1986). Between 1986 and 1990, the VVR steadily increased to 31.7. Personal theft victimization rates (TVRs) fluctuated between a low of 67.5 (1986) and a high of 97.3 (1977). During this period, Whites had higher victimization rates for simple assault, whereas Blacks had higher victimization rates for aggravated assault and robbery. Black females had a pattern of higher victimization rates for rape than did White females. Victimization levels and rates for larceny/theft with contact were higher for Blacks than Whites, and larceny/thefts without contact victimization were higher for Whites than Blacks. This pattern occurred regardless of gender. Rates of burglary incidents and motor vehicle thefts were much higher for Blacks and others than for Whites (Bastian, 1992).

More recent victimization trends indicate that American Indian violent victimizations between 1992 and 1996 were more than twice as high as the national rate. The aggravated assault rate was twice as high as that of Blacks and three times the national rate (Greenfield & Smith, 1999). Estimates of Hispanic victimizations between 1992 and 2000 fell from 63 to 28 per 1,000. Similar to other racial categories, simple assaults were most common, and males were victimized most often (Rennison, 2002b). The Hispanic VVR fluctuated more between 1993 and 2000; the highest rate was reported in 1994 (61.6) and the lowest rate in 2002 (23.6).

In 2001, Americans experienced approximately 24.2 million victimizations, 18.3 million property victimizations, and 5.7 million violent victimizations, a decrease from 2000 and a continuation of a downward trend that began in 1994 (Rennison, 2001a, 2002a). In 2002, violent and property victimizations continued to decrease (Rennison & Rand, 2003). For Blacks, the VVR has steadily decreased since 1993 (67.4) to 27.9 in 2002 (Harrell, 2007). In 2005, an estimated 23 million violent and property victimizations occurred (Catalano, 2006). Over time, violent victimization rates have decreased within demographic groups, although they increased for Blacks, Other race, Two or more Races, and Hispanics between 2004 and 2005 (see Table 2.6).

Average annual violent victimizations between 2001 and 2005 indicate that Blacks have higher VVRs for rape/sexual assault and robbery, although American Indian/Alaska Natives have the highest VVR (56.8) (see Table 2.7). More important, Americans report that they are victims of simple assaults more than any other violent crime regardless of race and ethnicity.

As previously stated, there were an estimated 25 million violent and property crime victimizations of persons over the age of 12 in 2006. Approximately 19 million were property crime victims, and 6 million were violent crime victims. Higher rates of violent victimizations are reported by males (26.5 per 1,000) compared with females (22.9) and 16- to 19-year-olds (52.3). Interestingly, only 48.9% of violent crimes and 37.7% of property crimes were reported to the police (Rand & Catalano, 2007). Homicide victimizations are discussed next.

Homicide Victimizations

Homicide victimizations are reported annually in the UCR Supplementary Homicide Reports and have been published in the NCVS as well. The BJS Web site has a section that describes homicide patterns and trends. According to the site, the homicide rate doubled during the mid-1960s and 1970s, peaked at 10.2 per 100,000 population in 1980, and decreased to 7.9 per 100,000 in 1985. It rose again toward the end of the 1980s and early 1990s, peaking again in 1991 at 9.8 per 100,000. During the 1990s, the rate declined sharply, reaching 5.5 per 100,000 in 2000 (see BJS Web site: www.ojp.usdoj.gov/bjs/homicide/hmrt.htm#longterm). For Whites, Blacks, and Others, homicide rates have declined to levels last reported in the decade of the 1960s (see Table 2.8) (J. Fox & Zawitz, 2003).

Most victims of homicides are males, although White females are more likely to be victims of intimate homicides (Fox & Zawitz, 2007). Black offenders are more likely than Whites to commit drug-related homicides, whereas Whites are more likely to commit murders in the workplace and have multiple victims (see Table 2.9).

Table 2.6 Violent Victimization Rates of Selected Demographic Categories, 1993–2005

Demographic characteristic of victim	Number of violent crimes per 1,000 persons age 12 or older													Percent change
	1993	1994	1995	1996	1997	1998	1999	2000	2001	2002	2003	2004	2005	
Gender														
Male	59.8	61.1	55.7	49.9	45.8	43.1	37.0	32.9	27.3	25.5	26.3	25.0	25.5	−57.4%*
Female	40.7	43.0	38.1	34.6	33.0	30.4	28.8	23.2	23.0	20.8	19.0	18.1	17.1	−58.0*
Race														
White	47.9	50.5	44.7	40.9	38.3	36.3	31.9	27.1	24.5	22.8	21.5	21.0	20.1	−58.0%*
Black	67.4	61.3	61.1	52.3	49.0	41.7	41.6	35.3	31.2	27.9	29.1	26.0	27.0	−59.9*
Other race	39.8	49.9	41.9	33.2	28.0	27.6	24.5	20.7	18.2	14.7	16.0	12.7	13.9	−65.1 *
Two or more races	–	–	–	–	–	–	–	–	–	–	67.7	51.6	83.6	–
Hispanic origin														
Hispanic	55.2	61.6	57.3	44.0	43.1	32.8	33.8	28.4	29.5	23.6	24.2	18.2	25.0	−54.7%*
Non-Hispanic	49.5	50.7	45.2	41.6	38.3	36.8	32.4	27.7	24.5	23.0	22.3	21.9	20.6	−58.4*
Annual household income														
Less than $7,500	84.7	86.0	77.8	65.3	71.0	63.8	57.5	60.3	46.6	45.5	49.9	38.4	37.7	−55.5%*
$7,500–$14,999	56.4	60.7	49.8	52.1	51.2	49.3	44.5	37.8	36.9	31.5	30.8	39.0	26.5	−53.0*
$15,000–$24,999	49.0	50.7	48.9	44.1	40.1	39.4	35.3	31.8	31.8	30.0	26.3	24.4	30.1	−38.6*
$25,000–$34,999	51.0	47.3	47.1	43.0	40.2	42.0	37.9	29.8	29.1	27.0	24.9	22.1	26.1	−48.8*
$35,000–$49,999	45.6	47.0	45.8	43.0	38.7	31.7	30.3	28.5	26.3	25.6	21.4	21.6	22.4	−50.9*
$50,000–$74,999	44.0	48.0	44.6	37.5	33.9	32.0	33.3	23.7	21.0	18.7	22.9	22.1	21.1	−52.0*
$75,000 or more	41.3	39.5	37.3	30.5	30.7	33.1	22.9	22.3	18.5	19.0	17.5	17.0	16.4	−60.3*

SOURCE: Catalano (2006, Table 4, p. 6).

NOTE: *1993–2002 difference is significant at the 95%-confidence level.s

Table 2.7   Number and Rate of Violent Victimization, 2001–2005

Race/Hispanic origin	Total violent crime	Rate per 1,000 persons age 12 or older	Assault		
			Robbery	Aggravated	Simple
Black/African American	28.7	1.7	4.3	7.7	14.9
White[a]	22.8	0.9	2.0	4.2	15.7
American Indian/ Alaska Native	56.8	0.9[b]	4.8[b]	11.6	39.5
Asian/Pacific Islander	10.6	0.5[b]	2.3	1.7	6.2
Hispanic/Latino	24.3	0.8	3.6	5.3	14.5

SOURCE: Harrell (2007, Table 2, p. 3).

a. Not Hispanic or Latino

b. Based on 10 or fewer sample cases.

Hate Crime Trends

The terms *hate* and *bias crimes* refer to offenses committed against individuals because of their race, religion, ethnicity, sexual orientation, or disability. As long as hate/bias statistics have been available, most offenses reported are motivated by anti-Black bias. The 1998 murder of James Byrd in Jasper, Texas, and the 2003 beating of Billy Ray Johnson of Linden, Texas, are examples of bias crimes. Anti-Black bias crimes outnumber offenses against other races as well as bias crimes motivated by victims' religion, sexual orientation, ethnicity, or disability (Federal Bureau of Investigation, 2006). According to the NCVS, half of victims of hate crimes reported perceiving race as the primary offender motivation (Harlow, 2005).

Table 2.10 presents offenses classified as "single-bias incidents" by race and ethnicity between 2003 and 2006. The total number of single-bias incidents fluctuated between 2003 and 2006. Between 2003 and 2006, incidents within racial categories fluctuated. Anti-Hispanic single-bias incidents steadily increased during this period perhaps due to anti-immigrant attitudes.

According to Perry (2002), hate crime is more likely than street crime to involve crimes against the person than crimes against property. In 2001, the most commonly reported offenses against persons were intimidation, simple assault, and aggravated assault. Although there were fewer crimes against property, destruction/damage/vandalism ranked second to intimidation in hate/bias crime incidents. Most hate crimes motivated by race occur at the victim's residence/home. Hate crime also is more likely to be interracial; the race of most known offenders is White (Federal Bureau of

Table 2.8   Homicide Victimization Rates per 100,000 Population by Race

	White	Black	Other
1976	5.1	37.1	4.9
1977	5.4	36.2	4.7
1978	5.6	35.1	4.0
1979	6.1	37.5	4.1
1980	6.3	37.7	5.7
1981	6.2	36.4	6.1
1982	5.9	32.3	6.5
1983	5.3	29.4	6.4
1984	5.2	27.2	5.5
1985	5.2	27.6	5.5
1986	5.4	31.5	6.2
1987	5.1	30.7	5.2
1988	4.9	33.5	4.0
1989	5.0	35.1	4.3
1990	5.4	37.6	4.2
1991	5.5	39.3	6.0
1992	5.3	37.2	5.4
1993	5.3	38.7	5.5
1994	5.0	36.4	4.6
1995	4.8	31.6	4.9
1996	4.3	28.3	4.1
1997	3.9	26.0	4.1
1998	3.8	23.0	2.9
1999	3.5	20.5	3.3
2000	3.3	20.5	2.7
2001	3.4	20.4	2.8
2002	3.3	20.8	2.7
2003	3.4	20.9	2.8
2004	3.3	19.7	2.4
2005	3.3	20.6	2.5

SOURCE: Fox and Zawitz (2007).

Table 2.9   Homicide Type, by Race, 1976–2005

	Victims			Offenders		
	White	Black	Other	White	Black	Other
All homicides	50.9%	46.9%	2.1%	45.8%	52.2%	2.0%
Victim/offender relationship						
Intimate	56.6%	41.2%	2.2%	54.4%	43.4%	2.2%
Family	60.7%	36.9%	2.4%	59.2%	38.5%	2.3%
Infanticide	55.9%	41.6%	2.5%	55.4%	42.1%	2.5%
Eldercide	69.2%	29.1%	1.6%	54.5%	43.8%	1.6%
Circumstances						
Felony murder	54.7%	42.7%	2.6%	39.1%	59.3%	1.6%
Sex related	66.9%	30.5%	2.5%	54.7%	43.4%	1.9%
Drug related	37.4%	61.6%	.9%	33.9%	65.0%	1.1%
Gang related	57.5%	39.0%	3.5%	54.3%	41.2%	4.4%
Argument	48.6%	49.3%	2.1%	46.8%	51.1%	2.2%
Workplace	84.6%	12.2%	3.2%	70.5%	26.7%	2.8%
Weapon						
Gun homicide	47.2%	50.9%	1.9%	41.9%	56.4%	1.7%
Arson	58.9%	38.1%	2.9%	55.7%	42.0%	2.3%
Poison	80.6%	16.9%	2.5%	79.8%	18.4%	1.8%
Multiple victims or offenders						
Multiple victims	63.4%	33.2%	3.3%	55.7%	40.8%	3.5%
Multiple offenders	54.8%	42.5%	2.7%	44.6%	53.0%	2.4%

SOURCE: Fox and Zawitz (2007).

Investigation, 2006). A study of hate crimes reported in the NIBRS, the Bureau of Justice Statistics (2001a) found that younger offenders were responsible for most hate crimes.

Conclusion

For more than 70 years, the UCR has been the primary source of crime statistics, and for more than 30 years, the NCVS has provided victimization data. Analyses of UCR and NCVS data are useful for understanding crime patterns and trends. After trending upward for several decades, arrests trended downward in the late 1990s, until 2001,

Table 2.10   Hate/Bias Crime Incidents Reported

	2003	2004	2005	2006
Total Single-Bias Incidents	3,844	4,042	3,919	4,000
Single-Bias Race				
Anti-White	830	829	828	890
Anti-Black	2,548	2,731	2,630	2,640
Anti-American Indian or Alaskan Native	76	83	79	60
Anti-Asian American or Pacific Islander	231	217	199	181
Anti-Multiple Races	–	182	183	229
Ethnicity				
Anti-Hispanic	426	475	522	576
Anti-Other Ethnicity/National Origin	600	497	422	408

SOURCE: Federal Bureau of Investigation, *Hate Crime Statistics,* Table 1, 2003–2006.

when they began to increase. There was fluctuation in arrests of persons within racial categories that is often overlooked. Although the data provide information about race and crime, racial categories in each data set are problematic. Jurisdictional differences in reporting from time to time and place to place also are difficult to determine. At best, we can conclude that Americans are arrested for a variety of offenses, including violent crimes, property crimes, alcohol-related offenses, and drug abuse violations. With few exceptions over time, Blacks continue to be arrested for murder more often than any other race, although reported murders and arrests have decreased (compared with earlier decades), and Black homicide victimizations are at their lowest levels. Fewer persons are arrested for violent personal crimes, although they tend to receive the most attention in the study of crime, especially in comparisons across racial categories. Victimization surveys provide support for patterns of arrests: Theft victimizations are more likely to be reported than rapes, aggravated assaults, and robberies. Among violent victimizations, simple assaults are reported most often. Perhaps the most surprising finding from the NCVS is the high victimization rates for Native Americans.

Decades of comparisons made between Blacks and Whites as arrestees and victims have resulted in several misperceptions about race and crime. In fact, comparisons between racial categories reveal more similarities than differences. Unfortunately, fixation on violent crimes has contributed to the racialization of crime; statistics are used to provide empirical support for the overrepresentation of most minorities as arrestees and victims. As a result, many Americans equate crime with minorities. Support for this idea can be easily found by asking students enrolled in criminal justice courses a few questions about persons arrested before presenting the material. Invariably, they believe that more Blacks than Whites are arrested, especially for drug offenses and rape.

Despite what they do tell us, what these two sources of crime statistics do not tell us is just as important. Arguably, they really do not tell us much about race and crime because we have vague categories to describe race and Latinos/Latinas are still included in the White category for arrest data. The data do not tell us about police bias and how that affects arrests. Most important, crime statistics tell us either nothing (UCR) or little (NCVS) about class and crime. For now, the extent of race and crime as reported in arrest and victimization data is useful, although not definitive. Because it appears that the limitations of these data sets will not be overcome in the near future, it is important to focus on the offenses that are most frequently reported in arrest and victimization data and move away from our fixation on violent crimes.

Finally, although Americans have experienced hate/bias crimes for more than 100 years, hate crime statistics are still in their infancy. Despite this, they reveal a dimension of race and crime conveniently ignored: Blacks are more likely to be the victims of bias crimes than any other group. Unlike most reported victimizations, bias crimes are often interracial. White violence such as lynching may not be a crime problem today, but assaults and intimidation against individuals based on their race are problems. Over time, these victimizations will be better understood when the NIBRS has more participants and hate/bias crimes are integrated into the NCVS.

Photo 2.2. Eighteen-year-old David Ritcheson, a young Latino man from Houston, survived a brutal racially motivated attack in 2006 and committed suicide in 2007 by jumping from a Carnival Cruise Ship.

Discussion Questions

1. Discuss the importance of crime and victimization data in the study of race and crime.

2. Discuss the definitional issues related to racial categories in crime and victimization data.

3. Discuss why racial and ethnic minorities are often viewed as criminals in our society.

4. Why do you think some minorities are disproportionately arrested and victimized?

5. Why is there less publicity about anti-Black hate crimes than about black arrests and other types of crime victimizations?

Internet Exercises

Visit the FBI Web site (www.fbi.gov) and view the most recent UCRs online. Select one racial group and describe the arrest trend for three to five offenses for 3 years.

Visit the Southern Poverty Law Center Web site (http://www.splcenter.org/) and view hate groups in your state.

Internet Sites

Bureau of Justice Statistics: http://www.ojp.usdoj.gov/bjs; Crime and Victim Statistics: http://www.ojp.usdoj.gov/bjs/cvict.htm

Bureau of Justice Statistics, Homicide in the United States: http://www.ojp.gov/bjs/homicide/homtrnd.htm

Federal Bureau of Investigation: http://www.fbi.gov

Theoretical Perspectives on Race and Crime 3

> *A wide variety of sociological, psychological, and biological theories have been proposed to explain the underlying causes of crime and its social, spatial, and temporal distribution. All of these theories are based on the assumptions that crime is accurately measured. But when variations in crime patterns and characteristics [are] partially attributable to unreliability in the measurement of crime, it is impossible to empirically validate the accuracy of competing criminological theories.*
>
> —Mosher, Miethe, and Phillips (2002, pp. 179–180)

Considering the historical and contemporary crime and victimization figures presented in Chapter 2, the logical next question is: What explains the crime patterns of each race? Based on this question, we have formulated two goals for this chapter. First, we want to provide readers with a rudimentary overview of theory. Second, we want to provide readers with a summary of the numerous theories that have relevancy for explaining race and crime. In addition to this, where available, we also discuss the results of tests of the theories reviewed. Last, we also document some of the shortcomings of each theory.

Decades ago, it was pretty standard to find criminology textbooks with whole chapters devoted to race and crime (Gabbidon & Taylor Greene, 2001). Currently, however, most texts cover the topic, but only in a cursory way, with either a few paragraphs or several highlight boxes. A rare exception to this is Knepper (2001), who returned to the practice of having a full chapter devoted to the subject. In general, because of the additional focus on race and crime, scholars have written more specialized books, such as this one, to more comprehensively cover the subject. But even in these cases, many authors devote little time to reviewing specific theories related to race and crime. In her 1993 tome on race and crime, *Unequal Justice: A Question of Color*, Mann provided one of the most comprehensive reviews of theories that have been applied to race and crime. More recently, Gabbidon's (2007a) *Criminological Perspectives on Race and Crime* considerably updated and expanded Mann's coverage and represents the first text solely devoted to examining how well criminological theories contextualize race and crime. We attempt to provide an overview of the vast number of major and lesser known theories that have been applied to understanding racial patterns in crime and victimization. Before we begin, however, we review the fundamentals of theory, noting what theory is, distinguishing the various types of theories, and discussing the usefulness of having theory.

What Is Theory?

According to Bohm (2001), "A theory is an explanation" (p. 1). Some theory can be found in practically everything we do. When it comes to explaining crime, just about everyone has an opinion. All of these insights, however, might not qualify as *scientific* theory. To qualify as a scientific theory, Curran and Renzetti (2001) noted that "a theory is a set of interconnected statements or propositions that explain how two or more events or factors are related to one another" (p. 2). Furthermore, theories are usually logically sound and empirically testable (Curran & Renzetti, 2001).

Theories can be further categorized by whether they are macrotheories, microtheories, or bridging theories (Williams & McShane, 2004). Macrotheories focus on the social structure and are generally not concerned with individual behavior; conversely, microtheories look to explain crime by looking at groups, but in small numbers, or at the individual level (Williams & McShane, 2004). Bridging theories "tell us both how social structure comes about and how people become criminals" (Williams & McShane, 2004, p. 8). Many of the theories reviewed in this chapter fit some of these criteria, whereas others do not, but in our view nonetheless provide useful insights into race and crime. Thus, we discuss some nontraditional approaches that have not been folded into the mainstream of scientific criminological theory. It is important to note here that this chapter does not review every criminological theory. Our aim was to simply examine some of those that have been applied to the issue of race and crime. Others, such as rational choice theory, might also have some relevancy, but were left out because there is limited scholarship that makes the connection between the perspective and racial disparities in crime and justice.

Theories are valuable for a number of reasons. Curran and Renzetti (2001) provided an important summary of the usefulness of theory:

> Theories help bring order to our lives because they expand our knowledge of the world around us and suggest systematic solutions to problems we repeatedly confront. Without the generalizable knowledge provided by theories, we would have to solve the same problems over and over again, largely through trial and error. Theory, therefore, rather than being just a set of abstract ideas, is quite practical. It is *usable knowledge.* (p. 2; italics original)

There are several paradigms within criminological theory that are reviewed here. We review biological approaches that look to physical features and/or genetic inheritance to explain crime, other theories that have their foundations in the American social structure, social processes or one's culture, and theories that have psychological foundations. In recent years, more theorists have also sought to integrate some of these approaches (Messner, Krohn, & Liska, 1989). As one might expect, many of these theories have been applied to explain race and crime. We begin with a review of biological explanations of crime.

Biology, Race, and Crime

The linking of biology and crime has its roots in Europe. Reid (1957) wrote that "[in] the year 1843 a Spanish physician Soler was [the] first to [mention] the concept of the born criminal" (p. 772). It was also Europe where phrenology, the study of the external shape of the head, was first popularized (Vold, Bernard, & Snipes, 1998). The publication of Darwin's *The Origin of the Species* (1859) and *Descent of Man* (1871) was also influential in this era.

Once the ideas became accepted, Cesare Lombroso, a doctor in the Italian Army in the 19th century and the so-called father of criminology, began studying army personnel from the southern portions of Italy where, in addition to being considered inferior beings, the citizens were thought to be "lazy, incapable, criminal, and barbaric" (Vold et al., 1998, pp. 42–43).

In Lombroso's first major work, *The Criminal Man* (1876/1911), he made clear the importance of race in explaining crime. He mentioned that some tribes in parts of India and Italy had high crime due to "ethnical causes" (Lombroso, 1876/1911, p. 140). He added that "the frequency of homicide in Calabria, Sicily, and Sardinia is fundamentally due to African and Oriental elements" (p. 140). When Lombroso took on the task of explaining criminality among women, he again saw race as being an important contributor to crime. In his view, "Negro" women and "Red Indian" women were seen as manly looking, which contributed to their criminality. His works were widely hailed and were soon translated into English. By the time Lombroso's works were translated into English, the notion of biological determinism had already taken hold on American shores.

As in Lombroso's work, racial and ethnic groups were the focus of ideas that inferior "stocks" were polluting society. The most virulent attacks were reserved for African Americans. Books such as Charles Carroll's (1900) *The Negro a Beast* spoke to the notion that African Americans were not human; they were more akin to apes. Relying heavily on biblical interpretations, Carroll sought to show why the White race was superior to the African American race. Around the same time, there was the thought that, because of their genetic inferiority, African Americans would eventually die off (Hoffman, 1896). Although these notions were vigorously challenged here and abroad, such ideas dominated the late 19th- and early 20th-century literature and gave rise to the racist eugenics movement. However, as noted in Chapter 1, with the increasing immigration to the United States, these ideas were also applied to the unwelcome new arrivals.

Early sociologists, such as Charles Henderson of the University of Chicago, picked up on the connection among biology, race, and crime and wrote,

> There can be no doubt that one of the most serious factors in crime statistics is found in the conditions of the freedmen of African descent, both North and South. The causes are complex. The primary factor is racial inheritance, physical and mental inferiority, barbarism and slave ancestry and culture. (Henderson, 1901, p. 247)

Although Henderson also discussed the importance of sociological factors, he clearly prioritized the role of genetic inheritance.

Analyses such as Henderson's predominated the first several decades of the 20th century. Noting the overrepresentation of African Americans and some immigrants in the crime statistics, observers continued to look to racial and ethnic diversity to explain these differences. The work of Harvard anthropologist Earnest A. Hooton took the lead in this area. His two 1939 books, *Crime and the Man* and *The American Criminal*, were based on his 12-year study of more than 17,000 subjects (14,000 male prisoners and 3,203 civilians as the control group) from 10 states. His study focused on the relationship between physiology and criminality. He listed a variety of general characteristics that were typical of criminals, many of which supported the earlier work of Lombroso. The study was sharply criticized by sociologist Robert Merton and anthropologist M. F. Ashley-Montagu for a number of reasons, including methodological and definitional problems. In terms of race, they noted that Hooton's work was contradictory; in many instances, when the data fit Hooten's suppositions, he stated that biology was the cause, but when the data did not fit, he suggested that socioeconomic conditions were relevant. Merton and Ashley-Montagu (1940) also argued that Hooton's analysis of prisoners did not account for selective enforcement on the part of the police, which would undoubtedly influence the demographics of those found in institutions.

In a short but influential work, *Race and Crime*, Dutch criminologist Willem Bonger (1943) provided a different perspective on the topic:

> Criminality is not a characteristic. It is neither a physical quality such as the possession of blue eyes, nor the spiritual one such as musicality. No one comes into the world with "criminality," in the way in which one is born with a certain color of eyes, and so forth. Crime is something completely different. (p. 27)

In his view, the notion of criminal and noncriminal races was ridiculous. On this point, he opined, "The truth is naturally, that crime occurs in all races, and, by the nature of things, is only committed by a number (generally very limited) of individuals in each race. In principle the races do not differ" (p. 28).

Crime and Human Nature

Because of the persistent criticism of the biological perspective, support for the ideology lay dormant until 1985, when Wilson and Herrnstein resurrected it with their publication of *Crime and Human Nature*. The work has been generally seen as taking an approach similar to the work of Lombroso, Hooton, and other early biological theorists (Lilly, Cullen, & Ball, 2001). In their chapter on race and crime, Wilson and Herrnstein (1985) pointed to constitutional factors that may contribute to the overrepresentation of Blacks in crime; such constitutional factors "merely make a person somewhat more likely to display certain behavior; it does not make it inevitable" (p. 468). Drawing on the foundations of Sheldon's work, Wilson and Herrnstein suggested that Black males tend to be more mesomorphic (muscular) than White males; in addition, because they have higher scores on the Minnesota Multiphasic Personality Inventory (MMPI) than Whites, this shows they are "less normal." Another constitutional factor mentioned by the authors is low IQ (see Wilson & Herrnstein, 1985). This connection is discussed further in the next section of the chapter.

As with their predecessors, Wilson and Herrnstein have had their critics. Most notably, there were concerns about the clarity of concepts and other measurement issues. Another concern related to their exclusive use of the theory to explain crime in the streets, not "crimes in the suites" (Lilly et al., 2001, pp. 212–213). This obviously speaks to race and crime because it is clear that these conservative thinkers have more interest in explaining crimes associated with racial minorities than those overwhelmingly committed by middle- and upper class Whites.

Intelligence, Race, and Crime

With the development and acceptance of intelligent tests, another linkage was developed: intelligence and crime (S. Gould, 1996). Much of the early literature suggested that criminals were of low intelligence or "feebleminded." This line of thinking was based on the early work of Richard Dugdale's 19th-century Jukes study, which chronicled the genealogy of a family that had experienced generations of immorality and criminality.

Building on the Jukes study, in the early 1900s, Henry H. Goddard studied the lineage of a family in New Jersey. Goddard found that one side of the family produced primarily descendants of superior intelligence, whereas the other side of the family produced offspring that were considered immoral, criminal, and alcoholics. Goddard's study was later found to be faulty because he had his assistant, Elizabeth S. Kite, conduct the research, and she failed to use an IQ test to determine feeblemindedness. Instead, she made her assessments based on physical appearance (Knepper, 2001). In addition, it was found that Goddard had altered some of the pictures in his books to make study participants look diabolical. Although these actions severely challenged Goddard's work, the notion of intelligence and crime had already become accepted. The notion of intelligence and crime had existed prior to the aforementioned studies, but the development of the IQ test gave proponents of the idea a tool to test their beliefs.

Because of a critical review of numerous studies on IQ and crime by Edwin Sutherland, as well as Simon Tulchin's (1939) classic *Intelligence and Crime*, intelligence-based theories disappeared from the criminological literature until the 1970s (Hirschi & Hindelang, 1977). At this time, two prominent criminologists, Hirschi and Hindelang, conducted a review of the literature on intelligence and crime. On the issue of race, they wrote, "There can be no doubt that IQ is related to delinquency within race categories" (p. 575). From their research, they concluded that students with low intelligence had difficulty in school and, as a result, were more likely to engage in delinquency—ergo, given that Blacks have traditionally scored lower on IQ tests, they are likely to commit more crimes.

The debate lingered with a few publications here and there, until Herrnstein and Murray (1994) published their controversial work, *The Bell Curve*. The book picked up where the debate left off. They suggested that low IQ contributed to a host of factors, including crime, poverty, illegitimacy, unemployment, welfare dependency, and others. How? Well, the authors present a few ways in which this connection materializes. First, they state that low IQ results in school failure, which tends to lead to crime. Second, they argue that low IQ leads to people being drawn to danger and having "an insensitivity to pain or social ostracism, and a host of derangements of various sorts" (Herrnstein & Murray, 1994, p. 240). Combined, these factors, in their minds, were precursors for a criminal career. Finally, the authors suggest that those with low IQs would have a hard time following ethical principles. According to their theory, people with low IQ might "find it harder to understand why robbing someone is wrong, find it harder to appreciate values of civil and cooperative social life, and are accordingly less inhibited from acting in ways that are hurtful to other people and to the community at large" (Herrnstein & Murray, 1994, pp. 240–241). Implicit in their thinking is that, because Blacks tend to have lower IQs, they are likely at greater risk for engaging in criminality.

Numerous shortcomings have been noted with the intelligence, race, and crime approach, however. First, there still remain questions as to what IQ tests really measure. Moreover, there have always been questions of cultural and class biases with IQ tests. An additional concern relates to the question that if a lack of intelligence is associated with crime, then what explains the fact that persons with high IQs commit white-collar and political crime (M. Lanier & Henry, 1998)? Finally, there is also some uncertainty about whether differences in IQ are genetic or related to one's environment (Onwudiwe & Lynch, 2000; Vold et al., 1998). One final biological approach reviewed is the *r/K* theory, which has garnered significant attention in recent years.

r/K Life History Theory

Possibly one of the most controversial theories of our time related to race in general and crime in particular is the *r/K* life history theory. Created by Harvard biologist E. O. Wilson to explain population growth and the decline in plants and animals, the theory has been adapted to humans by Rushton (1999), professor of psychology at Western Ontario University. The gene-based evolutionary theory links many of the differences among the races, including crime patterns, to migrations out of Africa.

Rushton agrees with the hypothesis that all humans came out of Africa. It is his contention, however, that there was a split of the population before humans left Africa and that this split is responsible for the current position of Blacks, Whites, and Asians. As he sees it, those who stayed in Africa (now referred to as Black people) were subjected to unpredictable droughts and deadly diseases, which caused them to die young (Rushton, 1999). Those who migrated to Eurasia (now referred to as Whites and Asians) had to deal with other concerns, such as "gathering and storing food, providing shelter, making clothes, and raising children during the long winters" (Rushton, 1999, p. 85). These tasks were more mentally demanding and, according to Rushton, required greater intelligence. Moreover, "They called for larger brains and slower growth rates. They permitted lower levels of sex hormones, resulting in less sexual potency and aggression and more family stability and longevity" (Rushton, 1999, p. 85).

At the heart of the *r/K* theory are reproduction, climate, and intelligence. Because Africans were faced with early death, they often had to bear more children to maintain their population, which left them unable to provide significant care for their offspring. Conversely, those falling under the *K*-strategy, Whites and Asians (Rushton acknowledges only three races: Negroid, Caucasoid, and Mongoloid), reproduced less and generally spent more time caring for their offspring.

Rushton's theory relates to race and crime in that aggression, impulsive behavior, low self-control, low intelligence, and lack of rule following are all associated with criminals, and, according to Rushton, those who fall under the *r*-strategy, namely, Black people. To support his approach, Rushton has conducted cross-national studies that looked at race and crime (see Rushton, 1995; Rushton & Whitney, 2002). Other scholars have also adopted some of Rushton's ideas in the areas of crime (Ellis, 1997; Ellis & Walsh, 1997, 2000; Walsh, 2004; Walsh & Ellis, 2003; J. Wright, 2008) and skin color and intelligence (Lynn, 2002; see Highlight Box 3.1).

As with all theories, there have been several notable criticisms of the *r/K* theory. First, Rushton generally ignores sociological factors. Most of his cross-national comparisons point strictly to numbers, without taking into account variables such as socioeconomic status, discrimination, and other important sociological variables. Second, in the 21st century, there are few "pure" races, especially in the United States, where, as noted in Chapter 1, White sexual aggression against Black females during the slave era produced countless mixed-race offspring. Therefore, the rigorous adherence to the Black–White–Asian split is problematic. Third, if Rushton's theory were true, what would explain White aggression as colonizers around the globe? In contrast to Rushton's theory, Bradley (1978) argued that, as a result of migration to colder regions, since the beginning of humanity, Whites have been the global aggressors. Finally, there have been concerns that Rushton's research returns us to the days of the eugenics movement.

In summary, many of the current biologically oriented theories either directly or indirectly point to some race and crime linkage. Nevertheless, for more than a century, opponents of such approaches have countered with alternative sociological perspectives, some of which are reviewed in the next section.

Highlight Box 3.1

Skin Color and Crime: Is the Connection Intelligence or Social Distance?

Theorists such as Rushton (1995) have adopted the r/K theory. As a part of the theory, they suggest that certain traits inherent in Black people make them less intelligent and they eventually end up engaging in crime. Drawing on the work of Rushton, Richard Lynn (2002) sought to determine whether skin color was related to intelligence. That is, if intelligence were related to race, then Whites would have higher IQs than light-skinned people, who would have higher IQs than medium-skinned people, and so on. He tested this proposition by using data from the 1982 National Opinion Research Center's (NORC) opinion poll. One of the few surveys to ask questions regarding skin color, the survey also included a 10-word vocabulary test, which Lynn used as an indicator of intelligence.

Describing the racial identification part of the survey, Lynn (2002) wrote, "First, respondents were asked if they would describe themselves as White, Black or other. Second, if they described themselves as Black they were asked whether they would describe themselves as very dark, dark brown, medium brown, light brown or very light. On the basis of their replies they were placed into one of these categories" (p. 370). Table 3.1 shows the results of the study. Lynn indicated that the results show that there is a strong correlation between light skin color and intelligence among African Americans. Considering that theorists such as J. Phillipe Rushton, Richard Herrnstein, and Charles Murray link low IQ to crime, the findings from Lynn's research likely serve as a bridge to the notion that the darker one is, the more likely one is to engage in crime.

Table 3.1   Mean Vocabulary Scores of Whites and Blacks, Analyzed by Self-Reported

Group	N	Mean	SD
Whites	1,245	6.18	2.06
Blacks	442	4.81	2.08
Very dark	42	4.43	1.98
Dark	104	4.09	1.68
Medium	204	5.01	2.15
Light	66	5.33	2.08
Very light	14	5.0	2.29

SOURCE: Lynn (2002, p. 371).

Tatum (2000) explored whether skin color is an important criminological concept. Tatum wrote, "Skin color (or skin tone) can refer to the prejudicial or preferential treatment of in-group members based on the lightness or darkness of their skin. Like race, skin color is also a measure of the similarity or dissimilarity that in-group members share with the dominant group" (p. 32). Incorporating the work of Daniel Georges Abeyie, Tatum looked at the concept of social distance. She defined *social distance* as "the degree of closeness or remoteness that the majority group desires in its interactions with members of a particular group" (p. 33). Tatum listed the following six factors that influence social distance:

1. Extent of physiognomic, phrenologic, and anatomic differences;

2. Perceived extent of "contributions" to the national development;

3. Perceived extent of cultural and social network differences between minority/out-groups and majority/in-groups;

4. Perceived threat to the social order;

5. Perceived criminality; and

6. Perceived "intrinsic worth." (p. 33)

Tatum (2000) noted that African Americans have traditionally ranked lower on these factors than other minority groups. This is where skin color also begins to matter. As the tone of one's skin color gets darker, Tatum posited, there is greater social distance from the majority group. She provided historical context by drawing on slave history, whereby light-skinned slaves received the "better" jobs in the house, whereas the darker-skinned slaves were relinquished to the lower-status position of field hands.

According to Tatum (2000), there are three ways that skin color is related to crime. First, people may discriminate against those with darker skin and perceive them as dangerous and untrustworthy. Subsequently, at every stage of the criminal justice system, a dark-skinned person would encounter skin color discrimination. Second, drawing on strain theory, if dark-skinned persons are discriminated against more than other persons for prospective job opportunities, they are likely to have a "higher probability of unemployment, lower occupational status and income, and [reside in] high-stress areas [which] theoretically suggests that these individuals have higher levels of frustration and a higher probability of criminal offending" (pp. 40–41). Finally, because of their low status in the pecking order, dark-skinned persons might develop low self-esteem and engage in higher levels of violence, substance abuse, and suicide.

Both perspectives look to skin color as having potential explanatory power when considering race and crime. Lynn (2002) looked to make biological connections between skin color and intelligence, which also presupposes that, in line with the r/K theory, darker-skinned persons would have lower intelligence and would therefore be engaged in more crime than those having lighter complexions. In contrast, Tatum (2000) sees skin color being connected to crime in a sociological way. That is, discrimination is the more pertinent explanation as to why people of darker hues potentially engage in more crime than people with lighter complexions. What do you think?

Sociological Explanations

Sociological explanations for crime in general have existed for nearly two centuries. Beginning with the early work of the cartographic school, led by Adolphe Quetelet, who some have said produced the first scientific work on crime (see Quetelet, 1833/1984), this approach looked to sociological factors to explain criminality (i.e., age, social class, poverty, education level, etc.). Several decades after the publication of Quetelet's work, as noted earlier, biological notions related to crime were being espoused in America. Numerous American scholars, however, challenged the biological approach using sociological analyses of crime problems. With the development of the first sociology department at the University of Chicago in 1892, and other such programs at universities across the United States, scholars saw this new discipline and a sociological approach as a means to solve some of the pressing issues particularly plaguing overcrowded northern cities.

In the late 1890s, Philadelphia was one of those cities looking for answers to its concerns regarding the burgeoning African American population. At the time, city officials sought out W. E. B. Du Bois to conduct a study of Philadelphia's notorious Seventh Ward. Du Bois conducted a comprehensive review of the ward, outlining the conditions in the area, and also pointing to several possible explanations for crime among African Americans. One of his explanations was as follows:

> Crime is a phenomenon of organized social life, and is the open rebellion of an individual against his social environment. Naturally then, if men are suddenly transported from one environment to another; the result is lack of harmony with the new conditions; lack of harmony with the new physical surroundings leading to disease and death or modification of physique; lack of harmony with social surroundings leading to crime. (Du Bois, 1899/1996, p. 235)

He felt that the mass migration from the South to the North produced problems of adjustment for African Americans, who were previously familiar only with southern life.

Du Bois' ideas were in line with the concept of social disorganization, which we discuss later. Like Quetelet earlier, to explain criminality in the Seventh Ward, Du Bois pointed to issues related to age, unemployment, and poverty. Du Bois, however, added the sociological variable of discrimination, noting that Blacks were arrested for less cause than Whites, served longer sentences for similar crimes, and were subject to employment discrimination (Gabbidon, 2007b; Taylor Greene & Gabbidon, 2000). Other early studies would echo similar sentiments on crime in the African American community (K. Miller, 1908/1969; R. Wright, 1912/1969), with some scholars going as far as to suggest that, because of the deep-seated societal discrimination contributing to crime in the African American communities, Whites were the "Ultimate Criminals" (Grimke, 1915).

Social Disorganization

Northern cities, such as Chicago, were also experiencing the same social problems as Philadelphia as a result of population booms caused by the mass immigration by racial and ethnic groups outlined in Chapter 1. With unparalleled philanthropic support from numerous foundations (Blumer, 1984), by the 1920s, the University of Chicago had put together a formidable cadre of scholars to investigate the social ills plaguing the city.

Together, these scholars combined their ideas to formulate what is now known as the "Chicago School."

The leaders of the school were Robert Park and Ernest Burgess. They viewed the city as an environment that functioned much like other ecological environments: It was formed based on the principles of invasion, dominance, and succession. In short, one group moves in, battles the previous group until they dominate the area, after which, to continue the cycle, it is likely that another group will invade the area and pursue dominance. This ecological approach was believed to explain the conflict that occurred in emerging cities across the United States. Moreover, it was Burgess (1925) who had earlier conducted a study that produced the notion that towns and cities "expand radially from its central business district—on the map" (p. 5). From this, he and Park produced their now famous map of Chicago. The map divided the city into several concentric circles or "zones," as described by Park and Burgess. Of the numerous zones, Zone 2 is of most significance to the theory. This area was referred to as "the capital zone in transition" or "the slums," which, according to the theory, is where most of the crime should take place. As predicted by the theory, the farther one moves away from this zone, the more crime decreases (Shaw & McKay, 1942/1969).

In the tradition of Quetelet's earlier work, two researchers, Clifford Shaw and Henry D. McKay, who worked at the University of Chicago's Institute for Juvenile Research, but were not faculty members, tested the theory by examining juvenile delinquency. To do so, they made use of 20 different types of maps. Each of the maps charted out different characteristics of Chicago's residents and delinquent youth. For example, there were maps that outlined neighborhood characteristics such as population fluctuations, percentage of families on welfare, monthly rents, percent foreign-born and Negro, and distribution of male delinquents (Shaw & McKay, 1942/1969). Their results were striking. As postulated by the theory, over several decades and with several changes in ethnic groups, Zone 2 had the most delinquency. Describing this dramatic finding, Shaw and McKay (1942/1969) wrote:

> . . . the proportions of Germans, Irish, English-Scotch, and Scandinavians in the foreign-born population in 8 inner-city areas underwent, between 1884 and 1930, a decided decline (90.1 to 12.2 per cent); while the proportion of Italians, Poles, and Slavs increased . . . the 8 areas maintained, throughout these decades, approximately the same rates of delinquents relative to other areas. (pp. 150–151)

In the end, the scholars concluded that the crime in these areas was caused by social disorganization. Social disorganization refers to areas characterized by the following conditions: (a) fluctuating populations, (b) significant numbers of families on welfare, (c) families renting, (d) several ethnic groups in one area, (e) high truancy rates, (f) high infant mortality rates, (g) high levels of unemployment, (h) large numbers of condemned buildings, and (i) a higher percentage of foreign-born and Negro heads of families (Sampson & Groves, 1989; Shaw & McKay, 1942/1969).

During the late 1930s and early 1940s, a 1923 graduate of the "Chicago School," Norman Hayner, while investigating crime in diverse communities populated by

Whites, Asian Americans, African Americans, and Native Americans, utilized social disorganization theory (Hayner, 1933, 1938, 1942).

Under the "Oriental crime" heading, Hayner (1938) compared the situation of the Japanese, Chinese, and Filipinos in the Pacific Northwest. Overall, he noted that each of the groups had little criminality in comparison with Whites. However, he decided to look at variations among the three Asian American populations. According to Hayner, the Japanese had low rates of crime and delinquency because of their closely integrated families, the efficiency and organization of the community, and the lack of acquaintance with American ways. The closely integrated families were characteristic of their home countries. The organization of the community reflected the concern of the community for their youth. At that time, they maintained language schools to keep youngsters busy after their regular school days. Community attitudes concerning crime were so strong that, according to Hayner's (1938) research, those caught engaging in crime would likely commit suicide before facing the community. Finally, some Japanese steered their youth away from American traditions because there was the belief that increasing Americanization would lead to increasing levels of crime and delinquency (Hayner, 1938). In general, both of the other two groups had more crime than the Japanese, which was explained as a result of their either becoming too Americanized or having a gender imbalance (significantly more males than females) often due to earlier government immigration restrictions.

Turning to Native Americans, Hayner examined their condition on three reservations in the Pacific Northwest. He sought to determine why these Native Americans had become engaged in more crime. The first of his explanations pointed to the increased mixing between Whites and Native Americans. In his words, "As the Indian becomes more like the White biologically and, as a result, associates more freely with him, his criminal behavior tends to approximate more closely the White patterns" (Hayner, 1942, p. 605). Hayner also pointed to the social disorganization that occurred among Native Americans because of social isolation, concerns about sustenance, and, in other situations, "some Indians, pauperized by too much money [from successful suits against the government] or unwisely administered relief; lack the incentive to work; others, including many boarding-school graduates, want to work but lack the opportunity" (p. 613).

CONTEMPORARY SOCIAL DISORGANIZATION THEORY

Since these early articles, scholars have continued to explore the viability of social disorganization to explain crime, particularly in urban areas. Sampson (1987) found a connection between Black male joblessness and economic deprivation and violent crime. This connection was an indirect one mediated by family disruption (i.e., female-headed households). Another important article by Sampson and Groves (1989) expanded the theory and found considerable support for it. Building on this prior research and the important research of William Julius Wilson (1987), Sampson and Wilson (1995) posited a theory targeted at explaining race and crime with structural and cultural constructs:

> [Our] basic thesis is that macro social patterns of residential inequality give rise to the social isolation and ecological concentration of the truly disadvantaged, which in turn leads to structural barriers and cultural adaptations that undermine social organization and hence the control of crime. This thesis is grounded in what is actually an old idea in criminology that has been overlooked in the race and crime debate—the importance of communities. (p. 38)

The theory, which is referred to as the racial invariance thesis, draws heavily on two of W. Wilson's (1987) concepts from *The Truly Disadvantaged*. The first, concentration effects, speaks to the fact that Whites and Blacks live in considerably different areas. In his research, Wilson found that many African Americans live in areas where there are significant concentrations of poverty. Once neighborhoods reach this point, working-class and middle-class African Americans abandon these areas.

This removes important "social buffers" (role models) who show neighborhood kids that there are successful people who go to work, day in and day out. When all the "social buffers" have abandoned a community, Wilson suggested that the remaining individuals are in a state of *social isolation*, which he defined as "the lack of contact or of sustained interaction with individuals and institutions that represent mainstream society" (p. 60). The notion of social isolation adds the cultural component to the theory. By not being exposed to mainstream individuals and institutions, socially isolated people tend to develop their own norms within these isolated areas. In a series of articles, Lauren Krivo and Ruth Peterson of Ohio State University have tested some of the ideas of Wilson (1987) and Sampson and Wilson (1995) and have found considerable support for them (see Krivo & Peterson, 1996, 2000; R. Peterson & Krivo, 1993, 2005). Returning to the perspective, Sampson and Bean (2006) have called for a revision of the theory to account for concentrated immigration and culture, both of which have profound implications for communities.

Scholars have also applied the theory to nonurban areas and with populations other than African Americans. Bachman (1991), for example, applied the theories of social disorganization and economic deprivation to explain Native American homicides. The results from her examination of 114 counties, which were located either all or partially on reservations, showed considerable support revealing, as in urban areas: "Both social disorganization and poverty contribute to high levels of lethal violence in reservation communities" (Bachman, 1991, p. 468). Recently, Lanier and Huff-Corzine (2006) also investigated whether social disorganization was applicable to Native American homicides. Their research found support for the theory in two areas. More specifically, female-headed households and ethnic heterogeneity were positively related to American Indian homicides. However, other aspects of social disorganization (level of poverty and residential mobility) were not found to be important variables.

Besides Native Americans, Martinez (2003) and Lee and Martinez (2002) have found support for aspects of the perspective in Latino communities. Velez (2006) argues that there is a lower level of social disorganization in Latino communities. She outlines four characteristics for this difference. First, she notes that there are lower levels of concentrated disadvantage in such communities. This includes things such as male joblessness and female-headed households (see Highlight Box 3.2 on female-headed households). Moreover, in contrast to conventional wisdom, she argues that more immigrants "provide protective mechanisms against crime" (p. 92). Third, she indicates that Latino communities have better relations with economic officials, the police, and local politicians, all of whom are key "players" in all communities. Finally, she argues that Latinos tend to benefit from living in communities that are close to advantaged Whites. Based on data from Chicago, she provides data that support her assertions (Velez, 2006).

Highlight Box 3.2

Long After the Alarm Went Off

By Ellis Case

Late in life, Sen. Daniel Patrick Moynihan complained that big ideas were no longer the special province of the Democratic Party. But in the '60s, the party had big ideas aplenty. And none were bigger than those championed by Moynihan, then an assistant Labor secretary under Lyndon Johnson. After reviewing reams of statistics, Moynihan came to a startling conclusion: the Negro family was collapsing. But with a faith characteristic of the time, he conjectured that government might be able to set things right.

The so-called Moynihan Report, written 40 years ago this month, was never meant for general distribution. Its language was notably impolitic—describing, among other things, families trapped in a "tangle of pathology." When the report came out, in the aftermath of a riot in Watts that left hundreds injured and 34 dead, it set off shock waves of its own. Moynihan was denounced as insensitive, perhaps even racist. Though the "pathology" concept had originated with Kenneth Clark, a distinguished black psychologist, that did not soften the critical blows. Moynihan's daughter, Maura, 6 at the time, recalls "seeing my poor father in utter despair." She thinks she remembers him crying.

Moynihan's report died a public death—a victim of ideological politics, misleading press coverage and the report's own loaded language. Yet the truth is that Moynihan was on to something—just not precisely what he thought he was. Indeed, some of Moynihan's concerns were not so different from those that recently launched comedian Bill Cosby on a family-values crusade.

Moynihan was actually telling two stories—one about the crushing, intergenerational impact of slavery and racism, which he got pretty much right, and the other about the transformation of the family, the scope of which he got wrong. His mistake was in assuming trends that he saw in black families were somehow peculiar to the black community and that the white family was a model of stability. What he was actually seeing, as historian Stephanie Coontz puts it, was "a rehearsal for something that was going to happen in the white community."

He was alarmed that nearly a fourth of black families were headed by females and that the illegitimacy rate was climbing. In his view, this "matriarchal structure" was a major problem because it put the "Negro community . . . out of line with the rest of American society." Moynihan could not foresee that the statistics, so alarming about blacks, would eventually describe reality for many whites. Nearly a fourth of non-Hispanic white children currently live in something other than a two-parent household. Roughly a third of American births are to single women, as are nearly a fourth of all non-Hispanic white births. White women under 25 are more likely to have a child out of wedlock than in.

As Coontz sees things, Moynihan erred not only in misdiagnosing the breadth of the phenomenon but in seeing pathology where it did not necessarily exist. A woman's decision to be alone is not always a bad thing. In her forthcoming book, "Marriage, a History," Coontz makes the case that many things we see as social improvements—from increased employment opportunities for women to widespread acceptance of the notion that spouses should be in love—inevitably have undermined the institution of marriage. Also, the very fact that many social indicators got better (crime rates went down, educational rates went up) even as single-parent families were on the rise shows that something more complicated was going on than Moynihan realized, Coontz contends.

No serious scholar argues that the trends Moynihan highlighted were essentially for the best. Single-parenthood, as he correctly surmised, tends to be associated with a host of bad things—including poverty and an increased likelihood of children's dropping out of school and into jail—especially if the single parent is un-educated and unemployed. Practically speaking, it's not so much single-parenthood that is the problem as the things that often come with it.

Moynihan can hardly be faulted for not foreseeing the implications of the data floating before his eyes. He sounded an alarm and made a passionate argument for true equality. Had more people paid attention, we might be farther along in figuring out how to solve the problems he sensed but only dimly understood.

Instead, Moynihan learned that some subjects were simply not to be broached—not by certain people, anyway. It's hardly surprising that five years after his report imploded, he advocated a period of "benign neglect"—an era that, with notable exceptions here and there, has essentially lasted for decades. By now, 40 years after Moynihan's ill-fated study, it should be clear to his critics and defenders alike that silence will not make difficult problems go away. Generally it just makes them that much harder to understand.

SOURCE: *Newsweek* (March 14, 2005, p. 37).

MASS INCARCERATION AND SOCIAL DISORGANIZATION

In the late 1990s, Todd Clear and Dina Rose articulated an expansion of social disorganization theory. Contrary to the punitive approach being heralded at the time, Rose and Clear (1998) posited that mass incarceration actually exasperated social disorganization in the most depressed communities. According to their thesis, this happens for three reasons. First, it impacts on the socioeconomic nature of the community. Second, because mass incarceration results in people leaving for prison and then being released from prisons, it increases the mobility in certain communities. Finally, mass incarceration increases the heterogeneity of communities. This occurs because you have offenders who spend time in correctional institutions where they learn new antisocial behaviors that they bring back to their communities (for a recent articulation of the perspective, see Clear, 2007; see also Western, 2006). Using data from Florida, they found considerable support for their theory (Clear, Rose, & Ryder, 2001; Clear, Rose, Waring, & Scully, 2003).

Collective Efficacy

A decade ago, Sampson, Raudenbush, and Earls (1997) sought to determine why urban communities differ in their levels of crime. From their research, they concluded that crime was related to the amount of collective efficacy found in a particular community.

They defined *collective efficacy* as "social cohesion among neighbors combined with their willingness to intervene on the behalf of the common good" (p. 918). In short, in the communities where residents do not retreat behind their locked doors and actively look out for one another, there is a diminished likelihood that they will have many of the ills found in similar urban areas. Since their work, other scholars have found some support for collective efficacy among African Americans (Simons, Gordon Simons, Burt, Brody, & Cutrona, 2005) and Native Americans (Abril, 2007). Other research has suggested that the impact of collective efficacy is not as significant in communities as more official strategies such as community policing (Xu, Fiedler, & Flaming, 2005).

Both social disorganization and collective efficacy generally speak to high-crime urban areas. Because not all African Americans live in high-crime urban areas, some have wondered if those in middle-class areas also encounter higher crime rates than those in similarly situated White areas. To investigate this question, Pattillo (1998) conducted participant observation and 28 in-depth interviews in one such area of Chicago. He found that "middle class Black areas tend to be nestled between areas that are less economically stable and have higher crime rates" (p. 751). In addition, many of those Black residents who make it to these middle-class areas are "unstable" middle-class residents and struggle to maintain their status. In some instances, they cross over the line into crime to do so. Therefore, Pattillo also found that such residents were "given a degree of latitude to operate in the neighborhood" (p. 770). Based on the premise of social organization, which, along with being goal oriented, "stresses the importance of kin and neighborly ties for the social control of crime and disorder," Pattillo showed how these communities maintain order while allowing "the integration of licit and illicit networks both working toward common goals, with variant strategies" (p. 770).

All in all, there has been considerable support for social disorganization theory. There have been several persistent criticisms of the theory, however. The most often cited weakness of the social disorganization perspective is the so-called ecological fallacy. This refers to the fact that the perspective is usually tested at the aggregate level, but researchers still use the data to make assertions about individuals. The theory also does not explain how certain groups, such as Asians and Jewish communities, maintained low levels of crime and delinquency, although they lived in areas that might be categorized as socially disorganized (Lanier & Henry, 1998). Moreover, although there were high levels of delinquency in the study areas, the theory does not explain why, in general, most juveniles in these areas do not become delinquent.

Culture Conflict Theory

Originally formulated by criminologist Thorsten Sellin in the late 1930s, culture conflict theory, according to Williams and McShane (2004), is heavily based on the work on Chicago School graduates Louis Wirth and Edwin Sutherland (who was to have collaborated with Sellin). The theory has several basic ideas. A central idea of the theory relates to the rules or norms within a culture. Sellin (1938) suggested that, over a period of time, certain behavior becomes accepted within a culture. Such behavior becomes so accepted that "the violation of [it] arouses a group reaction. These rules or norms may be called *conduct norms*" (p. 28; italics original).

Sellin's (1938) theory states that all societies have conduct norms, which vary from one culture to the next and could result in violations in one society not being a violation of conduct norms in another. Within each society, those in power can control the definitions of conduct norms and hence determine what behaviors become crimes. This leaves the potential for culture conflict. In general, Sellin pointed to three ways that conflicts between various cultural codes arise: (a) when the codes clash on the border of contiguous cultural areas; (b) when, as may be the case with legal norms, the law of one cultural group is extended to cover the territory of another; or (c) when members of one cultural group migrate to another.

Summarizing these ideas, Sellin (1938) formulated two types of culture conflicts. Of the first type, called *primary conflicts*, he noted: "[If] the immigrant's conduct norms are different from those of the American community and if these differences are not due to his economic status, but to his *cultural origin,* then we speak of a conflict of norms drawn from different cultural systems or areas. Such conflicts may be regarded as *primary* culture conflicts" (p. 104; italics original). Sellin described *secondary conflicts* as "conflicts of norms which grow out of the process of social differentiation which characterize the evolution of our own culture" (p. 105). As an example of the applicability of his perspective, Sellin used Native Americans as an illustrative population. He wrote:

> We need only to recall the effect on the American Indian of the culture conflict induced by our policy of acculturation by guile and force. In this instance, it was not merely contact with the White man's culture, his religion, his business methods, and his liquor, which weakened tribal mores. In addition, the Indian became subject to the White man's law and this brought conflicts as well, as has always been the case when legal norms have been imposed upon a group previously ignorant of them. (p. 64)

Although the theory clearly has relevance for Native Americans and the various ethnic immigrants who were arriving in America during the early part of the 20th century, in recent years, however, it has not received much attention. One exception to this trend was a recent study by Lee (1995), which looked at culture conflict and crime in Alaskan villages.

Based on comparative data on eight Yupiit Nation villages and eight non-nation villages, Lee (1995) explored the following proposition: "that colonization (invasion, dominance, and succession) results in conflict (disorder/ crime) related to the imposition of laws (social control) associated with the dominant group" (p. 181). Essentially, Lee wanted to find out whether crime was less prevalent in villages that maintained their traditional values. Moreover, Lee was interested in finding out whether villages that enforce alcohol laws have higher or lower alcohol-related incidents. On the first point, Lee found that "the rates for felonies and misdemeanors are lower in Nation villages with the exception of liquor violation, drunk in public, minor in possession, and protective custody" (p. 184). On the second point, the results were mixed; however, one thing was clear: The formal laws were "not keep[ing] the villagers from drinking or acting out while drunk" (p. 186). Overall, the study found support for culture conflict theory.

More recently, Abril (2007) has also found support for the theory. Nonetheless, criminologists have generally neglected the theory; they have borrowed some ideas from the theory and formulated related theories, such as strain theory, subcultural theory, and conflict theory. We look at these theories in the following sections.

Strain/Anomie Theory

In the same year as the publication of Sellin's work on culture conflict, another important theory was presented. The 1938 publication of Robert K. Merton's "Social Structure and Anomie" produced what is likely one of the most cited theories in criminology, strain or anomie theory (Lilly et al., 2001). The theory was influenced by the classic work of Emile Durkheim, who first made use of the word *anomie* in a criminological sense. According to Akers (2000), "Durkheim (1951[1897]) used the term anomie to refer to a state of normlessness or lack of social regulation in modern society as one condition that promotes higher rates of suicide" (p. 143). Merton's (1938) work showed that in every society, there are "culturally defined goals, purposes, and interest" (p. 672). He also suggested that there are generally "acceptable modes of achieving these goals" (p. 673). Turning to American society, Merton recognized that "the extreme emphasis upon the accumulation of wealth as a symbol of success in our own society mitigates against the completely effective control of the institutionally regulated modes of acquiring a fortune" (p. 675). In short, in pursuit of the "American Dream," some people turn to alternative means to secure this cultural goal. When applying the theory to race and crime, Merton recognized the special case of African Americans, writing,

> Certain elements of the Negro population have assimilated the dominant caste's values of pecuniary success and advancement, but they also recognize that social ascent is at present restricted to their own caste almost exclusively. The pressures upon the Negro which would otherwise derive from the structural inconsistencies we have noticed are hence not identical to those upon lower class Whites. (p. 680)

Merton (1938) understood that the strain experienced by African Americans was unlike any other in American society. Basically, no matter how much they sought to achieve the "American dream," they could never "legitimately" reach the status of Whites, so they maintained lower aspirations and were resigned to achieving a lower level of success and advancement. Such a situation likely contributed to a strain that resulted in some African Americans turning to crime.

Epps (1967) tested some of Merton's ideas regarding race and crime. Using a racially diverse sample (159 Whites, 111 African Americans, and 76 Asian Americans) of high school juniors from Seattle, Washington, Epps tested several hypotheses that looked at whether there were differences in delinquency and aspirations by race and class. He did find some differences between the occupational expectations and aspirations and the educational aspirations of the students. In all cases, Whites and Asian

Americans had higher aspirations and expectations than did African Americans. No support, however, was found for any of the other hypotheses.

More than 30 years after Epps's research, Cernkovich, Giordano, and Rudolph (2000) tested whether African Americans still subscribe to the American Dream and whether this is related to their involvement in criminal behavior. Making use of longitudinal data involving African Americans and Whites from private households and an institutional sample (both from Toledo, Ohio), the authors found

> that African Americans maintain a very strong commitment to the American dream. Blacks report higher levels of commitment to economic success goals than do their White counterparts and indicate that they are prepared to work harder and sacrifice more to realize them. Even though the young Black adults in our study report low incomes and are more likely to be unemployed than are Whites, they continue to maintain a very strong commitment to the American dream. (Cernkovich et al., 2000, pp. 158–159)

Their study, which also partially tested social control theory, found support for strain theory, but only in the case of Whites. That is, many of the variables used to test strain theory "were significant correlates of crime among . . . Whites in our sample but not among African Americans" (Cernkovich et al., 2000, p. 161), a finding that the authors could not explain, but curiously implied that the African American participants might not have been forthright with their answers—something that likely applied to all participants.

McCluskey (2002) also applied strain theory to Latinos. Using survey data from Denver and Rochester, she sought to determine whether strain theory was applicable to all ethnic groups. However, even when she took into account various aspects of Latino culture (e.g., family involvement, acculturation, and religiousness), her results indicated that "the adequacy of traditional strain theory in explaining Latino delinquency is relatively weak" (McCluskey, 2002, p. 198). Because strain was not applicable to all ethnic groups, she suggested that the creation of culturally specific models might be necessary.

LIMITATIONS OF STRAIN/ANOMIE THEORY

Most of the criticisms of strain theory have been leveled at Merton's original formulation of the theory. Bohm (2001), for example, noted that anomie theories have a middle-class bias; they presume that lower class individuals commit crimes in an effort to reach middle-class status. As was seen by some of the research reviewed, this is not always the case. Another persistent criticism is that the theories do not explain white-collar and government crimes. Given that people at this level have already achieved middle-class status, why, then, do they engage in crime? Even in its various incarnations, the theory is generally silent on this issue.

Bohm (2001) also suggested that the theory suffers from overprediction. As he put it, "If strain is caused by the inability to achieve the American dream and is as widespread as Merton implies, then there ought to be much more crime than occurs" (p. 80). Because of the shortcomings of strain/anomie theory, Agnew (1992) developed a revised version of the theory.

General Strain Theory

Robert Agnew renewed interest in strain theory by adding that the removal (or loss) of positive or introduction of negative stimuli into an environment can cause a strain such that, as with blocked opportunities, the removal or loss of positive stimuli from an individual can result in criminal behavior. As for the removal of positively valued stimuli, Agnew (1992) specifically pointed to the following: "loss of a boyfriend/girlfriend, the death of or serious illness of a friend, moving to a new school district, the divorce/separation of one's parents, suspension from school, and the presence of a variety of adverse conditions at work" (p. 57). Turning to the presentation of negative stimuli, Agnew pointed to the following: child abuse and neglect, criminal victimization, physical punishment, negative relations with parents, negative relations with peers, adverse or negative school experience, stressful life events, verbal threats and insults, physical pain, unpleasant odors, disgusting scenes, noise, heat, air, pollution, personal space violations, and high density.

Building on these ideas, Jang and Johnson (2003) used the National Survey of Black Americans (comprising a sample of 2,107 African American adults) to test whether Agnew's theory holds true for African Americans. In addition to testing core tenets of Agnew's work, they sought to determine whether African American religiosity, an area where research has consistently shown more commitment by African Americans than by other ethnic groups, has any impact in helping them cope when strain occurs. In contrast to the earlier research of Cernkovich et al. (2000), these authors found support for Agnew's modified version of strain theory, noting the following regarding the role of religiosity:

> We find that individuals who are religiously committed are less likely than those who are not to engage in deviant coping in reaction to personal problems because their religiosity buffers the effects of negative emotions on deviance as well as directly and indirectly (via outerdirected emotions) affects their coping strategies. (Jang & Johnson, 2003, p. 98)

Studies by Simons, Chen, Stewart, and Brody (2003), Eitle and Turner (2003), and Rocque (2008) also found some support for general strain theory. In the Simons et al. (2003) research study, the authors found that experiencing discrimination was a significant predictor of delinquency. Eitle and Turner's work revealed that disparities in crime commission were largely attributable to African Americans' increased exposure to stressors. Most recently, Jang and Johnson (2005) found additional support for their earlier research on the benefit of religiosity when coping with strain (see also Jang & Lyons, 2006). We now turn our attention to subcultural theory.

Subcultural Theory

In the 1950s, several theories were formulated that consider criminality tied to the development of subcultures among White middle-class youth. In *Delinquent Boys*, Albert Cohen (1955) argued that gang delinquency was associated with juveniles being unable to achieve status among their peers. When they are unable to meet established White middle-class standards, they establish their own values, which generally involves activities and behaviors that are in conflict with middle-class values.

While examining a diverse, lower class area in Boston, Walter Miller (1958) also formulated a subcultural theory. Referred to as the "focal concerns" theory, Miller's thesis was based on 3 years of field research in lower class areas of Boston. His focal concerns were considered values to which lower class residents adhered. These included trouble, toughness, smartness, excitement, fate, and autonomy. Trouble referred to youth engaging in risk-taking activities. Toughness represented the notion that one was fearless and could "handle oneself" in a physical encounter. Smartness referred to street smarts, which are valued in lower class communities. Excitement is the pursuit of thrill seeking. Fate represents the feeling that lower class youth believe that their lives are controlled by things over which they have no control. Autonomy was the final focal concern and represents the resentment that lower class youth have regarding the control others have over their lives (Miller, 1958).

Two years after Miller's work, Cloward and Ohlin's (1960) *Delinquency and Opportunity* pointed to the opportunity structure as the key to understanding gang activities. They suggested that when there are limited opportunities, youth join gangs with one of three orientations. Those who cannot find legitimate opportunities join criminal gangs whose aim is to make money through a variety of illegitimate avenues. If, however, there remain few illegitimate opportunities, the youth might join a "conflict" gang. Such gangs primarily engage in violent activities, doing whatever is necessary to maintain their status in the streets. Youth who end up in "retreatist" gangs are what Cloward and Ohlin refer to as "double failures." Because such youth did not make it in either legitimate or illegitimate opportunities, they retreat to drug usage.

The same year Miller published his theory, noted criminologist M. Wolfgang (1958) published *Patterns in Criminal Homicide*. This was significant because, as an outgrowth of this pioneering work, less than a decade later, he along with Franco Ferracuti formulated the subculture of violence theory, which has been used to explain homicide, particularly in the African American community. We review this theory next.

The Subculture of Violence Theory

As reflected in Chapter 2, African Americans and Latinos are overrepresented in the violent crime categories. Again, as reflected by the historical data, in the case of African Americans, this is nothing new. In the late 1950s, while studying homicides in Philadelphia, Wolfgang (1958) found high homicide rates among African Americans in

Philadelphia. In addition, Wolfgang found interesting results related to sex differences, victim–offender relationships, weapons involved, and motives for the homicides. From this research and that of his colleague Franco Ferracuti, who had also conducted homicide research in Italy, Wolfgang and Ferracuti (1967) formulated the subculture of violence theory. Their theory, which draws from several other criminological theories, consists of seven propositions.

These propositions speak to a range of factors that encapsulate the subculture of violence. Some of these factors include the fact that those invested in the subculture of violence are not violent all the time. Although the subculture is found in all age segments of society, it is found most in those in the late-adolescence to middle-age categories. Because those vested in the subculture do not see violence as an "illicit conduct," they have no feelings of guilt toward their actions (Wolfgang & Ferracuti, 1967).

Several authors have either critiqued or tested the theory as it relates to race and the commission of violent crimes. Hawkins (1983) provided one of the earliest and most comprehensive critiques of the theory. In doing so, he also provided an alternative perspective. We begin with a summary of his critique; then we turn to a brief overview of his alternative theory. Hawkins pointed to the following five major weaknesses of the theory:

(1) There is an extreme emphasis on mentalistic value orientations of individuals—orientations that in the aggregate are said to produce a subculture,

(2) The theory lacks empirical grounding and indeed is put in question by some empirical findings,

(3) Much of the theory has tended to underemphasize a variety of structural, situational, and institutional variables that affect interpersonal violence. For Blacks, these variables range from historical patterns developed during slavery to the immediate social context of an individual homicidal offense to the operation of the criminal justice system, past and present,

(4) Subcultural theory underemphasizes the effects of the law on patterns of criminal homicide, and

(5) There are other plausible ways apart from the inculcation of values by which the economic, political, and social disadvantages of American Blacks may produce high rates of homicide. (pp. 414–415)

Hawkins' (1983) alternative theory provided three propositions that were meant to address the holes in the subculture of violence theory. Proposition 1 states, "American Criminal Law: Black life is cheap but White life is valuable" (p. 415). Here, Hawkins believes that, based on history, Black lives have taken on less value than White lives; as a result, African Americans can kill other African Americans without fear of being punished. In line with this argument, Hawkins expanded the work of G. Johnson (1941) and presented a hierarchy of homicide seriousness, which punctuates the least and most serious types of homicides (see Figures 3.1 and 3.2). Hawkins' (1983) second proposition states the following: "Past and present racial and social class differences in

Figure 3.1   Johnson's Hierarchy of Homicide Seriousness.

Rating	Offense
Most Serious	Negro versus White, White versus White
Least Serious	Negro versus Negro, White versus Negro

SOURCE: Johnson (1941).

Figure 3.2   Hawkins' Hierarchy of Homicide Seriousness

Rating	Offense
Most Serious	Black kills White, in authority
	Black kills White, stranger
	White kills White, in authority
	Black kills White, friend, acquaintance
	Black kills White, intimate, family
	White kills White, friend, acquaintance
	White kills White, intimate, family
	Black kills Black, stranger
	Black kills Black, friend, acquaintance
	Black kills Black, intimate, family
	White kills Black, stranger
	White kills Black, friend, acquaintance
Least Serious	White kills Black, intimate, family

SOURCE: Hawkins (1983, pp. 420–421).

the administration of justice affect Black criminal violence" (p. 422). This proposition speaks to the lack of attention paid to prehomicide behaviors in the Black communities. Hawkins believes that, because various prehomicidal assaults in the African American community do not receive the attention they deserve, homicides that could be prevented are not. Such inattention is also a product of the poor relationship between African Americans and police agencies (Brunson, 2007; Jones-Brown, 2007; Stewart, 2007).

As a product of poor relations with the police, in some instances, response times are slower, and at some point, African Americans lose faith in the police and refuse to call on them for assistance in certain instances. Relatedly, once a homicide is committed and the police are called in, the lack of serious attention provides no deterrent effect to the community. The final proposition—that "economic deprivation creates a climate of powerlessness in which individual acts of violence are likely to take place" (p. 420) speaks to the association between socioeconomic disadvantage and violence, a c tion generally lacking in the subculture of violence theory, but that was incorp into Sampson's (1985) test of Wolfgang and Ferracuti's version of the theory. Sampson tested the theory, looking at disaggregated homicide rates for 55 of the largest American cities. According to Sampson, if the theory were correct, he would find that "Black offending rates should be related positively to percent of Black violent crimes, independent of other structural characteristics, particularly poverty and inequality" (p. 52). Using a variety of sophisticated methods, no support was found for the theory.

During the 1990s, the theory also was tested to determine its applicability to Black women. Mann (1990a) examined homicide data from six major cities and found Black women to comprise 77% of the female murderers. However, after taking all factors into consideration, she concluded, "These women are not part of a 'subculture of violence' but of a 'subculture of hopelessness.' Their fierce independence, their tendency to batter or to kill when battered and their almost insurmountable economic obstacles represent a constant struggle" (Mann, 1990a, p. 198). When Ray and Smith (1991) took up the subject the following year, they noted that if there is a "subculture of violence" among African American females, there must also be one among White females who had identical offending patterns, primarily committing homicides against males of the same race with whom they have a close relationship.

During the mid-1990s, Harer and Steffensmeier (1996) did find support for the subculture of violence theory. Their research made use of prison misconduct data and examined the theory as applied to institutional violence. They found that Blacks were twice as likely to be found guilty of violent misconduct as opposed to Whites, even after controlling for standard variables. Even so, they pointed out that the differences also could be attributable to racial discrimination in the disciplinary process, something their research did not take into consideration.

Cao, Adams, and Jensen (2000) tested the theory using General Social Survey (GSS) data from 1983 to 1991 (excluding 1985). Focusing on all core elements of the theory, the authors found that, in contrast to the theory, "Whites are found to be significantly more vocal than Blacks in expressing their support for the use of violence in defensive situations, with the effects of other factors held constant" (Cao et al., 2000, p. 54). Finally, the authors concluded with this:

> Based on our data and analyses, there is enough evidence to conclude that Blacks in the general U.S. population are no more likely than Whites to embrace values favorable to violence. Our findings thus repudiate the idea that the causes of Black crime are rooted in unique aspects of Black culture. (Cao et al., 2000, p. 58)

They suggested that, given the limited support for the theory, for scholars to continue to promote it as an explanation for racial differences in violence implies that all African Americans are violent, something that is "unfair and potentially racist in nature" (Cao et al., 2000, p. 58). More recent research also has found little support for the perspective (Chilton, 2004; Pridemore & Freilich, 2006).

Other criticisms also have been leveled at the theory. Covington (2003) noted that supporters of the theory "fail to explain how [the] Black subculture of violence came to be more combative than the White subculture of violence" (p. 258). Psychologists also have argued that Wolfgang and Ferracuti (1967) "ignore the psychological underpinnings of [the] subculture" (Poussaint; cited in Covington, 2003, p. 259).

THE CODE OF THE STREETS

A recent subcultural theory approach that has some connections to several of the approaches previously reviewed is the "code of the streets" (Anderson, 1994, 1999; see Highlight Box 3.3). Based on his research in Philadelphia, Elijah Anderson, an urban ethnographer, published a highly acclaimed article, "The Code of the Streets," which focused on interpersonal violence in an impoverished Philadelphia neighborhood and how residents in the area adopted "the code of the streets" to survive. Anderson (1994) believes that, "at the heart of the code is the issue of respect—loosely defined as being treated 'right,' or granted deference one deserves" (p. 82). In such an environment, something that has little meaning to one person might be interpreted as "dissing" by someone else and result in a confrontation that could lead to violence. Being able to defend oneself is also an important part of the code. Within such depressed neighborhoods, Anderson suggested that there are "decent" and "street" families. Decent families "tend to accept mainstream values more fully and attempt to instill them in their children" (pp. 82–83). Such families are also strict and teach their children to respect authority and act in a moral way. In addition, they are not seriously tied to the code.

Highlight Box 3.3

Code of the Street Is Keeping Crime Alive

By Elijah Anderson

As of 7 p.m. yesterday, 95 people had been killed in Philadelphia in 2005, a large proportion of them in impoverished neighborhoods.

The factors that give rise to inner-city street crime and violence are many and complex, but spring mainly from the circumstances of life among the ghetto poor. Among them are the lack of jobs that pay a living wage and the stigma of race; an ad hoc financial system borne of the lack of economic resources, combined with the fallout of drug trafficking and rampant drug use; lack of faith in police "protection" or fair treatment by the criminal justice system; and the resulting alienation and lack of hope for the future. Essentially, many community residents believe there are two different systems of law, one for black people and one for whites.

In this environment lacking so many elements found in civil society elsewhere, an oppositional culture, that of "the street," whose norms are often consciously opposed to those of mainstream society, has filled the vacuum. Compounding the situation is the ready access to guns.

Large numbers of African Americans live in racialized poverty and second-class citizenship. The extent to which this is true in some objective sense is less important than that it is what so many black people believe. Black people live in areas of concentrated urban poverty to which the wider system has abdicated its responsibility, or so many residents believe. Even if this is not objectively true, there is enough evidence for many of them to be convinced.

Many black males, by the time they reach post-adolescence, have had some contact with the criminal justice system, through "misbehavior" in school or assault or petty crime, and they have acquired some kind of record, the result being, in effect, the criminalization of inner city childhood. Moreover, the inner-city youth culture strongly encourages drug experimentation. By the time they

(Continued)

(Continued)

apply for a job, young men are often disqualified from low-income service jobs by background checks that include a police check and a urine test. Rejected for employment, they are left without money or resources and have even more limited possibilities. Too often, they resort to the underground economy of hustling, drugs and street crime.

Despite a real scarcity of economic resources in the inner-city poor community, to make ends meet, people engage in numerous everyday exchanges—bartering, lending, as well as illegal enterprises such as the drug trade. These are performed without the benefit of civil law.

Strikingly, there is a profound lack of faith in the police and the criminal justice system. The policing mechanism that thus most often matters is street justice, essentially an eye for an eye and a tooth for a tooth. If someone takes advantage of you, you must get even. Not to be even is to be vulnerable to further advances. When debts are made, they must be repaid. No one must be allowed to get away with anything.

The police are most often viewed as representing the dominant white society and not caring to protect inner-city residents. In some of the most isolated communities, when called, they may not respond, which is one reason many residents feel they must be prepared to take extraordinary measures to defend themselves and their loved ones against those who are inclined to aggression. Lack of police accountability has, in fact, been incorporated into this local status system: The person who is believed capable of "taking care of himself" is accorded a certain respect.

Simply living in such an environment places young people at special risk of falling victim to aggressive behavior. Although there are forces in the community that can counteract the negative influences—by far the most powerful being a strong, loving, "decent" family committed to middle-class values—both of these two orientations, decent and street, socially organize the community. Above all, this dual environment means that even youngsters whose home lives reflect mainstream values—and the majority of homes in the community do—must be able to handle themselves in a street-oriented situation.

In the code of the street, a set of informal rules governs interpersonal public behavior, including violence. The rules prescribe both a proper comportment and the proper way to respond if challenged. They supply a rationale that allows those who are inclined to aggression to precipitate violent encounters in an approved way.

In this setting, there is tremendous jealousy and envy. For a person to "have something" can be seen as disrespecting another; it can be confused with making the other feel small. So if another person has money, material things, or even simply good looks—anything that can be taken as a social good—such possession can be cause for an altercation to settle things.

On the streets, the distinction between street and decent is often irrelevant; everybody knows that if the rules are violated, there are penalties. Knowledge of the code is essential for operating in public. Families with a decency orientation often reluctantly encourage their children's familiarity with it.

At the heart of the code is the issue of respect—loosely defined as being treated "right" or granted the "props" one deserves. However, in the troublesome public environment of the inner city, as people increasingly feel buffeted by forces beyond their control, what one deserves in the way of respect becomes more and more problematic and uncertain. This, in turn, further opens the issue of respect to sometimes intense interpersonal negotiation. In the street culture, especially among young people, respect is viewed as almost an external entity that is hard-won but easily lost, and so must constantly be guarded, if not regularly updated.

The person whose very appearance—including his or her clothing, demeanor, and way of moving—deters transgressions feels that he or she possesses, and may be considered by others to possess, a measure of respect. With the right amount of respect, for instance, one can avoid "being bothered" in public. If one is bothered, not only may he be in physical danger but he has been disgraced or dissed. Many of the forms that dissing can take might seem petty to middle-class people (maintaining eye contact for too long, for example), but to those invested in the street code, these actions become serious indications of the other person's intentions.

Thus the street code has emerged where the influence of the police ends and personal responsibility for one's safety is felt to begin. It is exacerbated by the proliferation of drugs and easy access to guns. On Saturdays, the gun sellers roam the inner-city neighborhoods, selling guns out of the trunks of their cars to anyone with the money. At night, in some of the most disenfranchised neighborhoods, random gunshots are heard as children try out their guns.

This volatile situation results in the ability of the street-oriented minority (or those who effectively "go for bad") to dominate the public spaces, including streets and school playgrounds and hallways.

In these neighborhoods, there exists a crisis of civil law, and when civil law is weakened, street justice emerges to fill the void. And people die in disproportionate numbers.

Elijah Anderson, author of "Code of the Street," is the William K. Lanman, Jr., Professor of Sociology at Yale University.

SOURCE: *Inquirer* (www.Philly.com), editorial posted on Sun. April 10, 2005.

In contrast, Anderson (1994) described "street families," who loosely supervise their children and in many cases are unable to cope with them. Unlike the decent families, "They believe in the code and judge themselves and others according to its values" (Anderson, 1994, p. 83). Subsequently, their lives "are marked by disorganization" (p. 83). In such families, children learn early on that they must fend for themselves. This produces a cycle in which they also become vested in the code and take to the streets to prove their "manhood," which involves securing pretty women, being able to defend themselves, and being able to support themselves "by any means necessary."

In recent years, there has been some support found for Anderson's ideas focusing on Blacks (Baumer, Horney, Felson, & Lauritsen, 2003; Brezina, Agnew, Cullen, & Wright, 2004; Chilton, 2004; Stewart, Simons, & Conger, 2002; Stewart & Simons, 2006), Hispanics (Lopez, Roosa, Tein, & Dinh, 2004), and, more recently, young Black women (Brunson & Stewart, 2006). Other recent studies have also noted the role of rap in music in the perpetuation of the code of the streets (Kubrin, 2005). In contrast to these positive findings, Stewart, Schreck, and Simons (2006) recently found limited support for the perspective. In line with the theory, they postulated that those who adhered to the code of the streets would reduce one's likelihood of being victimized. However, their research revealed the opposite: Adherents to the code of the streets reported *higher* levels of victimization (see also McGee, 1999; McGee, Barber, Joseph, Dudley, & Howell, 2005; Stewart, Schreck, & Brunson, 2008).

Besides the need for nationwide replications of the theory, there have been other concerns expressed about the viability of Anderson's ideas. Commenting on one of the life histories presented in Anderson's work, J. Miller (2001) wrote that, based on the way Anderson described the person's prison experience, it could be that the prison, not the streets, is the more powerful contributor to the development of the code of the streets. Elaborating on this point, he wrote,

> I do not feel that Professor Anderson gives enough weight to the influences of prison on the code of the streets. It is no accident that most of the known violent gangs in California developed in the institutions of the California Youth Authority or the California prisons. Leadership is confirmed by a stint in prison. The walk, the "pose," the language, the argot, the dress, the focus of one's eyes, and the studied indifference all bespeak prison. (J. Miller, 2001, p. 157)

Wacquant (2002) provided a more expansive critique of Anderson's work, pointing to the "loose and over expansive definition of the code of the streets" (p. 1491). Another point of concern is that "there is considerable confusion as to the origins and vectors of the code of the streets" (p. 1491). Wacquant further observed,

> Because he starts from an overly monolithic vision of the ghetto and conflates folk with analytic concepts, Anderson cannot relate the *moral distinctions* he discovers in it to the internal *social stratification*. He thus boxes himself into a culturalist position with deeply disturbing political implications insofar as they render ghetto residents responsible for their own plight through their deviant values or role ineptness. (p. 1500)

In general, a common shortcoming of subcultural theories is that they ignore criminality in the middle and upper classes (Hagan, 2002). In addition, as noted in the critiques of Hawkins and Cao and his colleagues, tests of the theory (specifically the popular subculture of violence theory) have found minimal support. A final persistent criticism of subcultural theories is that, in most instances, they speak only to male criminality (Lilly et al., 2001).

One of the most popular theories used to explain racial differences in offending is conflict theory. Our discussion of the theory is presented next.

Conflict Theory

Conflict theory likely represents the most popular theoretical framework used to explain race and crime. The theory, which has seeds in many of the ones previously discussed, has some of its origins in Germany. Specifically, the works of German scholars Karl Marx, George Simmel, and Max Weber have been credited with providing the impetus for the theory. According to Lilly et al. (2001), "Theories that focus attention on struggles between individuals and/or groups in terms of power differentials fall into the general category of *conflict theory*" (p. 126; italics original). In short, when applying conflict theory to race and crime, one would look to whether the enforcement of laws and the distribution of punishment are done in a discriminatory manner. Although social class and gender also would be important to investigate, the way in which the White power structure administers justice would be of central concern to conflict theorists.

CONFLICT THEORY, RACE, AND CRIME

An early observer of race and crime, W. E. B. Du Bois studied under Weber and produced one of the earliest works to incorporate a conflict analysis (Gabbidon, 1999, 2007b; Taylor Greene & Gabbidon, 2000). In 1901, he published an article on the convict-lease system (for more discussion on this system, see Chap. 8), which spoke to the conflict perspective. Du Bois (1901/2002) traced the history of the system, whereby immediately after the passage of the Thirteenth Amendment, states leased convicts out to private landowners, who no longer had the free labor of African American slaves.

Du Bois wrote about how states strategically enacted various laws (referred to as the "Black codes") to snare Blacks into the criminal justice system so they could be returned to the labor force, which helped maintain the power and privileged status of southern White landowners. In the article, Du Bois (1901/2002) also rebutted the biological theorists of his day by declaring,

> Above all, we must remember that crime is not normal; that the appearance of crime among Southern Negroes is a symptom of wrong social conditions—of a stress of life greater than a large part of the community can bear. The Negro is not naturally criminal; he is usually patient and law-abiding. If slavery, the convict-lease system, the traffic in criminal labor, the lack of juvenile reformatories, together with the unfortunate discrimination and prejudice in other walks of life, have led to that sort of social protest and revolt we call crime, then we must look for remedy in the sane reform of these wrong social conditions, and not in intimidation, savagery, or legalized slavery of men. (p. 88)

By this time, as reviewed earlier, Du Bois had already made significant statements on crime, pointing to discrimination, segregation, lynching, and the attitudes of the courts as explanations for African American criminality (Gabbidon, 2001; Taylor Greene & Gabbidon, 2000). Other prominent scholars would find considerable support for Du Bois' ideas (Myrdal, 1944; Sellin, 1928, 1935; Work, 1900, 1913). In each case, the authors wrote of the discrimination and economic conditions that were contributing to African American involvement in the criminal justice system, matters that directly speak to conflict theory.

It would be some time, however, before the formal articulation of conflict theory (also referred to as critical criminology) and a little longer before it incorporated race as a central component. The development of conflict theory over the last 40 years is often credited to the writings of Chambliss (1964, 1969), Turk (1969), and Quinney (1970). Much of these writings were class-based analyses that suggested that capitalism, class structure, and the manipulation of laws were significant contributors to crime, and, as such, changing the structure of society would go a long way toward eliminating crime.

In addition to these scholars, in his classic work *Crime and Privilege*, Krisberg (1975), while articulating a critical perspective (referred to then as "New Criminology"), clearly added the dimension of race to the theory by integrating the history of criminal justice practices used to control oppressed groups and also highlighting the prison writings of George Jackson, Angela Davis, and other high-profile African American prisoners of the early 1970s. Notably, building on the work of Blauner (1972), Krisberg (1975) devoted a whole section of the work to race privilege, which in recent years has been translated into the notion of "White privilege" (see McIntosh, 2002). This notion of White privilege within criminal justice translates into more focus on "crimes in the streets," as opposed to "crimes in the suites." Such actions criminalize the actions of other races and poor Whites while minimizing or looking past the crimes of Whites in power.

Over the years, in several editions of his classic text *The Rich Get Richer, and the Poor Get Prison* (2004), Reiman has spoken of this in terms of white-collar crimes, environmental crimes, and other corporate crimes that kill thousands of people, who are primarily poor and American minorities, but rarely result in anyone being severely punished.

Hawkins (1987) further expanded the conflict model by examining it in terms of race, crime, and punishment. He emphasized the need to consider race discrimination in conflict theory. According to Hawkins, at the time, other considerations usually lacking in conflict theory included victim characteristics, region, and accounting for race-appropriate behaviors. Whereas the first two characteristics are self-explanatory, for the latter, Hawkins noted that anomalies found in some studies do not take into account behaviors that are generally committed by one race, which, when committed by another, result in a punishment that seems out of line. Finally, Hawkins also suggested that too often conflict theorists do not consider the power threat approach of Blalock (1967). The approach, which some have called a "power threat version of conflict theory" (Ellis & Walsh, 2000, pp. 384–385), argues that once a majority population sees a minority group encroaching on spheres traditionally reserved for majority group members, they respond in a number of ways, including additional social control (Hawkins, 1987). This usually comes in the form of increased investments in police forces. According to past and recent scholarship, there is support for the "power threat" thesis (see D'Alessio, Eitle, & Stolzenberg, 2005; Jackson, 1989; R. King, 2007; Sharp, 2006).

Around the same time of Hawkins' important research, Wilbanks (1987), a professor of criminal justice at Florida International University, published his controversial work, *The Myth of a Racist Criminal Justice System*. In contrast to conflict theorists, who argue that discrimination represents a significant factor when seeking to explain why minorities are overrepresented in the criminal justice system, Wilbanks argued that, although he believed there was some discrimination in the criminal justice system (using the analogy of having a few bad apples in a barrel), contrary to what was being espoused in much of the race and crime literature, he opined,

> I do not believe that *the system* is characterized by racial prejudice or discrimination against Blacks; that is, prejudice and discrimination are not "systematic." Individual cases appear to reflect racial prejudice and discrimination by the offender, the victim, the police, the prosecutor, the judge, or prison or parole officials. But conceding individual cases of bias is far different from conceding pervasive racial discrimination. (pp. 5–6; italics original)

Wilbanks' perspective became known as the "no discrimination thesis" (NDT).

Highlight Box 3.4

Walker, Spohn, and DeLone's (2004) Discrimination-Disparity Continuum

Building on the "no discrimination thesis" (NDT) and "discrimination thesis" (DT) articulated by Wilbanks and Mann, Samuel Walker, Cassia Spohn, and Miriam DeLone created the "discrimination-disparity continuum," which was presented in the first edition of their book, *The Color of Justice* (1996). The continuum provides a useful framework for the NDT/DT debate. As they see it, discrimination in the criminal justice and juvenile justice systems can fall somewhere along their continuum (see Figure 3.3). At the one end of the continuum is systematic discrimination, whereby there discrimination takes place "at all places of the criminal justice system, at all times, and at all places" (Walker, Spohn, & DeLone, 2007, p. 19). At the other end of the continuum, there is pure justice, which suggests that there is no discrimination in the criminal justice system and the overrepresentation of minorities in the criminal justice system is likely a product of offending patterns. In between these poles, you have institutionalized discrimination, contextual discrimination, and individual acts of discrimination. Walker et al. (2007) defined institutional discrimination as "racial and ethnic disparities in outcomes that are the result of the application of racially neutral factors, such as prior criminal record, employment, and demeanor" (p. 19).

Contextual discrimination relates to discrimination that occurs only in certain contexts. The examples presented by the authors include discrimination found in "certain regions, particular crimes, or special victim-offender relationships" (Walker et al., 2007, p. 19). For individual acts of discrimination, the authors suggested that this is when there are simply individuals within the criminal justice and juvenile justice systems—not whole agencies—engaged in discrimination.

After presenting their continuum, the authors surveyed the current race and crime literature and concluded that although the criminal justice and juvenile justice systems were once characterized by systematic discrimination, that is no longer the case. Their review of the literature suggests that the two justice systems are now characterized by contextual discrimination (see Walker et al., 2007, p. 420).

Exercise: Pick a topic related to race and crime (e.g., the death penalty, racial profiling, etc.) and find three refereed journal articles on the subject. After reviewing the articles, see whether, as a whole, you also find that contextual discrimination pervades your selected area of race and crime.

Wilbanks' book and its perspective initiated a series of debates between Wilbanks and Coramae Richey Mann. In contrast to Wilbanks' position, Mann (1990b) felt,

> The racism in the criminal justice system has become institutionalized in the same way that it has in other organizational segments of the nation such as education, politics, religion, and the economic structure; and the barrel *is* rotten. (p. 16; italics in original)

Mann's perspective became known as the "discrimination thesis" (DT). Although the debates became heated, the two had brought to the fore an issue that lay below the surface among criminologists for many years. In 1993, Mann responded with her contemporary classic *Unequal Justice: A Question of Color.*

Although the debate cooled after the publication of her book, the level of discrimination in the criminal justice system continued to be a central focus of race and crime

Figure 3.3   Discrimination-Disparity Continuum

Systematic Discrimination	Institutionalized Discrimination	Contextual Discrimination	Individual Acts of Discrimination	Pure Justice

SOURCE: Walker, Spohn, & DeLone (2007).

researchers (Walker et al., 2004; see Highlight Box 3.4 and Figure 3.3). Moreover, although Wilbanks never produced the second edition he planned to write (Wilbanks, 1987, p. x), other scholars have continued in his tradition (see, e.g., Delisi & Regoli, 1999; DiLulio, 1996; MacDonald, 2003, 2008).

Along with Hawkins' concern about the oversimplification of the theory, a few other shortcomings have been noted with conflict theory. Bohm (2001) noted that the perspective does not take into account individual differences. That is, not all people who are oppressed or discriminated against will respond the same way. Finally, some have suggested that, in some of its forms, the theory is not testable. A perspective related to conflict theory that has been applied to race and crime is the colonial model.

The Colonial Model

The colonial model has its foundations in the work of psychiatrist and activist Frantz Fanon (Tatum, 1994). Although Fanon used the model to examine the relations between Blacks and Whites in colonial settings, Blauner (1969) and Staples (1975), leaning heavily on intellectuals of the Black power movement, such as Stokely Carmichael and Charles Hamilton, were among the first to substantively apply the theory to crime. Applying the perspective to the conditions of African Americans, Blauner provided the following definition of *colonialism*:

> Colonialism traditionally refers to the establishment of domination over a geographically external political unit, most of them inhabited by people of a different race and culture, where this domination is political and economic, and the colony exists subordinated and dependent on the mother country. Typically the colonizers exploit the land, the raw materials, the labor, and other resources of the colonized nation; in addition a formal recognition is given to the difference in power, autonomy, and political status, and various agencies are set up to maintain this subordination. (p. 395)

Blauner (1972) also generally applied the model to Native Americans. In the work *Gringo Justice*, Mirande (1987) reviewed the historical treatment of Mexican Americans by the criminal justice system and formulated a theory of "gringo justice," integrating the colonial model and conflict theory. Although African Americans were not colonized in the sense that Native Americans or Mexican Americans were, according to Tatum (1994), internal colonialism, which is "when foreign control of a state or territory is eliminated and the control and exploitation of subordinate groups passes to the dominant group within the newly created society" (p. 41), produces many of the same characteristics of the more traditional colonization process. Such characteristics include "a caste system based in racism, cultural imposition, cultural disintegration and recreation and members of the colonized being governed by representatives of the dominant power" (p. 41). Such characteristics within a society leave the colonized with feelings of alienation, which results in either crime and delinquency or the desire to assimilate or protest.

All articulations of the theory note the important role that agents of the criminal justice system (or "internal military agents," as they are called by Staples, 1975) play in maintaining order in a colonial society. In the words of Blauner (1969),

> The police are the most crucial institution maintaining the colonized status of Black Americans. . . . Police are key agents in the power equation as well as the drama of dehumanization. In the final analysis they do the dirty work for the larger system by restricting the striking back of Black rebels to skirmishes inside the ghetto, thus deflecting energies and attacks from the communities and institutions of the larger power structure. (pp. 404–405)

R. Austin (1983) was one of the first to empirically test the theory. Using violence rates before and after the decolonization of the Caribbean island of St. Vincent, he sought to determine whether crime rates declined following the removal of British colonial rule. Although he did find that crime rates declined after the end of colonial rule, this did not hold true when he examined data related to murder and manslaughter. Here, Austin noted that the increasing availability of guns might have played a role in this finding.

Nearly a decade ago, Tatum (2000) provided one of the more comprehensive tests of the theory. In her test of the theory, she formulated several propositions related to the model, including the connections among race, class, and oppression; how race and class are associated with the availability of social support; and issues related to alienation. Relying on survey data from African American, Mexican American, and White juniors and seniors at two high schools in a major southwestern urban area, she found limited support for the model.

The colonial model has applicability for racial groups who have been subjected to colonization (most notably, Native Americans, African Americans, and Mexican Americans). There have been mixed results when the theory has been tested, and there need to be more direct tests of it. Tatum (1994) also noted several additional concerns with the theory. First, as reflected in other structural models, she noted that two people can be exposed to the same oppression yet respond differently; in such instances, the model does not account for the different adaptations. Second, as with conflict theory, the model is difficult to test. Another weakness of the model is that it does not adequately address class issues (Tatum, 1994).

Criminologist Agozino (2003) also has considered colonialism in his groundbreaking work *Counter-Colonial Criminology: A Critique of Imperialist Reason*. In the work, he argued that "criminology is concentrated in former colonizing countries, and virtually absent in the former colonized countries, because criminology is a social science that served colonialism more directly than many other social sciences" (p. 1). More specifically, Agozino focused on

> how imperialism used criminological knowledge and how it can be seen as a criminological project—imprisonment with or without walls, a widening of the net of incarceration, and how the close kinship between the two fields of knowledge and power, criminology and imperialism, served both. (p. 6)

He also highlighted that the discipline of criminology originated "at the height of European colonialism" (p. 6). As a product of these origins, he noted that "criminology is dominated by scholars in former colonial centres of authority," which has led to what he considers "theoretical underdevelopment through the concealment of the bloody legacy of colonialist criminology" (p. 6). Although on the surface his ideas might seem "controversial," it is clear that Agozino's work provides a critical new direction for race and crime theorists.

In general, however, the impact of colonialism on countries around the globe has been neglected too long by criminologists. Notably, scholars have begun to revisit the role of colonialism in crime and justice (see Bosworth & Flavin, 2007; Saleh-Hanna, 2008).

Integrated and Nontraditional Theories on Race and Crime

STRUCTURAL-CULTURAL THEORY

In the 1980s, Oliver (1984) proposed that, to explain Black male criminality, one needs to use an integrated theory combining structural conditions of African Americans and their cultural adaptations to such conditions. In one of his early articles, he explored Black males and their "tough guy image" or, as he called it, the "Black compulsive masculinity alternative" (Oliver, 1984). Because of racial oppression, Oliver believes that Black males exhibit masculine behavior that places an overemphasis on "toughness, sexual conquest, manipulation, and thrill-seeking" (p. 199).

Oliver has argued that Black males act this way for two reasons. First, "lower-class Black males who adopt the compulsive masculinity alternative do so in order to mitigate low self-esteem and negative feelings which emerge as a consequence of their ability to enact the traditional masculine role" (p. 199). The second reason relates to the notion that males who adapt the masculine approach pass it on to other males. In later publications, Oliver applied his theory to sexual conquest and the adaptation of an Afrocentric perspective to ameliorate social problems in the African American community (Oliver, 1989a, 1989b), and he also has examined violence among African Americans in barroom settings (Oliver, 1994). More recently, he has continued to refine his perspective (see Oliver, 2003, 2006).

One limitation of Oliver's perspective relates to the central role of low self-esteem. There has been some debate as to whether low self-esteem is really the central problem contributing to social problems among African Americans generally and African American males in particular (see Ross, 1992). Covington (2003) also has argued that Oliver's approach labels activities "race specific" that Whites also engage in. For example, many of the functions that bars serve for African Americans serve the same functions for Whites. Finally, Covington noted that in one of his studies, Oliver's "sample of African-American participants in violent transactions report that many of their fights seem to have been precipitated for non race-specific reasons that apply equally well to violent Whites" (p. 266).

ABORTION, RACE, AND CRIME

In a highly controversial paper, Donohue and Levitt (2001) proposed that more than 50% of the crime drop in the 1990s could be attributed to the 1973 *Roe v. Wade* Supreme Court decision that legalized abortion. They pointed to three important factors that support this thesis. First, they noted that the decline in crime coincided with the landmark decision and the period when those who would have been born would have reached their peak years of criminal activity (ages 18–24). Second, they suggested that the states that legalized abortion 3 years before the *Roe v. Wade* decision experienced earlier crime drops than the remaining states. Finally, they pointed to the fact that states that have the highest abortion rates also have had the largest declines in crime (Donohue & Levitt, 2001).

At the core of the theory are two premises. First, abortion reduces the pool of individuals who would later engage in crime. Second, the theory relates to race and crime in that, in this view, abortion is not random. According to their thesis, those likely to have abortions include unwed women, teenagers, and Blacks. As such, because of a host of challenges often faced by such parents, they would be less than ideal parents and place their children "at risk" for criminal activity. Donohue and Levitt's thesis, which has been widely disseminated in the scholarly community as well as Levitt and Dubner's (2005) best-selling book, *Freakonomics*, has garnered significant attention.

Among scholars, there has been vigorous debate about the veracity of their thesis, with some scholars supporting the thesis (Barro, 1999), and other researchers (some of whom conducted re-analyses of Donohue and Levitt's data) having found little to no support for the thesis (Chamlin, Myer, Sanders, & Cochran, 2008; Foote & Goetz, 2006; Hay & Evans, 2006; Joyce, 2004a, 2004b; Lott & Whitley, 2007). More recently, scholars have tried to apply the abortion and crime thesis to England and Wales and found no support (Kahane, Paton, & Simmons, 2008). Some noted scholars, such as Al Blumstein, have suggested that, although creative, the thesis does not give enough "attention to other factors, such as the decline in crack cocaine street dealing, the booming economy, and the efforts of police to keep guns away from juveniles" ("Renowned Criminologist Eschews Alarmist Theories," 1999, p. B5). Overall, although Donohue and Levitt (2004, 2006) have vigorously defended their perspective, there has not been a major shift toward using their thesis to explain the significant crime dip of the 1990s.

CRITICAL RACE THEORY

In addition to theories based on biological, sociological, or other traditional perspectives, critical race theory (CRT), which emanated from the critical legal studies movement during the 1970s (K. Russell, 1999) and closely aligns with radical criminology (Delgado & Stefancic, 2001, p. 113), represents another perspective that has proved useful for contextualizing race and crime. Founded by Derrick Bell, Richard Delgado, and other legal scholars, in recent years, it also has become more widely known in social science circles. The perspective has two goals. The first is to understand how the law is used to maintain White supremacy and continue to oppress people of color. The second is countering or stopping the use of the law to maintain White supremacy (K. Russell, 1999). It is here where critical race theorists have expressed concern about laws (e.g., three strikes and you're out) and practices (e.g., racial profiling, wrongful convictions, etc.) that directly impact on racial and ethnic minorities.

In addition to the aforementioned goals, there are several tenets of the perspective. First, racism is ever present in American society and is, thus, a daily occurrence in American society. The second tenet is referred to as "interest convergence," or the notion that Whites benefit (materially and in other ways) from racism, so they "have little incentive to eradicate it" (Delgado & Stefancic, 2001, p. 7). Third, critical race theorists believe race to be socially constructed, manufactured classifications. Here, critical race theorists are particularly concerned about the racialization of groups. Specifically, they express concern about "the ways the dominant society racializes different minority groups at different times, in response to shifting needs such as the labor market" (Delgado & Stefancic, 2001, p. 8). Critical race theorists also believe that, because of their distinct histories and experiences, racial and ethnic minorities have a "unique voice of color" perspective to offer society.

There have been a few persistent criticisms of the theory. First, because much of the work is based on storytelling and personal narratives, which move away from "objective" or "value-free" analyses, some critics have concluded that the perspective is unscientific. Also, according to K. Russell (1999), some have argued that "CRT amounts to academic whining about women and minorities" (p. 183). Even with these criticisms, the perspective has become a standard legal theory, especially among women and minority legal scholars.

Conclusion

Just as there is little consensus among criminologists about the causes of crime (Ellis & Walsh, 1999), there is even more debate about which theory best explains racial patterns in crime and victimization. However, one thing is apparent from the coverage in this chapter—numerous theories have been applied to the question of race and crime (see Table 3.2). In the beginning, scholars turned to the biology of African Americans, Native Americans, and Asian Americans to answer this question; however, over the years, this has changed. The decline in popularity of the biological approach gave rise to the sociological approach. Beginning with scholars such as Du Bois, the sociological approach continues to be a mainstay of those interested in studying race and crime. Subcultural approaches seem to have also maintained their place in the race and crime literature. Conflict theory now represents one of the more popular theoretical frameworks when studying race and crime. In addition, scholars are beginning to reexamine the role of colonization in race, crime, and justice.

It is interesting to note that, with the return of the biological-sociological debate in the form of Rushton's *r/K* theory, criminology has come full circle. Moreover, with the ongoing Human Genome Project, scholars have become even more interested in the role that genetics might play on human behavior and also in explaining racial disparities in offending. When one reviews the various theories, it seems safe to say that, although the research methodologies have become more sophisticated, many of the same ideas presented about race and crime 100 years ago remain popular today. As was the case a century ago, among racial and ethnic groups, African Americans remain the focal attention of theories related to race and crime. Chapter 4 looks at the police and their historical and contemporary roles in handling race and crime.

Table 3.2 Theoretical Contexts of Race and Crime

Theory	Context
Biological	White superiority, genetic inferiority, low IQ, physical characteristics, evolutionary factors
Sociological	Social conditions, social structure, heterogeneity, mobility, mass migration, impact of mass incarceration on communities
Culture Conflict	Cultural differences, violation of social norms within a community
Subcultural	Street culture, code of honor rooted in violence
Conflict	Racial discrimination in criminal justice
Colonial/Countercolonial	Role of imperialism, criminology's role in imperialism
Structural-cultural	Social structure, culture, and black masculinity
Legalized Abortion	Unborn "at-risk" children
Critical Race Theory	Law maintains White supremacy

Discussion Questions

1. Why are theoretical perspectives so critical for race and crime researchers?

2. How do biologically based theories explain racial differences in crime trends?

3. How do sociologically based theories explain racial differences in crime trends?

4. What is colonialism? How does it explain the disproportionate involvement of racial and ethnic minorities in certain criminal offenses?

5. In your view, which of the theories in the chapter best explains racial disparities in the criminal justice system?

Internet Exercise

Look at the index offenses presented at www.fbi.gov/ucr/ucr.htm and discuss which of the theories presented in this chapter can provide an explanation for homicide trends by race.

Policing 4

Police have historically enforced laws that are today widely regarded as violations of the human rights of racial and ethnic minorities. One may argue that the people who drafted these laws were a product of a specific time in history; however, such arguments are based on the assumption that human rights are relative and cannot be applied across time and place. We reject this assumption.

—Barlow & Barlow (2000, p. 53)

Police are on the front line of the criminal justice system. This means that a citizen's introduction to the administration of justice usually begins with contact with a police officer. Most Americans have positive attitudes toward the police, although minorities are less likely to view the police as favorably as Whites do. For example, in 2003, while 92% of Whites had either a great deal or some confidence in the police, only 80% of minorities and 73% of Blacks did (Maguire & Pastore, 2003). In 2007, reported confidence in the police (i.e., a great deal or some) decreased in all groups to 90% of Whites, 75% of minorities, and 65% of Blacks (Pastore & Maguire, 2007). Unfortunately, recurring incidents of police brutality, racial profiling, and, more recently, the use of tasers continue to erode confidence in the police. In their book entitled *Police in a Multicultural Society*, the authors of the previous quote note the long history of conflict between police and communities of racial and ethnic minorities (Barlow & Barlow, 2000). Yet as Russell-Brown (2004) noted, "The historically oppressive relationship between police and African-Americans is lost in contemporary discussions" (p. 66).

Today, police agencies are more diversified, professional, and technologically advanced than at any other time in our history. There are more minority, women, educated, and community policing officers. Many departments have a presence on the World Wide Web, utilize computer mapping, and have access to the National DNA Index System. Despite these advancements, the terrorist attacks on September 11, 2001, have led to what Forst (2003) has described as the "Terrorist Era of Policing," characterized by increased federalism, loss of civil liberties, greater reliance on private security, and a propensity toward errors of justice. The 2001 USA PATRIOT Act and its reauthorization signed into law in 2006 was enacted to counter terrorism, although it is viewed by some as infringing on the rights of citizens. During this era, there is increased concern about privacy and due process for all Americans, as well as closer scrutiny of "new" minority groups, including Middle Easterners and Muslims, and those who are identified as belonging to groups perceived as threatening (although they may not be), such as Sikhs. Aside from shifting the focus to "new" minority groups, it is unclear how police relations with "traditional" minority groups will unfold during this era. Will the shift in focus to terrorism, terrorists, and home-land security improve on or exacerbate the strained relationship between police and minority groups?

The goals of this chapter are to provide an overview of policing in America, present the history of race and policing minority groups in America, and examine several contemporary issues, including citizen satisfaction with the police, police deviance, and policing innovations. This chapter focuses primarily on race and policing in municipal (local) police agencies for two reasons. First, police histories usually focus on cities like Boston (MA), New York City (NY), Chicago (IL), and Charleston (SC). Second, most police are employed at the local level in cities and counties. We forgo an extensive overview of policing, which can be found in most law enforcement and criminal justice textbooks. Instead, our overview describes major components of the policing industry in America and provides recent employ-ment data for minorities. Our history of policing presents developments since the colonial era and relevant historical information on policing Native Americans, African Americans, Asian Americans, Latinos, and immigrant Whites. By examin-ing the past, we are able to better understand the complexities of police and com-munity relations today.

Overview of Policing in America

Police are the most visible symbol of governmental authority in our country. There are numerous federal, local, and state agencies, which vary in size and are bureaucratic and quasimilitary in structure. In addition, there are Native American (tribal) police agencies, special police agencies with limited jurisdiction (e.g., transit police), and private police. The roles and functions of police agencies are specified in federal and state statutes and local ordinances. In municipal agencies, the police role consists primarily of maintaining the social order, preventing and controlling crime, and enforcing the law. Police are responsible for social control, have a considerable amount of discretion (decisional latitude), and are authorized to use force in the performance of their duties. Public police agencies rely heavily on the citizenry and work closely with other local, state, and federal police agencies in task forces to address specific problems, including gangs, homeland security, immigration fraud, narcotics law enforcement, and terrorism.

Research on employment patterns and practices in police agencies tends to focus on large metropolitan agencies where minorities have made substantial gains. In large cities, minority representation increased from 29.8% in 1990 to 38.1% in 2000. Hispanic representation increased the most (from 9.2% to 14.1%), followed by Blacks (18.4% to 20.1%) (Reaves & Hickman, 2002b). Minority officers outnumber White officers in some agencies, although the ratio of minority police officers to minority group members in large cities has increased only slightly, from .59 in 1990 to .63 in 2000. This means that there were 63 minority officers for every 100 minority citizens. The ratio for Blacks increased to .74, for Latinos to .56, and for other minorities to .37 (Reaves & Hickman, 2002b). Although many officers are employed in large cities, 63% of local police departments employ only between 2 and 24 sworn officers (Reaves & Hickman, 2002a). Minority representation in smaller police departments as well as in state and federal agencies is still relatively low. Minorities comprised 22.7% of local sworn police officers and 17.1% of sworn officers in sheriffs' departments (Hickman & Reaves, 2003; Reaves & Hickman, 2003). Sometimes relying on percentages masks the actual representation of minorities in policing. For example, in 2000, 22.7% of the 440,000 sworn officers in local police agencies represents about 101,000 officers, and 17.1% of police in sheriffs' agencies represents almost 30,000 sworn officers (nationwide).

In 2004, there were more than 12,700 local agencies, 3,000 sheriffs' offices, 49 state agencies, 1,481 special jurisdiction agencies, and 513 other agencies (such as constables) in the United States. The number of sworn personnel in state and local agencies increased to 732,000, a 3.4% increase since 2000. Local municipal agencies employed approximately 447,000 sworn officers, and sheriffs employed about 175,000 sworn officers. In 2004, the 49 state police agencies employed 58,190 full-time sworn officers, a 3.3% increase since 2000 (Reaves, 2007).

At the federal level, passage of the Homeland Security Act in 2002 created the Department of Homeland Security (DHS) in order to provide a more integrated approach to security in the United States. The legislation transferred either all or part of 22 federal agencies to DHS, making it the largest employer of federal law enforcement officers, and created the Transportation Security Agency (Reaves & Bauer, 2003). Prior to the creation of the Department of Homeland Security in 2002, the Department of Justice (DOJ) and Department of Transportation were the largest employers of federal officers with arrest and firearm authority. The Immigration and Naturalization Service (INS) had the largest minority percentage (46.75%), comprised primarily of Latinos (38.1%). Other DOJ agencies, including the Federal Bureau of Investigation (FBI) (16.8%), Drug Enforcement Administration (17.7%), and U.S. Marshals Services (17.6%), had lower minority representation.

The 2004 Census of Federal Law Enforcement Officers identified 105,000 full-time personnel authorized to arrest and carry firearms. The U.S. Customs and Border Protection (CBP) and U.S. Immigration and Customs Enforcement (ICE) employ more than 38,000 officers. Most federal officers are White males, and about one third are members of a racial or ethnic minority group (see Figure 4.1) (Reaves, 2006).

Changes in police hiring practices during the latter part of the 20th century contributed to current representation of minorities in policing, especially in local agencies. Despite considerable progress, minorities are still underrepresented in law enforcement.

Figure 4.1   Gender and race of full-time federal officers with arrest and firearm authority, September 2004

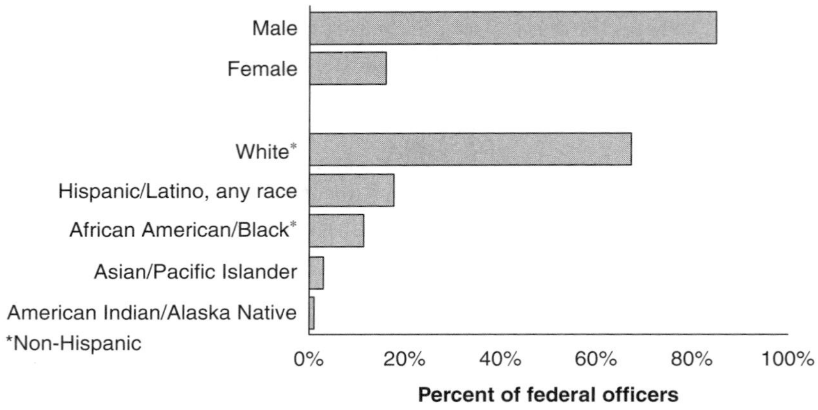

SOURCE: Federal Law Enforcement Officers, 2004 (Reaves, 2006, p. 5).

Efforts to make policing a more diverse profession often were met with resistance by police officers and the citizenry throughout American history.

Historical Overview of Race and Policing

Historical accounts of the police often exclude the horrendous treatment of minorities by the police, but rather focus on the need for social control of these "threatening" groups. Facts about policing minorities and some immigrants in the 18th, 19th, and 20th centuries are often omitted from criminal justice and law enforcement textbooks. It is impossible to separate the history of policing minority groups from the general history of the experiences of minorities in the United States, which were presented in Chapter 1. Similar to their treatment by most segments of our society, Native Americans, African Americans, Asian Americans, Latino/Hispanic Americans, and some White immigrants were treated differently from the so-called law-abiding native Whites by the police. Here, we provide (a) a brief summary of the history of policing in America, beginning with the colonial period to the present; and (b) a discussion of the historical treatment of race/ethnic groups by the police. It is important to remember that historical research on some groups is still limited and that policing practices varied considerably from one time to another and from one locale to another.

According to Walker and Katz (2002), "American policing is a product of its English heritage" (p. 24). During the colonial era, there were no formal police departments. Like their British counterparts, the colonists initially made policing the responsibility of every citizen and later utilized the sheriff, constable, and watch (Uchida, 1997). During this time, there were regional differences in policing. For example, in some places like Boston (MA) and New Amsterdam (NY), night watches were established, whereas in the South, slave patrols were more common. Over time, social disorder and crime increased so much, especially in northern cities, that the early forms of policing proved ineffective and were gradually replaced with more formal police agencies patterned after the British model (Fogelson, 1977; Lane, 1967; Monkkonen, 1981; Richardson, 1970). Unlike the British police, American police agencies were decentralized, locally controlled, and influenced by politics.

According to Emsley (1983), "The history of American police during the nineteenth century is the history of separate forces in separate cities" (p. 101). For example, White immigrants dominated police forces in some northern cities, and slave patrollers in the South were usually poorer Whites. Slave patrols relied on a more military style of policing. The first African American police were "free men of color" who served in police organizations in New Orleans, Louisiana, between 1805 and 1830. For the most part, they were responsible for slaves (W. Dulaney, 1996). D. Johnson (1981) mentioned that social tensions among immigrants, Blacks, and native Whites often resulted in conflicts and crime. Examples of these conflicts include verbal and physical abuse of abolitionists, race riots, labor strikes, and draft riots (Barlow & Barlow, 2000). Outside the urban areas, especially on the frontier, policing was characterized as vigilante in nature.

Although paid police forces that emerged in the 1800s were viewed as better than the night-and-day watches of the colonial era, social disorder and crime continued to be challenges. This led to the formation of private police, such as the Pinkerton Agency, and the reorganization of public police. Law enforcement textbooks allude to two important reform efforts. The first, during the late 1800s, attempted to reduce political control of the police. The next occurred during the early 20th century and focused on developing more professional police forces (McCamey, Scaramella, & Cox, 2003; Thurman, Zhao, & Giacomazzi, 2001; Walker & Katz, 2002). These reforms are important because the police were abusing their powers and resorting to unsavory practices such as corruption and the "third degree" (discussed later). Although the reform movement in the beginning of the 20th century was significant, much-needed personnel, policy, and training reforms did not occur until much later. During the 1960s, police were closely scrutinized following a great deal of civil unrest and several U.S. Supreme Court decisions, including *Mapp v. Ohio* (1961), *Escobedo v. Illinois* (1964), and *Miranda v. Arizona* (1966). Between 1970 and the present, policing has changed dramatically due to the convergence of several previously mentioned factors, including the hiring of minorities and women, technological developments, community policing, and the research revolution.

During the late 1980s, many police leaders and scholars attended the Executive Session on Policing at Harvard University's John F. Kennedy School of Government. Two participants, Kelling and Moore (1988), identified three eras in the history of policing often cited in many law enforcement textbooks: political (1840s to early 1900s), reform (1930s to 1960s), and community problem-solving (1970s to present). Williams and Murphy (1990) viewed Kelling and Moore's analysis as incomplete, noting:

> The fact that the legal order not only countenanced but sustained slavery, segregation, and discrimination for most of our nation's history—and the fact that the police were bound to uphold that order—set a pattern for police behavior and attitudes toward minority communities that has persisted until the present day. . . . The existence of

this pattern . . . meant that, while important changes were occurring in policing during our Nation's history, members of minority groups benefited less than others from these changes. (p. 2)

The following sections describe how the police (and other governmental officials) interacted with minority groups to uphold the legal and social order throughout American history.

NATIVE AMERICANS

In the foreword to F. Cohen's (1971) classic book entitled *Handbook of Federal Indian Law*, Robert L. Bennett and Frederick M. Hart described the complexities of understanding Indian law due to the numerous treaties, statutes, judicial decisions, and administrative rulings that emerged during the 1800s and 1900s. F. Cohen noted further that "Indian country" must be viewed temporally because federal and tribal law had changed over time. For example, although the act of May 19, 1796, set the boundary between Indian country and the United States, by 1799, the boundary was changed to give federal courts jurisdiction over U.S. citizens who committed crimes against Indians. The act of March 3, 1817, further extended federal law to prevent "White desperados" from escaping federal and state law (F. Cohen, 1971, p. 6). Indian country and/or territory were defined as "country within which Indian laws and customs and federal laws relating to Indians are generally applicable" (p. 5).

Following the Civil War, the United States minimized the political autonomy of tribal leaders and encouraged government representatives to deal with individual Indians and families (Tyler, 1973). As early as 1878, Congress established the U.S. Indian Police, which by 1881 comprised 40 agencies, 162 officers, and 653 privates (Tyler, 1973). By the 1880s, the West was becoming rapidly developed and settled by Whites, which precipitated the demand for the acquisition of Indian land and resources. According to Tyler, there were chiefs of Indian police on the reservations who often mediated between White man's law and Indian custom. At the same time, the decimation of tribal leadership often resulted in the breakdown of social control. In 1887, the General Allotment or Dawes Act permitted tribal lands to be allotted to individual Indians, making them landowners and farmers. Although not applicable to all tribes initially, over time, the Dawes Act was extended to the Five Civilized Tribes in Indian Territory (Cherokees, Choctaws, Creeks, Chickasaws, and Seminoles) by the Curtis Act of 1898. By the turn of the century, the allotments were primarily an opportunity for the U.S. government and White settlers to secure both land and natural resources.

Interestingly, the FBI's first major homicide investigation involved Native Americans. During the 1800s, oil was discovered on the land of the Osage Indian tribe of Oklahoma. Beginning in the late 1800s, the reservations began yielding barrels of oil. By the early 1900s, the tribe was extremely wealthy considering that "[i]n 1923 alone, the Osage Tribe received $27 million. In two decades, the Osage would receive more money from oil than all the old west gold rushes combined had yielded" (Hogan, 1998, p. 28). Eventually, this made them the target of elaborate schemes to defraud them of their wealth. After asking for assistance from the government, during the

1920s, the FBI infiltrated the tribe and uncovered a major scheme that involved inter-marriage and murder. In the end, the FBI solved the case, but had local officials make the arrests "since Special Agents of the FBI did not have the power of arrest (or the authority to carry firearms) at the time" (Hogan, 1998, p. 194).

On August 15, 1953, Public Law 280 transferred federal responsibility for criminal and civil jurisdiction, including law enforcement duties, to six states (Alaska, California, Minnesota, Nebraska, Oregon, and Wisconsin) and made it optional for sev-eral other states to assert jurisdiction (Luna-Firebaugh, 2003; Tyler, 1973). Although P.L. 280 did not terminate tribal jurisdiction, it was the belief that states and sheriffs were now responsible for law enforcement. When it was widely recognized that states were not providing adequate policing services, some tribal governments sought to reestablish and strengthen tribal police services. In 1974, almost a century after the federal government first established police in Indian country, police protection on reservations was described as follows:

> On reservations where State laws apply, police activities are administered in the same manner as elsewhere. On reservations where State laws do not apply, tribal laws or Department of the Interior regulations are administered by personnel employed by the Bureau of Indian Affairs, or by personnel employed by the tribe, or by a combination of both. (Bureau of Indian Affairs, 1974, p. 27)

In February 1973, a group of young Indians seized the village of Wounded Knee to protest conditions on the Pine Ridge reservation. Federal agents were called in to end the takeover, which lasted 71 days. Afterward, relations between the FBI and the American Indian Movement, a more radical American Indian rights group, continued to be tense. By 1976, the Bureau of Indian Affairs and/or tribes maintained law enforcement services on 126 reservations assigned to 61 Indian agencies. These agencies were responsible for around 200,000 square miles, 416,000 Indians, and numerous other non-Indians (Bureau of Indian Affairs, 1976).

In the 1990s, more than 56 million acres of land were owned and policed by tribal nations in the United States. "Indian country" includes reservations in 34 states, most of which are located west of the Mississippi River (Wakeling et al., 2001). Tribal police have responsibility for large land areas and often share criminal jurisdiction with fed-eral and state agencies, "depending upon the particular offense, the offender, the victim and the offense location" (M. Hickman, 2003, p. 4). According to the *Census of State and Local Law Enforcement Agencies, 2004*, there were 154 tribal police agencies with 2,490 sworn personnel (Reaves, 2007). As previously mentioned, tribes have assumed responsibility and control of policing on many reservations. The typical department is small, and therefore police protection is considerably lower than in other urban and rural agencies. The workload of officers in Indian police departments has increased as a result of several factors, including more reliance on the police (instead of traditional methods) and more crime and emergencies (Wakeling et al., 2001).

Today, there are several administrative arrangements for policing Native Americans that include cross-deputization agreements with several entities. Table 4.1 presents several tribal law enforcement agencies in Arizona, Michigan, New Mexico, Oklahoma, and South Dakota that illustrate the complexities of tribal law enforcement.

Table 4.1   Selected Tribal Law Enforcement Agencies Cross-Deputization and Arrest Powers

| | | Tribal law enforcement agency | | | | | | |
| | | Cross deputization agreements with | | | | | Arrest authority over | |
State	Tribal name	BIA	Neighboring tribes	Neighboring non-tribal authorities	Federal law enforcement other than BIA	Recognized by state to have peace officer authority	Tribal members off the reservation	Non-Indians on the reservation
Arizona	Navajo Nations	Yes	No	Yes	Yes	Yes	No	Yes
	White (River) Mountain Apache Tribe of the Fort Apache Reservation (Arizona)	Yes	Yes	Yes	No	Yes	No	Yes
Michigan	Saganaw Chippewa Indian Tribe of Michigan	Yes	No	Yes	No	Yes	Yes	Yes
New Mexico	Jicarilla Apache Nation, New Mexico (formerly the Jicarilla Apache Tribe of the Jicarilla Apache Indian Reservation)	No	No	Yes	No	Yes	Yes	No
Oklahoma	Cherokee Nation	No	No	Yes	No	Yes	Yes	Yes
	Choctaw Nation of Oklahoma	Yes	No	Yes	No	No	No	No
South Dakota	Standing Rock Sioux Tribe of North and South Dakota	No	No	Yes	No	Yes	No	Yes

SOURCE: Census of Tribal Justice Agencies in Indian Country (2002, Table 3, pp. 6–12).

The Relocation Program of the Bureau of Indian Affairs transplanted Native Americans to cities in the early 1950s. By 1973, more than 100,000 Native Americans had relocated to urban areas, where they were faced with social, economic, and psychological challenges (Fixico, 2000). Historical research does not focus on either police–Indian relations in urban areas or crime among Indians who relocated, although considerable attention is devoted to the problem of alcohol-related crimes both on and off the reservation.

AFRICAN AMERICANS

African Americans have posed a unique problem for America since their arrival as slaves in the 16th century. By the early 18th century, most of the early colonies regulated the movement and activities of free and enslaved Negroes by enacting special codes to totally control them. During the antebellum period, slaves were unprotected from crimes by slave owners, including murder, rape, assault, and battery (Kennedy, 1997). According to Dulaney (1996), by the mid-1700s, "patterrollers," or slave patrols, the first distinctively American police system, existed in every southern colony. Slave patrols carried out numerous duties, including checking passes of slaves leaving plantations, routinely searching slave quarters for stolen property, and administering whippings (Websdale, 2001). Relatedly, slave revolts were of increasing concern and, along with slavery, mandated a brutal policing mechanism to both protect Whites and dehumanize Blacks (T. Jones, 1977).

During the 1800s, in the North and South, both before and after Emancipation, free blacks were treated like slaves. Their movements were regulated: They were prohibited from entering several states and had to be employed to remain in states; in Virginia, an emancipated slave had to leave the state within 12 months or forfeit his freedom (Taylor Greene & Gabbidon, 2000). Kennedy (1997) describes the unwillingness of Southern Whites to recognize Blacks' rights to protection and the failure of local authorities to restrain or punish violence against them.

African Americans were excluded from policing in most cities until the mid-1800s. Black police served in Washington, D.C., as early as 1861 (Kuykendall & Burns, 1980). The first wave of Black police officers in the United States appeared during Reconstruction. Several cities in the North and South employed Negroes, including Chicago, Philadelphia, and Columbia and Charleston, South Carolina, and several cities in Mississippi and Texas. However, by the 1890s, most Negroes had been eliminated from police agencies in the South (Dulaney, 1996). At the same time, lynching was a serious problem. Some police participated in lynchings or silently supported attacks against African Americans by White mobs (Russell-Brown, 2004).

A similar problem occurred in numerous cities during race riots in the 19th and early 20th centuries. Race riots were clashes between Blacks and White citizens, in which the police were accused of indifference, brutality, and leaving the scene. With few exceptions, police did not protect Blacks from these attacks on their persons and property, and sometimes they even participated in these events. Police officers played a pivotal role in race riots in East St. Louis (1917), Chicago (1919), Tulsa (1921), Detroit

(1943), and Washington, D.C. (1919). In the 20th century, efforts were made to protect Blacks in some instances, and federal troops were sometimes called in to assist in restoring order (Grimshaw, 1969).

In 1938, Gunnar Myrdal undertook a comprehensive study of the Negro in the United States. His findings are reported in *An American Dilemma*, where he states,

> The average Southern policeman is a promoted poor White with a legal sanction to use a weapon. His social heritage has taught him to despise the Negroes, and he has had little education which could have changed him. . . . The result is that probably no group of Whites in America have a lower opinion of the Negro people and are more fixed in their views than Southern policemen. (Myrdal, 1944, pp. 540–541)

There were approximately 50 Negro policemen by 1940. At the time, three fourths of the Negro population lived in the South. Alabama, Georgia, Louisiana, Mississippi, and South Carolina, the states with the largest Negro populations, had no Negro policemen (Rudwick, 1960). Substantial gains were made by Blacks in law enforcement as a result of the civil rights movement of the 1960s, although it would take numerous court battles to integrate police departments, and the problems of underprotection and police violence persisted. Since the 1970s, there has been an increase in the number of Black police officers and police administrators. Black police organizations and police leaders have been instrumental in calling attention to police use of deadly force, brutality, and racial profiling, issues of concern to Black Americans (Taylor Greene, 2004). Nevertheless, the ratio of Black officers to Black citizens is still low, even in larger cities. It is important to note that employment gains by Blacks in law enforcement have caused some citizens to complain that their treatment by Black officers often is just as problematic as the historical treatment by White officers.

ASIAN AMERICANS

As discussed in Chapter 1, many Chinese immigrated to the United States as laborers. In 1876, there were 151,000 Chinese in the United States, and 116,000 were in the state of California. Like others, many Chinese immigrated to California in search of gold, although they were also known to serve as cooks, laundrymen, and servants (Norton, 2004). Early laborers were indentured servants who financed their passage by agreeing to work for merchant creditors (Courtwright, 2001). Around 1880, El Paso, Texas, became a point of entry for Chinese laborers from Mexico after the United States made entry for Chinese workers illegal. Often referred to as "coolies," the Chinese were involved in laying tracks for the Southern Pacific Railroad or remained in El Paso after completing railroad tracks that originated in the West (Dickey, 2004; Institute of Texan Cultures, 1998).

The fact that many Chinese immigrants were illegal did not seem to matter as much as their opium smoking and opium dens. Opium smoking was common among Chinese immigrants and confined to them between 1850 and 1870. Soon after, opium and opium dens gained popularity with the underworld of gamblers, prostitutes, and other criminals (Courtwright, 2001). In 1883, several dens were operating in El Paso and were frequented by members of both the lower and upper classes. Many believe that the opium dens were eventually targeted by the police and other governmental officials over concerns that Whites were mixing with Chinese, viewed by some as "polluting" (Institute of Texan Cultures, 1998). In the 1870s and 1880s, federal, municipal, and state governments passed laws that penalized opium smoking. According to Courtwright (2001), "Dens patronized by Whites were the most likely to be raided" (p. 77). By 1909, federal legislation banned the importation of opium for smoking by stipulating that it could be imported only for medicinal purposes.

After the anti-opium smoking legislation, imports decreased and coincided with a decrease in the size of the Chinese population that was also related to more restrictive immigration policies. During most of the 20th century, Asian Americans were more likely to remain in cultural enclaves and less likely to be involved in crime in their communities. Thus, although Asian gangs in California and other jurisdictions are still involved in drug distribution (heroin, ice), prostitution, and other illegal activities, the fact that most Asian Americans are insulated from police attention is due, at least in part, to their cultural values, ethnic isolation, and language barriers. More recent Southeast Asian immigrants and some second-generation immigrants target other Asian immigrants, Asian Americans, and businesses (www.knowgangs.com/gang_ resources/sea/southeast_asian_gangs_001.htm). These individuals, often belonging to gangs, are more likely to come to the attention of police today.

LATINOS

Research on the experiences of Mexican Americans in the Southwest and Los Angeles is more readily available than studies of the historical experiences with the police of other Latino groups such as Puerto Ricans, Dominicans, Colombians, and others. Trujillo (1974/1995) described how American capitalists who colonized the Southwest during the 1850s and 1860s created a cheap labor force and perpetrated an atmosphere of violence against Chicano immigrants. He maintained that repression of Chicanos by the police, military, and vigilantes led to the formation of protective guerilla units and bandits that are misrepresented in American history. Brutality against Mexican Americans included lynchings and murders. The number of brutal incidents in California and Texas prompted the Mexican ambassador to formally protest the mistreatment of Mexicans in 1912 (Kanellos, 1977). According to Chabrán and Chabrán (1996), Latino migrant workers in the Southwest received the harshest treatment. Because many of them were illegal immigrants, they were not as protected by the American legal system.

In the 1931 Wickersham Commission Report on Crime and the Foreign Born, three chapters specifically address the Mexican American immigrant. The focus is primarily on arrests, although there is some information on discriminatory treatment by the police. For example, P. Taylor (1931) found several individuals who noted how the police in Chicago and Gary, Indiana, were more likely to arrest Mexican Americans than Poles for drunkenness.

By the 1940s, police–Chicano relations in Los Angeles, where many Mexican Americans resided, had deteriorated to what Escobar (1999) described as extreme hostility and suspicion toward each other. Chicano youth were viewed as a social threat. On June 3, 1943, several sailors stated that they were attacked by a group of Mexicans, which marked the beginning of what is referred to as the "zoot suit" or "sailor" riots. The sailors retaliated by going to East Los Angeles and attacking Mexican Americans, especially those wearing zoot suits. The riots continued until June 7, when the navy finally put an end to the wanton attacks by declaring Los Angeles off limits for military personnel. The police and sheriff arrested several sailors, although none were charged with crimes. In contrast, hundreds of Chicano youth were arrested without cause (Suavecito's, 2004).

During the latter part of the 20th century, Latinos, like African Americans, continued to have strained relations with the police. The problems of illegal immigration, human smuggling, and drug smuggling, especially in the southwest region of the country, have increased the amount of contact between Latinos and state, local, and federal police agencies. Although their entry into the ranks of policing is not as dramatic as that of African Americans, their presence, especially in communities with large Latino populations, is more noticeable.

Photo 4.1   A group of Hispanic men stand chained together on their way to court in June 1943.

WHITE IMMIGRANTS

Throughout American history, the majority of immigrants have been of European descent. During the 1800s, immigration increased considerably as waves of Germans, Irish, and the "new immigrants" (Italians, Poles, Greeks, Portuguese, Jews, and others) arrived in the United States. These new immigrants were viewed less favorably than those who had arrived prior to 1850. This was due, at least in part, to beliefs that "incoming migrants from eastern and central Europe brought many negative social characteristics often attributable to their race," a term used rather carelessly to refer to nationality and/or culture (Graham, 2004, p. 25). By the late 1800s, immigration was seen by Progressive Era reformers as a threat to democracy, social order, and American identity (Graham, 2004). Poverty, illiteracy, unemployment, and underemployment in the ethnic ghettos meant that various types of criminal enterprises were common, including gambling, prostitution, theft, and con men (Fogelson, 1977).

The police experiences of White immigrants were drastically different from those of Native Americans, African Americans, Asian Americans, and Latinos. First, it was easier for foreign-born Whites to gradually blend into American society. Second, many White immigrants were able to secure jobs as police officers. Third, White immigrants often were assisted by the police in their transition to life in a new country. As early as the 1850s, the Boston police provided firewood and other necessities to Irish immigrants. Decades later, one function of the Boston police department was to distribute free soup in station houses, especially during the winter (Lane, 1967). Police also assisted parents looking for lost children (Monkkonen, 1981) and accommodated overnight lodgers (Lane, 1967; Monkkonen, 1981).

By the turn of the 20th century, tension between upper- and middle-class native-born and foreign-born Whites often resulted in more aggressive policing that emphasized law enforcement more than service. In some cities, policing of White immigrants depended on whether the police were controlled by political machines friendly to either immigrants or Progressive Era reformers seeking to restrict the immorality of immigrants (Brown & Warner, 1992). Ironically, because policemen were recruited from the lower and lower middle classes, they "had little or no inclination to impose the morality of the upper middle and upper classes on the ethnic ghettos" (Fogelson, 1977, p. 38). Later in the 20th century, serial killers, including Theodore Bundy, David Berkowitz, and John Wayne Gacy; White supremacists Randy Weaver and the Ruby Ridge incident; and domestic terrorists Timothy McVeigh and Terry Nichols, who bombed the Alfred P. Murrah Federal Building in Oklahoma City, were key factors in the development of criminal profiles and revisions to training and policy, especially at the federal level.

It is easy to lose sight of the fact that, throughout history, most persons who have been arrested (i.e., have had contact with the police) have been White. Historically, interactions between the police and Whites were less controversial than encounters between police and minority groups. Compared with early periods in American history, policing minorities has gradually improved as a result of the civil rights movement, federal legislation, affirmative action policies, human rights advocates, community policing, minority political empowerment, minority ascendancy in police organizations, policing research, and increased media attention focusing on police behaviors. Despite improved police–community relations, there are several contemporary issues regarding police and minority groups, which are discussed next.

Contemporary Issues in Race and Policing

As noted at the beginning of the chapter, the majority of Americans view the police favorably, although the percentages recently have decreased. Nevertheless, there is still variation in citizen satisfaction with the police, and several problematic police behaviors continue to capture the attention of all Americans. For example, minorities continue to be disproportionately the victims of police use of force and racial profiling. It is also unclear how recent innovations in policing impact minorities, including community policing, problem-oriented policing, and zero-tolerance policing. These issues are discussed next.

CITIZEN SATISFACTION WITH POLICE

Police need the cooperation of the citizenry to effectively perform their duties. Citizens who are dissatisfied with the police and are distrustful are less likely to partner with them to prevent and control crime. They also are unlikely to aspire to law enforcement careers (Ross, 2001). Race is one of the strongest predictors of attitudes toward the police (Weitzer, 1999). Efforts to gauge public opinion about the police include the measurement of several related concepts, including perceptions of satisfaction and confidence. Citizen satisfaction research assesses perceptions about different aspects of policing, including community policing (Skogan, Steiner, DuBois, Gudell, & Fagan, 2002; S. Smith, Steadman, Minton, & Townsend, 1999; Wilder, 2003), police misconduct (Sigelman, Welch, Bledsoe, & Combs, 1997; Weitzer, 1999), racial profiling (Weitzer & Tuch, 2002), and complaint procedures (Walker, 2001).

During the 1960s, when the National Opinion Research Center (NORC) conducted a national survey for the President's Commission on Law Enforcement and the Administration of Justice, they found that 67% of the general public responded that their police did an "excellent" or "good" job, and 77% responded that police did a "pretty good" to "very good" job of protecting people in their neighborhoods. Non-Whites (primarily Blacks) gave a rating of "very good" half as often as Whites and a "not so good" rating twice as often. Blacks were also more likely to view the police as discourteous (President's Commission on Law Enforcement and the Administration of Justice, 1967). In 1979, according to the National Crime Survey, whereas 54% of White victims of crime evaluated their local police as "good," only 25% of Blacks and 19.2% of Hispanic victims did (Carter, 1983).

The Gallup Organization also collects public opinion data. Citizens are queried on several topics, including their confidence in the criminal justice system, the police, honesty and ethical standards of the police, and perceptions of when it is acceptable for police to strike a citizen. The findings appear annually in the *Sourcebook of Criminal Justice Statistics*. Since 1998, the majority of survey respondents report a great deal/quite a lot of satisfaction with the police. In 2000, the Gallup Poll Social Audit found that Black Americans' general perception of fair treatment decreased to 36%, down from 49% in 1997 and lower than ratings of fairness in the 1960s, which ranged from 41% to 44% (Newport & Ludwig, 2000). More troubling is the fact that 55% to 64% of Blacks believe they are not treated as fairly as Whites by police (Newport & Ludwig, 2000). Table 4.2 presents trends in the percentage of respondents agreeing with the statement that they have a great deal/quite a lot of satisfaction with the police between 1998 and 2007. Although there was fluctuation during the 10-year period, the percentage of Whites, non-Whites, and Blacks reporting confidence was lower in 2007 than it was in 1998. Between 2005 and 2007, with only a few exceptions, confidence decreased nationally, as well as by sex, education, and community (see Table 4.2).

Table 4.2   Reported Confidence in the Police by Demographic Characteristic, United States, 1998–2007

Question: "I am going to read you a list of institutions in American society. Please tell me how much confidence you, yourself, have in each one – a great deal, quite a lot?"

	1998 Great deal/quite a lot	1999 Great deal/quite a lot	2000 Great deal/quite a lot	2001 Great deal/quite a lot	2002 Great deal/quite a lot	2003 Great deal/quite a lot	2004 Great deal/quite a lot	2005 Great deal/quite a lot	2006 Great deal/quite a lot	2007 Great deal/quite a lot
National	58%	57%	54%	57%	59%	61%	64%	63%	58%	54%
Sex										
Male	59	56	54	56	61	60	64	65	58	58
Female	57	57	54	58	57	61	64	61	58	51
Race										
White	61	59	57	59	63	65	70	66	63	60
Non-White[b]	40	43	38	44	43	43	43	53	40	32
Black	34	40	38	38	31	43	41	49	28	22
Age										
18 to 29 years	46	55	43	45	56	61	61	52	55	55
30 to 49 years	59	51	50	59	55	59	62	66	60	56
50 to 64 years	61	59	60	60	60	63	65	63	57	50
50 years and older	64	62	63	60	64	64	68	64	59	52
65 years and older	68	56	67	60	69	66	71	66	62	56
Education										
College post grad	55	57	54	57	62	67	66	72	65	59
College graduate	67	50	58	61	69	70	72	70	64	57
Some college	57	58	52	56	56	64	61	61	59	58
No college	57	57	54	57	56	53	64	59	53	48

(Continued)

113

Table 4.2 (Continued)

	1998 Great deal/quite a lot	1999 Great deal/quite a lot	2000 Great deal/quite a lot	2001 Great deal/quite a lot	2002 Great deal/quite a lot	2003 Great deal/quite a lot	2004 Great deal/quite a lot	2005 Great deal/quite a lot	2006 Great deal/quite a lot	2007 Great deal/quite a lot
Income										
$75,000 and over	–	59	56	66	68	70	69	72	65	59
$50,000 and over	59	58	56	62	65	68	70	63	69	51
$30,000 to $49,999	58	57	53	58	58	64	60	64	58	55
$20,000 to $29,999	65	53	55	57	51	48	57	48	45	44
Under $20,000	51	54	50	48	49	49	60	53	47	51
Community										
Urban area	50	52	50	51	53	59	60	61	55	52
Suburban area	60	60	57	58	60	64	64	66	62	59
Rural area	63	57	52	62	60	55	69	61	56	47
Region										
East	55	54	56	53	63	58	62	61	62	52
Midwest	59	64	51	57	56	62	68	64	58	59
South	56	56	54	57	53	64	63	65	58	53
West	62	52	54	59	64	57	65	62	56	53
Politics										
Republican	68	67	61	67	73	70	79	78	68	70
Democrat	54	50	52	57	51	58	59	60	54	44
Independent	55	54	50	47	52	55	55	52	55	52

SOURCE: Table constructed from the *Sourcebook of Criminal Justice Statistics.*

[a] Includes Black respondents.

In addition to public opinion polls, there is a considerable amount of research on citizen satisfaction with police in the context of their experience with police and neighborhood conditions (quality of life) (Reisig & Parks, 2001; Wilder, 2003). Direct and indirect contact with the police have been found to impact citizen satisfaction with the police differently, as do perceptions of crime, fear of crime, and neighborhood conditions. Some researchers have explored the intersection of race and class and found that both factors are important (Weitzer, 1999; Weitzer & Tuch, 2002). Weitzer (1999) found that perceptions of the extent of police misconduct were inversely related to neighborhood class and that middle-class Blacks experienced more suspicion and mistreatment outside their own neighborhoods. The lack of satisfaction and trust in the police does not bode well for citizens either working with the police or entering the law enforcement profession. One explanation for minorities' (especially Blacks') lower opinions and levels of trust of the police is related to their overrepresentation as victims of varying forms of police deviance.

POLICE DEVIANCE

Police deviance is a broad term used to refer to "police officer activities that are inconsistent with the officer's official authority, organizational authority, values, and standards of ethical conduct" (Barker & Carter, 1991, p. 4). Use of excessive force (including both police homicides and police brutality), corruption, perjury, sex, sleeping and drinking on duty, discrimination, and failure to enforce the law are examples of police deviance (Barker & Carter, 1991). Police deviance has been known to exist since the 1800s (see, e.g., Fogelson, 1977; Richardson, 1970; Sherman, 1974). When President Herbert Hoover appointed a commission to investigate the shortcomings of the administration of justice, the third degree, a mostly secret and illegal practice, was a common form of police deviance. The *Report on Lawlessness in Law Enforcement* stated,

> After reviewing the evidence obtainable the authors of the report reach the conclusion that the third degree—that is, the use of physical brutality, or other forms of cruelty, to obtain involuntary confessions or admissions—is widespread. (National Commission on Law Observance and Enforcement, 1931b, p. 4)

Deviant police practices continued to receive considerable attention throughout the 20th century. During the civil unrest of the 1960s, following several questionable police homicides during the 1970s and 1980s, and after the Rodney King beating in 1991, police abuse of their authority precipitated research and legislative changes. Another egregious incident of police deviance occurred in 1999 in Tulia, Texas, when 46 people (the majority of whom were Black) were arrested by Officer Tom Coleman and charged with being cocaine dealers. Many of them went to prison and later were pardoned by Governor Rick Perry after an appellate judge accused Coleman of lying, falsifying evidence, and racism (Drug Policy Alliance, 2008). In 2001, more than a dozen police departments, including Cincinnati, New York City, Detroit, New Orleans, and Tulsa, were under investigation by the U.S. Department of Justice (DOJ), Office of Civil Rights (OCR), to determine whether they engaged in a pattern and/or practice of either racial discrimination or brutality (Roane, 2001). The DOJ OCR continues to provide technical assistance to police agencies to remedy complaints of police misconduct and discrimination. Recent investigations include departments in Miami (FL), Warren (OH), Alabaster (AL), and Franklinton (LA).

Goldkamp (1982) identified two perspectives that help us to understand why minorities are overrepresented as victims of police use of force. Belief Perspective I points to differential law enforcement and the possible effects of prejudice and discrimination. Belief Perspective II posits that minorities are involved in crimes that increase their likelihood of victimization by the police. The National Organization of Black Law Enforcement Executives (NOBLE) offers some support for Belief Perspective I. They believe that justice is a system of people influenced by biases and stereotypes that are often incorporated into police decision making and actions. They have noted that the use of excessive force, brutality, and shootings of unarmed minority suspects and undercover officers are "symptoms and manifestations of biased-based policing" (National Organization of Black Law Enforcement Executives, 2001, p. 4). Another explanation of how discrimination manifests itself in police behavior is referred to as "reasonable racism." Kennedy (1997) explained,

> Most courts . . . have authorized police to use race in making decisions to question, stop, or detain persons so long as doing so is reasonably related to efficient law enforcement and not deployed for purposes of racial harassment. (p. 141)

More troubling is the fact that "reasonable" racial discrimination continues to be viewed as "a practical aspect of good law enforcement" (Kennedy, 1997, p. 141). He went on to state,

> Reasonableness, then, is not a definite, arithmetic objective quality that is independent of aims and values. It is a concept that is considerably more subtle, complex, malleable, and mysterious than the simplistic model of human decision making relied upon by those who accept at face value the "reasonableness" and "rationality" of conduct that not only expresses controversial moral and political judgments but that also expresses deep seated, perhaps unconscious, affections, fears, and aversions. (pp. 144–145)

Police discrimination is not as blatant and overt today. For example, police use of deadly force is no longer as extensive as it was before the U.S. Supreme Court decision in *Tennessee v. Garner* (1985), which found shooting of fleeing felons unconstitutional. Ironically, the police officer and victim in this case were both Black. Before the Court ruled in the *Garner* decision, researchers and practitioners were questioning the disproportionate number of minorities shot and killed by the police (see, e.g., Fyfe, 1981, 1982; Sherman, 1980; Takagi, 1974).

Goldkamp's Belief Perspective II has some support as well. The disproportionate involvement of minorities (especially Blacks) in arrests for violent crimes, as discussed in Chapter 2, leads to more contact with the police. The Death in Custody Reporting Act of 2000 mandates that all states collect and report data on arrest-related deaths. The first reported findings provide some support for Belief Perspective II; between 2003 and 2005, 2,002 deaths were reported. Of these, 1,095 arrest-related homicides were by police officers, and 75% involved arrests for a violent crime. In 80% of the homicides, a weapon was used by the arrestee. Relatedly, 16 of the 24 Conducted-Energy Devices (CEDs, Tasers) arrest-related deaths were persons arrested for violent crimes (Mumola, 2007).

An examination of police officers feloniously killed in the line of duty also provides support for Belief Perspective II. In 2006, 48 officers were killed; 12 died during arrest situations, 10 were ambushed, 8 were responding to disturbance calls, and 8 were killed during traffic pursuits or stops. Thirty-eight of the officers killed were White, 5 were Black, 1 was American Indian/Alaskan Native, and 1 was Asian/Pacific Islander. Among the 55 alleged assailants, 29 were Black and 25 were White (Federal Bureau of Investigation, 2006).

Opponents of Goldkamp's second perspective argue that many victims of police use of excessive force are not engaged in violent criminal acts when they are beaten or killed by police. Before Rodney King refused to stop his car, was he involved in a violent crime? When Abner Louima was savagely sodomized and beaten by New York City police officers, was he involved in a crime? Another victim of excessive force, Amadou Diallo, a Guinean immigrant, was suspected of rape and having a gun when he was shot and killed by New York City police in 1999, although neither was true. More recently, in December 2007, Sean Bell was killed by police in New York City after leaving a Queens strip club. Russell-Brown (2004) has succinctly noted,

> There is no evidence that the level of police assaults against Blacks increases and decreases on the basis of the rate of Black offending. Further, according to this reasoning, the majority of Blacks are required to pay the debt incurred by the small percentage of the group who are offenders. (p. 60)

Photo 4.2 A candle-light memorial dedicated to shooting victim Sean Bell (seen in picture with his fiancee Nicole Paultre).

Earlier in this chapter, some historical evidence of police brutality was provided (Myrdal, 1944; National Commission on Law Observance and Enforcement, 1931b). The International Association of Chiefs of Police has been developing a use-of-force database since the 1990s, although it is not representative of either police agencies or the population. Since 1995, 564 agencies have submitted data, and in 2001, about 225 agencies provided data (International Association of Chiefs of Police, 2003). If the limited data available are accurate, police brutality is rare, involves few officers, and occurs most often during arrests (Adams, 1999; International Association of Chiefs of Police, 2003). It is estimated that only 1% or approximately 446,000 police-citizen encounters involved either the threat or use of force (Adams, 1999). Between 1999 and 2000, the most common form of force was physical, followed by chemical and firearm force (International Association of Chiefs of Police, 2003). Most officers who did use force were White, and about 53% of victims were minorities.

More recently, the DOJ document entitled *Citizen Complaints About Police Use of Force* (Hickman, 2006) provides data from large state and local agencies collected in 2002. More than 26,000 complaints were received, most of them in large agencies. Only 8% of the complaints were sustained and warranted disciplinary action. It was estimated that the rate of police use of excessive force is about 1 incident per 200 full-time sworn officers. Like crime and victimization statistics presented in Chapter 2, this data source has methodological problems related to reporting and participation by agencies. One problem is the inclusion of only large state and local agencies. Another is the collection of data by police agencies.

Interestingly, police officers' attitudes toward treatment of minority and poor citizens also vary by race. Weisburd, Greenspan, Hamilton, Williams, and Bryand (2000) reported considerable variation in White, Black, and other minority officers' perceptions of treatment of Whites versus Blacks and use of physical force. Forty-seven percent of Black officers felt that police officers treat Whites better, compared with 11.2% of Whites and 21% of other minority officers. Forty-eight percent of Black officers felt that police are more likely to use physical force against Blacks and other minorities than against Whites, compared with 4.5% of Whites and 10% of other minorities. In a recent study in the Commonwealth of Virginia, Ioimo, Tears, Meadows, Becton, and Charles (2007) found that 21% of police officers believed officers practice biased-based policing, 15% reported witnessing such behaviors, and many believed it was a serious or somewhat serious issue. Biased-based policing and the historical pattern of police discrimination against minorities may help us to understand the contemporary practice referred to as *racial profiling*.

RACIAL PROFILING

Criminal profiling was popularized in the mid-1990s when John E. Douglas, a former FBI special agent, published his biography, *Mind Hunter: Inside the FBI's Elite Serial Crime Unit* (Douglas & Olshaker, 1996). Criminal profiles are a useful tool for identifying suspects and solving crimes, especially serious crimes. Today, profiling is not the exclusive domain of investigators and crime labs; it is also a tool of narcotics law enforcement and traffic enforcement. Police officers often use characteristics of a crime or group of crimes to develop a criminal profile (Harris, 2002). Profiling on the highways and in the streets is much more dangerous because it is based on beliefs about those who might commit crimes (predictive), rather than those who have committed them (descriptive) and is less formal (based on empirical support) and more informal (based on personal experiences) (Harris, 2002).

Racial profiling is a prime example of how far we have not come in police–community relations. It is not new, just a new name for an old problem (del Carmen, 2008; Meeks, 2000; National Organization of Black Law Enforcement Executives, 2001). During the 1990s, "driving while Black/brown" (DWB) was the most visible form of racial profiling, although other forms exist, including "walking and shopping while Black" (see Highlight Box 4.1) (Gabbidon, 2003; Gabbidon, Craig, Okafo, Marzette, & Peterson, 2008; Gabbidon & Higgins, 2007; Higgins & Gabbidon, 2008; Russell, 2002) and profiling in airports. Public opinion polls about racial profiling indicate that most Americans believe it exists (Barlow & Barlow, 2002; Gallup, 2004; McMahon, Garner, Davis, & Kraus, 2002), although there is less agreement about its definition. Racial profiling refers to "any action that results in the heightened racial scrutiny of minorities—justified or not" (Russell-Brown, 2004, pp. 98–99). This definition recognizes racial discrimination by store clerks, governmental officials, and police. In recent years, profiling has become a subspecialty within policing and has produced countless articles and books on the topic (for a recent summary of the extant literature, see Withrow, 2006).

Highlight Box 4.1

Study finds racial profiling of shoppers is real, but it goes unreported

Wednesday, October 10, 2007
   By Teresa F. Lindeman, Pittsburgh Post-Gazette
   Stacy Innerst/Post-Gazette
   Consumers who feel they've been singled out as potential shoplifters because of the color of their skin rarely make official complaints, choosing instead to shrug it off and share their painful stories with family and friends.

   "They have normalized the treatment—and accept it as a fact of life," said Shaun L. Gabbidon, a professor of criminal justice at Penn State's Harrisburg campus who has been looking into consumer racial profiling for several years. He spoke yesterday at the University of Pittsburgh's Center on Race and Social Problems.

   Dr. Gabbidon's latest research on the subject involved a telephone survey done last fall of almost 500 people, both blacks and whites, in the Philadelphia area. Forty-three percent reported they had the experience of being treated differently while shopping because of their race.

   He's convinced that, like racial profiling in traffic stops, the problem is real, but he is interested in trying to learn more about how consumers feel about the experience and how they respond. The results of the survey indicated that they don't like it but they tend to let it go.

(Continued)

(Continued)

Despite feeling angry, shocked, sad or embarrassed, 82 percent of those who said they had been racially profiled while shopping told the survey takers they never reported the experience to anyone other than their family and friends. About half still made a purchase at the store, said Dr. Gabbidon, who collaborated on the research with George Higgins from the University of Louisville.

Why buy despite the chilly reception? The offended shoppers sometimes wanted to prove they really could afford to shop in that store. Almost 40 percent even returned there to shop again.

That isn't likely to send the right message, said Dr. Gabbidon, who advocates alerting retailers about the problem. Otherwise, in his view, nothing will change. Eventually, if enough concerns are raised, a merchant worried that sales might be hurt will respond.

He became interested in the topic nearly two decades ago when he took a job after college as a store detective in a Macy's department store near Baltimore. Before long, he noticed he was getting a lot of calls to check on minority shoppers who didn't seem to have done anything suspicious. It didn't get any better after he was promoted to a store in Tysons Corner, Va. One clerk even called for him to check out himself.

Those surveyed last fall reported more than 50 percent of the employees who appeared to judge them based on race were white. But more than 20 percent were black, 11 percent were Asian-American and 5 percent were Hispanics, proving that the practice isn't limited to one group.

The majority of those doing the profiling were store clerks. "People bring their biases to work with them," said Dr. Gabbidon, who would like to see more research proving thieves come in all colors. Retailers, who really do lose a lot of money to criminals every year, might want to step up programs that train employees to focus on behavior, not race.

The situation has improved since the days when black customers weren't allowed to try on clothes in some stores, he said. Yet, even if consumers don't make a fuss about having someone follow them around stores when another shopper of a different skin color might be given free rein, he said the treatment creates unnecessary stress in people's lives.

Dr. Gabbidon hopes to learn more on the subject through future research that might look at other communities around the country or interview store clerks to study the problem from their side of the counter. He also plans to study data collected by the Pennsylvania Human Relations Commission, a state organization charged with investigating complaints of discrimination and monitoring bias-related crimes.

SOURCE: Lindeman (2007).

Here, we use Northeastern University Law Professor Deborah Ramirez's definition, as cited by National Organization of Black Law Enforcement Executives (2001):

> Racial profiling is any police-initiated action that relies on the race, ethnicity, or national origin of an individual rather than the behavior of an individual or information that leads the police to a particular individual who has been identified as being, or having been, engaged in criminal activity. (p. 5)

Fredrickson and Siljander (2002) argued that racial profiling does not exist because race is just one of several factors considered in the process of police profiling. Rather, they posited that criminal profiling may sometimes be racially biased and discriminatory. Harris (2002) credited lawyers in the New Jersey case of *State v. Soto* (1996) with demonstrating that at least in New Jersey criminal profiling had become racial profiling.

The drug courier profile is an earlier version of what is now referred to as "racial profiling during traffic stops." It remains unclear whether DWB traffic stops are driven either by honest efforts to control drug trafficking or the result of biased and racist beliefs about minority involvement in the drug trade, or are a more sinister and less recognized opportunity for police to seize drug assets. In 1996, in their controversial *Whren v. United States* decision, the U.S. Supreme Court exacerbated the DWB problem by granting police officers the power to stop persons suspected of drug crimes under the pretext of probable cause for a traffic violation. Whren and Brown, two Black males, argued that the police officers were racist, and that if they had been White, the traffic stop would not have occurred. Cooper (2001/2002) pointed out that the more troubling aspect of the *Whren* decision was the Court's ruling that a police officer's state of mind is irrelevant if probable cause for his or her action exists. For many observers, by ignoring the importance of an officer's racial prejudices, *Whren* opened the door to racial profiling during traffic stops (Russell, 2002). Since *Whren*, the Supreme Court has granted police even greater powers over drivers and passengers (American Civil Liberties Union, 1999).

Harris (2002) described the experiences of Blacks and Hispanics of varying classes with the police. During the 1980s and 1990s, African American motorists complained of being targeted and stopped on the New Jersey Turnpike; as a result, a group of lawyers filed a lawsuit on their behalf. The judge in the previously mentioned *Soto* case concluded that New Jersey State Police had targeted African Americans and that police administrators knew about the practice. Two years after *Soto*, in 1998, four Black men in a van traveling to North Carolina Central University for basketball tryouts were stopped by two troopers on the New Jersey Turnpike, and three of them were shot. This incident brought even more national attention to DWB (Harris, 2002). In 1999, the New Jersey Attorney General appointed a task force to study data collected by the New Jersey State Police and found race was a factor in vehicle stops (Russell-Brown, 2004). By 1999, the American Civil Liberties Union (ACLU) had filed lawsuits challenging racial profiling in eight states (American Civil Liberties Union, 1999).

Statewide studies and other racial profiling/traffic stop research provide overwhelming support for the overrepresentation of Blacks and Latinos in traffic stops (Becker, 2004; Dedman, 2004; Durose, Smith, & Langan, 2007; Engel & Calnon, 2004; Greenleaf, Skogan, & Lurigio, 2008; Growette Bostaph, 2008; Harris, 2002; Higgins, Vito, & Walsh, 2008; Institute on Race and Poverty, 2003; Lundman & Kaufman, 2003; Racial Profiling Data Collection Resource Center at Northeastern University, 2003; Russell-Brown, 2004; Warren, Tomsakovic-Devey, Smith, Zingraff, & Mason, 2006). In early 2008, 28 states had mandated the collection of data (Racial Profiling Data Collection Resource Center at Northeastern University, 2008). An analysis of traffic stops in New Jersey was an important part of the *Soto* decision. A pattern of racial profiling was also found in an analysis of traffic stops in Maryland (Harris, 2002). More recently, Becker (2004) analyzed racial profiling in the context of drugs found in vehicles stopped and searched in Maryland. The study reported inappropriate profiling of Hispanics and Blacks for any drugs and inappropriate profiling of Whites for large amounts of hard drugs (Becker, 2004). Findings from the 1999

NCVS on contacts between the police and the public indicated that Blacks were more likely than Whites to be stopped one or more times and that Black and Hispanic citizens felt that police acted improperly during the stop (Harris, 2002; Lundman & Kaufman, 2003). A more recent analysis of this data confirmed that young Black and Hispanic males are at greater risk of coercive action by police (Engel & Calnon, 2004).

In Minnesota, law enforcement officers were found to stop minorities, including Black, Latino, and American Indians, at higher rates than Whites. A pattern of stop-and-search disparities was found in almost every participating jurisdiction, including suburban and central cities, suggesting that racial/ethnic bias played a role in traffic stops statewide (Institute on Race and Poverty, 2003). Profiling of Hispanic and Black motorists was also found in a study conducted in the state of Kansas (Racial Profiling Data Collection Resource Center at Northeastern University, 2003). A state study of racial profiling in Massachusetts also found that minority drivers are disproportionately ticketed and searched in dozens of communities, including Boston (Dedman, 2004). In a survey of African American police officers in the Milwaukee Police Department, Barlow and Barlow (2002) found that 69% of their respondents felt they had been victims of racial profiling. The Racial Profiling Data Collection Resource Center at Northeastern University (RPDCRC) provides a state-by-state list of jurisdictions collecting racial profiling data. Twenty-five states have legislated collection of data, 21 states voluntarily collect data, and 4 states (Hawaii, North Dakota, Mississippi, and Vermont) and Puerto Rico do not require collection.

At the federal level, the Bureau of Justice Statistics collected data on police–citizen contacts in 2005 in a supplement to the NCVS that included 63,943 respondents ages 16 or older (Durose, Smith, & Langan, 2007). An estimated 19% of U.S. citizens (over 16) had face-to-face contact with a police officer in 2005. About 41% of all contacts were for traffic stops, and most drivers were stopped for speeding. Of those stopped, Black drivers were more likely to be arrested, and Hispanics were more likely to receive a ticket (see Table 4.3). The rate of drivers stopped was similar for Whites, Blacks, and Hispanics in 2002 and 2005, although minorities were more likely to be searched by police (Durose, Smith, & Langan, 2007).

One explanation for the continued overrepresentation of Black men in racial profiling incidents is the prevailing stereotypes about criminals rooted in inaccurate and discriminatory information. Several factors, including perceptions of Black crime, media images of Blacks as criminals, and racial hoaxes, may contribute to stereotypes of young Black men and justify racial profiling (Welch, 2007). A recent study of federal litigation involving minority plaintiffs (mostly Blacks and Hispanics) who allege they were racially profiled during a traffic stop did reveal the presence of criminal activity in more than half the cases (Gabbidon, Marzette, & Peterson, 2007). Given this finding, the authors reported that nearly 70% of the plaintiffs lost their case. The study also revealed that non-class action racial profiling litigation, at the federal level, inexplicably declined after peaking in the early part of the 2000s. The authors surmised that the heightened awareness surrounding racial profiling and the banning of the practice in police departments across the country might have accounted for the decline (Gabbidon et al., 2007).

One issue that is absent from the racial profiling research is whether traffic stops are used as a pretext for curtailing illegal immigration. Some states have enacted legislation requiring state and local law enforcement officers to act as federal immigration

Table 4.3 Enforcement Actions Taken by Police During Traffic Stops in 2005 by Demographic Characteristics

Characteristic of stopped driver	Total	Arrested	Ticketed	Issued a written warning	Given a verbal warning	No enforcement action was taken
Total	100%	2.4	57.4	9.1	17.7	13.5
Gender						
Male	100%	3.2	59.2	7.6	16.5	13.4
Female	100%	1.1	54.4	11.4	19.5	13.5
Race/Hispanic origin						
White[a]	100%	2.1	56.2	9.7	18.6	13.4
Black/African American[a]	100%	4.5	55.8	8.4	13.7	17.6
Hispanic/Latino	100%	3.1	65.0	5.9	14.5	11.6
Other[a,b]	100%	1.9[c]	63.6	7.0	16.0	11.5
Age						
16–19	100%	2.2[c]	60.7	8.7	19.2	9.2
20–29	100%	4.8	58.8	8.1	18.0	10.4
30–39	100%	1.9	60.8	8.1	15.3	13.9
40–49	100%	1.6	56.4	10.0	17.6	14.5
50–59	100%	1.3[c]	52.4	9.7	19.0	17.5
60 or older	100%	–	50.4	12.2	19.2	18.2

SOURCE: Durose, Smith, & Langan (2007, Table 8, p. 6).

NOTE: Total includes estimates for persons identifying with two or more races, not shown separately. Data on whether drivers were arrested, ticketed, or issued a verbal or written warning were known for 96.3% of cases. Detail may not add to total because of rounding.

– Less than 0.5%.

[a]Excludes persons of Hispanic origin.

[b]Includes American Indians, Alaska Natives, Asians, Native Hawaiians, and other Pacific Islanders.

[c]Estimate is based on 10 or fewer sample cases.

agents (*Minnesota Lawyer*, 2008). Activists and police agencies have taken varying positions on the role of state and local officials, which some believe is the responsibility of the federal government. Nevertheless, immigration status checks have become routine during traffic stops nationwide. At issue is whether an aggressive strategy of verifying residency during traffic stops results in racial profiling (S. Williams, 2007; Young, 2007). A recent article in the *St. Louis Post-Dispatch* highlighted the arrests of 52 illegal immigrants during routine traffic stops in the first 5 weeks of Missouri's Governor Blunt's get-tough policy. Immigration rights groups encourage police departments to take a hands-off immigration policy during traffic stops (see Highlight Box 4.2).

Highlight Box 4.2

Police urged not to check legal status: Activists want immigration standing off-limits in stops; some chiefs agree.

Byline: Scott Williams

Oct. 8—Christine Neumann-Ortiz had heard enough stories about immigrants facing deportation after traffic stops or random encounters with police, so the immigrant rights leader went right to the source of concern.

Neumann-Ortiz is asking local police departments for new policies to prevent officers from questioning people about their immigration status during unrelated investigations.

Some departments are going along.

A policy Milwaukee police officials adopted recently prohibits officers from asking immigration questions or alerting federal authorities to suspected illegal immigrants, with some exceptions.

Neumann-Ortiz's group, Voces de la Frontera, is asking other departments to follow suit, suggesting that enforcing federal immigration rules not only distracts police from investigating local crime but also drives a wedge between law enforcement and minorities.

"It creates kind of a poisonous culture," she said.

Others object to her view, saying that illegal immigration is too big a problem for any law enforcement agency to abdicate its role.

Racine Ald. Greg Helding said his community, for one, has no intention of adopting the policy sought by Voces de la Frontera. "It's just not going to happen," he said.

Helding questioned whether requiring police officers not to enforce certain laws would be legal. He said he would rather see Racine ask the federal government for special police authority to pursue immigration cases and start deportation proceedings.

Waukesha County considered seeking that special authority earlier this year, but officials dropped the matter after community leaders privately raised concerns about a potential police crackdown.

Anselmo Villarreal, executive director of the Waukesha agency La Casa de Esperanza, said that although he does not plan to pursue a hands-off-immigration policy with police, he warned authorities earlier that equipping officers with special immigration power would jeopardize good relations with the minority community.

"I truly believe that's not the solution. It's just going to create more problems," Villarreal said.

As illegal immigration has grown into a national issue the past couple of years, stepped-up enforcement efforts have produced deportation stories that rankled immigrant rights groups.

In 2004, a Waukesha County mother of two was sent back to India after she got a flat tire and police officers discovered that she had entered the United States illegally 12 years earlier.

Neumann-Ortiz recounted a situation in Illinois where a man was murdered and his grieving family was questioned at the murder scene about their immigration status.

Illegal immigration is generally treated as a civil offense—not a crime—and enforcement historically has been handled by the federal government with little local police involvement.

Milwaukee-based Voces de la Frontera decided to pursue a hands-off policy with area police departments after a police raid at a Whitewater factory where 25 illegal immigrants from Mexico were arrested in August 2006.

The Whitewater Police Department was one of the first departments to change its policies.

Whitewater Police Chief Jim Coan said he agreed to stop collecting Social Security numbers from traffic offenders and others because of concerns in the minority community that such information was used to trace immigration records. Not having Social Security numbers sometimes can make other police work more difficult, Coan said.

"It was a tradeoff that we were willing to make," he said. "It's a very polarizing issue—no question about that."

Milwaukee police spokeswoman Anne E. Schwartz released a copy of her department's policy and confirmed that it was updated in April, but she declined to comment further.

According to the policy, Milwaukee officers can question a person's immigration status or alert federal authorities only in cases of violent crimes, suspected terrorism, street gang crimes or other limited cases.

Neumann-Ortiz acknowledged that public sentiment in many circles favors tougher enforcement of immigration laws, and said federal reform is needed so that the nation has fewer illegal immigrants.

Yet the current push for local enforcement, she said, has created an environment in which Hispanics and other minorities are being subjected to racial profiling. She said her group's efforts to promote hands-off policies will defuse the situation.

"It really is about a separation of roles," she said. "It has become a new kind of battle front."

Do you think local police should ask for drivers' immigration status during traffic stops?

SOURCE: *Milwaukee Journal Sentinel* (Oct. 8, 2007).

Racial profiling research has methodological problems, including defining racial profiling, collecting data/measuring racial profiling, and selecting study variables (Schafer, Carter, & Katz-Bannister, 2004). Methods of data collection include officer-reported data, existing data sources, observational research, and citizen surveys (Schafer et al., 2004). Some researchers have focused on law-violating behavior, such as speeding and hit rates related to transporting contraband, and outcome tests that compare the hit rates or success of searches across racial groups (Engel & Calnon, 2004). One fundamental problem with racial profiling research is the heavy reliance on information provided by police officers, who are usually at the center of racial profiling controversies. Last, there is no required reporting to a centralized federal agency such as the FBI. As noted in Chapter 2, even national compilations of statistics are problematic. It is encouraging that questions on traffic stops are part of the National Crime Victimization Survey (NCVS). The Racial Profiling Resource Center identifies several benefits of collection data, including building trust and respect, providing information, and shaping officer training (Racial Profiling Data Collection Resource Center at Northeastern University, 2008). Collecting data also is important because it may, in and of itself, deter racial profiling by requiring that police officers record race/ethnicity-related data about traffic stops.

Engel and Calnon (2004) concluded that "targeting drivers solely or even partially on the basis of their race/ethnicity is not an effective, efficient, or responsible policing strategy at the national level" (p. 82). They suggested that profiling strategies may be institutionalized in police organizations and that the myth about profiling effectiveness may be hard to overcome. As more and more states make their data-collection analyses available, a better understanding of the extent and nature of profiling might emerge if methodological issues can be addressed. Although we cannot accurately assess the extent of police brutality and racial profiling, we can improve police accountability for incidents that do occur.

POLICE ACCOUNTABILITY

Police deviance is an inevitable outcome of the right to use force that Americans grant police. We entrust the police with the ability to use their power legitimately and want to hold them accountable when they do not. There are numerous mechanisms designed for police accountability: Citizens can file complaints with local, state, and federal agencies; bring civil lawsuits; and participate in civilian review boards or other external oversight agencies. Departments can conduct internal investigations, hold officers criminally liable for wrongdoing, use early warning systems, and implement training and policy directives. As mentioned earlier in the chapter, the DOJ can investigate a department's patterns and practices of either racial discrimination or brutality. Police officers who use excessive force are also liable under the U.S. Code, Title 42, Section 1983. Citizens are often dismayed over inadequate responses to complaints and the failure of internal investigations to find against police officers. In many cases of abuse, lawsuits are filed; if won, the citizens pay because the funds come from insurance coverage, not police departments. Most police departments view their internal affairs division as the most capable mechanism for investigating alleged police misconduct, and they resist citizen involvement in policing the police. This only exacerbates the problem of distrust because citizens who lack confidence in the police will be leery of internal investigations that find no wrongdoing in what appear to be cases of abuse.

To increase police accountability and bridge the gap between police and minorities, more effective measures must be implemented. Walker (2001) suggested that the monitoring function of some oversight agencies could be undertaken by citizens and police departments. Community outreach, customer assistance, policy review, and auditing complaint investigations are components of monitoring that may promote police accountability and change organizations. Police departments could also make the complaint review process more transparent by posting information on their Web pages. Some agencies have video equipment installed in police cars that can deter misconduct. In addition, many agencies are improving accountability by utilizing early warning systems that identify officers with a pattern of citizen complaints.

Crank (1998) presented one cautionary note, which he referred to as the "paradox of accountability":

> Administrative and citizen-based efforts to control accountability of individual officers will result in increased strengthening of the police culture and diminish the ability of administrators to hold individual officers accountable for their behavior. (p. 236)

Crank (1998) identified racism as one of many themes of solidarity in the police culture. He noted that there is evidence of police racism that is best understood if Americans look inside themselves because "the police do what we want them to do. They can't change because we won't" (p. 214). Despite this potential dilemma, police accountability is necessary. Many believe that some recent police innovations indirectly improve police accountability.

POLICE INNOVATIONS

There have been many police innovations in recent decades, including information systems, less-than-lethal weapons, community policing (COP), problem-oriented policing (POP), quality-of-life or zero-tolerance policing (ZTP), and multicultural/diversity training for officers. Here, we focus on three of these innovations: COP, POP, and quality-of-life policing or ZTP. These strategies have conceptual differences, although in practice, they are designed to prevent and control crime. For two decades, proponents of COP have touted its benefits over the traditional policing strategy. COP is a proactive approach that strives to bring police and citizens closer together in their efforts to reduce crime and disorder and solve related problems. Many Black police administrators, including Lee P. Brown, Hubert Williams, Rueben Greenberg, and other members of NOBLE, were early proponents of COP. In the 1980s, Brown and Williams were among the first in a cadre of progressive chiefs of police to successfully implement foot patrol programs in an effort to reduce the fear of crime in Houston, Texas, and Newark, New Jersey, respectively. COP was attractive to many progressive police administrators because it could be tailored to fit the needs of a particular neighborhood or jurisdiction. Other administrators and police officers resisted COP because it required a completely different police role; crime control was replaced with addressing not only crime, but also problems of disorder.

COP received a major impetus when the Crime Control Act of 1994 made federal funds available for hiring COP officers. At the time, although many agencies had adopted COP programs, most others were not convinced of its utility. During the past decade, research on COP addressing numerous topics and varying methodologies, including descriptive studies of implementation, organizational issues, evaluative studies, and analyses of citizen and officer perceptions of COP, informs us about its progress and obstacles. Today, most COP programs have a problem-solving component that originated in POP.

POP, as conceived by Goldstein (1979), differs from (early) COP programs in that it is a reactive approach that initially did not require citizen input. According to Piquero and Piquero (2001), POP requires that the underlying causes of repeated calls for service be identified and addressed. POP, sometimes referred to as "problem-solving policing," uses a process known as scanning, analysis, response, and assessment (SARA) to identify and respond to problems (Dempsey, 1999). Early and more recent problem-solving strategies often target public housing developments and other high-crime neighborhoods (Braga et al., 1999; Greene, Piquero, Collins, & Kane, 1999; Mazerolle & Terrill, 1997). More recently, the Indianapolis Directed Patrol Experiment and the Boston Gun Project have used POP approaches.

Before Mayor Rudolph Giuliani garnered national attention following the September 11, 2001, terrorist attack on the World Trade Center, he had established a reputation in law enforcement for instituting ZTP in New York City in the early 1990s. ZTP is based on "broken windows theory" and focuses on crime and disorder by utilizing aggressive policing tactics to restore order. In New York City, ZTP targeted minor crimes such as urinating in public, prostitution, and public drunkenness, although it is not viewed as improving police accountability.

Politicians in New York City claimed that ZTP and computerized statistics (COMPSTAT) drastically reduced crime (Walker & Katz, 2002). Shortly thereafter, many other cities adopted some version of ZTP. Unfortunately, ZTP is based on principles that often alienate the public, especially those who live in neighborhoods plagued by problems of disorder and crime.

COP, POP, and ZTP have had varied responses in different places, and the verdict is still out as to whether they improve police–community relations. Because there is little research that separates COP and POP, the focus here is COP. Some municipal agencies have excelled at adopting a philosophy of COP that is emphasized in their organizations' mission statements and values, although they have had more limited success in gaining support from minority groups. Kusow, Wilson, and Martin (1997) noted that citizen satisfaction with the police, as well as community trust and confidence, are prerequisites to successful COP programs. The NCVS surveyed residents in 12 cities with COP and found that, although 90% of Whites were satisfied with police who served their neighborhoods, only 76% of Blacks and 78% of other minorities (excluding Latinos) were satisfied (S. Smith et al., 1999). An evaluation of the Chicago Alternative Policing Strategy (CAPS), a community policing/problem-solving approach, found that Latinos were less likely to be familiar with it. After CAPS was implemented, most Whites perceived improvements in the quality of policing, although Latinos and Blacks were more skeptical. Latinos were not as aware or as participatory in beat meetings as others. Minorities also perceived little improvement in police misconduct after the implementation of CAPS (Skogan et al., 2002). Websdale (2001), in a case study of COP in Nashville, Tennessee, concluded that residents felt it did not make much difference that the crime problems remained the same at night and that community interest dissipated over time. Weisburd et al. (2000) reported that nationally Black officers were more likely than Whites and other minority officers to believe that COP could reduce the use of excessive force. Although COP continues to be an important policing innovation, priorities at the federal, state, and local levels have shifted to homeland security, immigration, and terrorism.

Conclusion

Race and policing is an important topic in the study of crime. This chapter provided a brief overview of policing and traced its history, paying specific attention to policing racial and ethnic groups. The quality of policing has improved dramatically during the past 200 years. Countless individuals and organizations have labored to ensure that citizens are provided with police who understand their role to protect and serve. New recruits receive more training and are better educated than ever before. Officers who successfully complete the police academy are more representative of minorities in their communities. The changing face of police officers did not occur overnight and was initially resisted by the police establishment. Today, there is still a need for affirmative action at all levels of government, and racial conflict within police departments as well as between police and the citizenry continues.

Despite progress, police deviance and misconduct are ongoing problems in policing. Cases of police brutality and racial profiling continue to occur and thus contribute to lower confidence and less trust in the police by Blacks, other minorities, and some Whites. Recent efforts to curtail illegal immigration have led to more allegations of racial profiling during traffic stops. The data collected on racial profiling attest to differential treatment of minorities (Blacks and Hispanics). In spite of this, the U.S. Department of Justice is considering a policy that appears to legalize investigations of Americans based on their race and ethnicity that would be used to develop profiles and strengthen the intelligence capabilities of the FBI (Jordan, 2008). Although police brutality and use of excessive force incidents are thought to be rare, there is still no reliable data set. New policing strategies like community policing have helped improve police and minority group relations in some jurisdictions, although ZTP seems to be more detrimental as a result of aggressive policing.

Discussion Questions

1. Why has the public's confidence in the police decreased in the recent past?

2. Why do you think minorities are often involved in brutality and profiling incidents?

3. How reliable do you think police–citizen contact data are?

4. Is racial profiling of minorities the same as profiling criminal suspects?

5. How can you tell whether community policing is a major focus of the police agency from which you come?

Internet Exercise

Use the RPDCRC to compare the most recent reports for two states in two different regions: http://www.racialprofilinganalysis.neu.edu.

Internet Sites

American Civil Liberties Union: http://www.aclu.org/PolicePractices/PolicePracticesMain.cfm

Racial Profiling Data Center: http://www.racialprofilinganalysis.neu.edu

Arrest Related Deaths in the United States, 2003–2005: http://www.ojp.usdoj.gov/bjs/abstract/ardus05.htm

Courts 5

My experience in the criminal court is that the colored defendant, even in bailable cases, is unable to give bail. He has to stay in jail, and therefore his case is very quickly disposed of by the prosecutor. Defendants locked up are usually tried first. The colored man is more apt to be out of work than the White man, and that is a possible reason for the large number of arrests of Negroes. His sphere is very limited, and if there is any let up in the industry that is involved in that sphere, he is a victim. I have often wondered if you could change the skin of a thousand White men in the city of Chicago and handicap them the way the colored man is handicapped today, how many of those White men in ten years' time would be law-abiding citizens.

—Judge Kickham Scanlan
(cited in Chicago Commission on Race Relations, 1922, p. 356)

As one of the key components of the criminal justice process, the courts represent an entity in which nationally billions of dollars are spent each year. Moreover, according to the Bureau of Justice Statistics, in 2000, these courts dealt with more than 1 million adults who were convicted of felonies (Durose & Langan, 2007), most of whom were handled in state courts. Furthermore, 38% of those convicted in these courts were Black (Durose & Langan, 2007). Given these statistics, prior to and following Judge Scanlan's telling comments, the question of the courts and their treatment of persons of color have piqued scholarly interest. It is also important to note here the central role that class and social status plays in the justice received in the courts (see Highlight Box 5.1).

Highlight Box 5.1

Paris Hilton Is Ordered Back to Jail

By SHARON WAXMAN
Published: June 8, 2007

LOS ANGELES, June 8—The national obsession with celebrity collided head-on with the more serious issue of the equal application of justice on Friday, as a judge sent the socialite Paris Hilton back to jail some 36 hours after she was released for an unspecified medical problem.

Judge Michael T. Sauer ordered Ms. Hilton to serve out the remainder of her sentence in a county lock-up after the city attorney, whose office had prosecuted the heiress, filed a petition asking that the sheriff's department be held in contempt or explain why it had released her with an ankle monitor on Thursday, after she had served just five days.

Ms. Hilton had been sentenced to 45 days in jail for violating the terms of her probation in an alcohol-related reckless driving case. With time off for good behavior, she had been expected to serve 23.

Ms. Hilton, 26, wearing no makeup and with her hair disheveled, sobbed and screamed, "Mom, this isn't right," as she was taken from the packed courtroom by deputies.

It was a rare moment in this star-filled city, where badly behaving celebrities can seemingly get away with anything—or at least D.U.I. But Ms. Hilton, for all her money and celebrity, seems to have been caught between battling arms of the legal justice system here, with prosecutors and Judge Sauer determined to make a point by incarcerating her, only to have the sheriff's office let her go.

"She's a pawn in a turf fight right now," said Laurie Levenson, a law professor at Loyola Law School Los Angeles. "It backfired against her because she's a celebrity. She got a harsher sentence because she was a celebrity. And then when her lawyer found a way out of jail, there was too much public attention for it to sit well with the court."

The struggle between the judge and the Los Angeles Sheriff's Department, which runs the jail, incited indignation far beyond the attention normally paid to a minor criminal matter.

Judicial and police officials here said they were inundated with calls from outraged citizens and curious media outlets from around the country and beyond. The Rev. Al Sharpton, the civil rights activist and media fixture, decried Ms. Hilton's release as an example of "double standards," where consideration was given to a pampered rich girl that would never be accorded an average inmate.

Even the presidential candidate John Edwards found himself drawn into the debate. When asked about Ms. Hilton's release on Thursday he said, "Without regard to Paris Hilton, we have two Americas and I think what's important is, it's obvious that the problem exists."

California has been struggling to comply with a federal order to ease overcrowding in its jails and prisons, and Los Angeles County Sheriff Lee Baca has for the past year implemented a program of early release. But that has frustrated prosecutors who believe that early release undermines their efforts to punish those found to have broken the law.

At a news conference on Friday, Mr. Baca said, "The special treatment appears to be her celebrity status. She got more time in jail." Under the normal terms of the early release program, he said, Ms. Hilton would not have served "any time in our jail."

The city attorney whose office prosecuted Ms. Hilton's case, Rocky Delgadillo, said preferential treatment had led to her being sent home with an ankle bracelet. In the original order sentencing Ms. Hilton to jail, the judge had specifically stated that Ms. Hilton would not be allowed a work furlough, work release or an electronic monitoring device in lieu of jail time.

"We cannot tolerate a two-tiered jail system where the rich and powerful receive special treatment," Mr. Delgadillo said after learning of the release.

In a news conference on Friday afternoon, Mr. Baca said that Ms. Hilton "had a serious medical condition," though he declined to say what it was. He said, "This is evidence that this lady has severe problems." But, he added, "The criminal justice system should not make a football out of Miss Hilton's status."

In a scene that seemed a strange parody of O. J. Simpson's low-speed chase more than a decade ago, news cameras from across the country followed a police cruiser containing Ms. Hilton as it drove slowly down from her home to Superior Court Friday .

The issue became non-stop fodder for news channels like CNN and Fox News, crowding out news on immigration legislation or the latest from Iraq, as legal experts debated how rare the decision was to release her, and whether doing so neutralized, negated or otherwise neutered the judge's original order.

Amid the debate over serious questions of equal justice under the law came speculation over the nature of Ms. Hilton's "medical situation," which Mr. Baca gave as the reason for her release. On television, commentators questioned whether she was a suicide risk or if she was eating properly in jail.

Judge Sauer had ordered the hearing for 9 a.m. When Ms. Hilton did not appear, apparently believing that she could participate by telephone, he sent sheriff's deputies to escort her from her home.

When she arrived and the hearing began, the judge said he had received a call on Wednesday from an undersheriff informing him that Ms. Hilton had a medical condition and that the sheriff's office would submit papers to the judge to consider releasing her early. The judge said the papers describing a "psychological" problem never arrived and, every few minutes during Friday's court session, interrupted the proceedings to state the time and note that the papers had still not shown up.

In ordering her return to jail, Judge Sauer said that there were adequate medical facilities within the system to deal with Ms. Hilton's problems.

Ms. Hilton was not the only high-profile defendant whose celebrity prompted a raised eyebrow from a judge this week. Also on Friday, the judge who sentenced I. Lewis Libby Jr. to prison this week issued an order dripping with acid sarcasm after receiving a supporting brief from a dozen prominent legal scholars, including Alan M. Dershowitz of Harvard and Robert H. Bork, the former Supreme Court nominee.

The judge, Reggie B. Walton of the Federal District Court in Washington, said he would be pleased to see similar efforts for defendants less famous than Mr. Libby, formerly the chief of staff to Vice President Dick Cheney.

"The court trusts," he wrote, in a footnote longer than the order itself, that the brief for Mr. Libby "is a reflection of these eminent academics' willingness in the future to step up to the plate and provide like assistance in cases involving any of the numerous litigants, both in this court and throughout the courts of our nation, who lack the financial means to fully and properly articulate the merits of their legal positions."

"The court," he added, "will certainly not hesitate to call for such assistance from these luminaries."

Maria Newman and Adam Liptak contributed from New York. Ana Facio Contreras contributed from Los Angeles.

SOURCE: Waxman (2007).

Photo 5.1   Paris Hilton at the Los Angeles Municipal Court

Considering the significance of the courts in the race and crime discourse, our objectives for this chapter are threefold. First, we provide an overview of how the American courts operate. This is followed by a brief historical overview of race and the American courts. Last, we examine some contemporary issues related to race and the courts. Our primary focus here is on the various aspects of the court process and whether discrimination (race or gender) remains a problem. In addition, we examine drug courts, which, over the past decade, have served as a way to handle the overflow of drug cases. Because racial minorities are frequently diverted to drug courts, after reviewing their structure and philosophy, we examine some recent evaluations.

Overview of American
Courts: Actors and Processes

Like so many other facets of American life, the American court system owes much to the English justice system (Chapin, 1983). In general, as American society gained its independence, the courts became more complex with the development of the federal court system and the Supreme Court (Shelden, 2001). Today, because of these early changes, the U.S. court system is referred to as a "dual-court system." There are both state and federal courts, each with trial courts at the lowest level and appellate courts at the top of the hierarchy. The federal court system starts with U.S. magistrate courts, "who hear minor offenses and conduct preliminary hearings" (Shelden & Brown, 2003, p. 196). U.S. district courts are trial courts that hear both civil and criminal cases. Positioned above the U.S. district courts is the U.S. Court of Appeals, which is the final step before reaching the Supreme Court. The Supreme Court represents the highest court in the land and has the final say on matters that make it to that level. The American court system (both state and federal) involves several actors and processes. Subsequently, over time, some of these actors and processes have come under scrutiny for a variety of reasons, including race and class-related concerns (Reiman, 2007). As for actors, the court is generally comprised of three main figures: the judge, the prosecutor, and the defense attorney. Although the system is theoretically based on the ancient system of "trial by combat," in practice, it has been suggested that these main figures are part of the "courtroom work group" who actually work together to resolve matters brought before the court (Neubauer, 2002).

There are several processes involved when one is navigating through the court system. First, there is a pretrial process, where the decision is made whether to move forward with a particular case. If the case is moved forward, the question of pretrial detention and bail is decided next. Other processes in the court system include the preliminary hearing, grand jury proceedings, and the arraignment, where a defendant first enters his or her plea. Although plea bargaining determines the outcome in most of the cases, if the case happens to go to trial, there are other processes that move the case along. Once the decision to go to trial is made, unless the case is going to be decided solely by a judge (referred to as a "bench trial"), the next process would be jury selection. Following the completion of jury selection, the trial begins. If the defendant is found guilty, the sentencing phase begins (this phase is discussed in detail in Chap. 6).

Photo 5.2   The U.S. Supreme Court

A NOTE ON THE PHILOSOPHY, OPERATION, AND STRUCTURE OF NATIVE AMERICAN COURTS

Because of the unique history and worldview of Native Americans (Tarver, Walker, & Wallace, 2002), much of the previous dialogue does not apply to Native American courts. As such, we provide a brief overview of the extent, philosophy, operation, and structure of their court system. Recent figures reveal that there are more than 200 courts in Native American jurisdictions. These courts have "approximately 200 judges, 153 prosecutors and 20 peacemakers" (J. Smith, 2009, p. 3). Even so, because of limited resources, "nearly half of all tribes rely on state courts for judicial service" (J. Smith, 2009, p. 3). The philosophy of tribal courts certainly follows a different paradigm than Anglo-American justice systems (see Table 5.1). Tarver et al. (2002) noted that the indigenous justice paradigm follows an approach that includes spirituality and oral customs, which are not welcome in traditional American courts.

As was previously mentioned in Chapter 4, since the enactment of Public Law 280 in 1953, the jurisdiction for criminal and civil cases on tribal lands in several states was turned over to local and state governments (Tarver et al., 2002). Other jurisdictions were also impacted by this new law, which "gave to state and local police the enforcement authority in Indian communities, and cases would be adjudicated in state courts" (Tarver et al., 2002, p. 88). The complexity of jurisdictional issues on tribal land is illuminated by Table 5.2.

Although almost no research examines whether American Indians are discriminated against in their own court system, it remains important to provide an overview of how their courts operate. To do this, we focus on the courts of the Navajos, the second largest American Indian tribe (U.S. Bureau of the Census, 1995). Tso (1996) noted that over the years, Navajo courts have been structured like Anglo courts. There are several judicial districts; as is seen in Table 5.2, the jurisdiction is determined based on the parties involved (i.e., Indian or non-Indian). In general, however, district courts are courts of general civil jurisdiction and of limited criminal jurisdiction. Civil

Table 5.1   Comparison of American and Indigenous (Native) Justice Paradigms

American Justice Paradigm	Indigenous (Native) Justice Paradigm
Vertical power structure	Circular structure of empowerment
Communication is rehearsed	Communication is fluid
Written statutory law derived from rules and procedures	Oral customary law learned as a way of life
Separation of church and state	The spiritual realm is invoked in ceremonies and with prayer
Time-oriented process	No time limits on the process

SOURCE: Tarver et al. (2002, p. 89).

Table 5.2   Indian Country Jurisdiction in Criminal Cases

Suspect	Victim	Jurisdiction
Indian	Indian	Misdemeanor: Tribal jurisdiction Felony: Federal jurisdiction No state jurisdiction No federal jurisdiction for misdemeanors
Indian	Non-Indian	Misdemeanor: Tribal jurisdiction Felony: Federal jurisdiction No state jurisdiction
Non-Indian	Indian	Misdemeanor: Federal jurisdiction Felony: Federal jurisdiction Normally not state jurisdiction (the U.S. Attorney may elect to defer prosecution to the state) No tribal jurisdiction
Non-Indian	Non-Indian	Misdemeanor: State jurisdiction Felony: State jurisdiction Normally U.S. Attorney will decline prosecution No tribal jurisdiction
Indian	Victimless	Misdemeanor: Tribal jurisdiction Felony: Federal jurisdiction
Non-Indian	Victimless	Misdemeanor: Usually state jurisdiction Felony: Usually state jurisdiction Normally U.S. Attorney will decline prosecution

SOURCE: Tarver et al. (2002, p. 89).

jurisdiction extends to all persons residing within the Navajo Nation or who cause an act to occur within the nation. The limitations of criminal jurisdiction are determined by the nature of the offense, the penalty to be imposed, where the crime occurred, and the status and residency of the individual charged with an offense (Tso, 1996).

There are also children's courts within each district. These courts hear "all matters concerning children except for custody, child support and visitation disputes arising from divorce proceedings, and probate matters" (Tso, 1996, p. 172).

The Navajo also have a Supreme Court and Peacemaker Courts. The Navajo Nation Supreme Court "hears appeals from final lower court decisions and from certain final administrative orders" (Tso, 1996, p. 173). Peacemaker Courts rely on mediation to resolve some lesser matters.

Judiciary selections are screened by the Judiciary Committee of the Navajo Tribal Council. Using the Navajo Tribal Code as its basis for selection, the committee selects those who are most qualified according to the code. In Navajo courts, where the Navajo Tribal Code prevails, people can represent themselves; however, only members of the Navajo Bar Association can participate in the courts. These include actual lawyers who have attended law school and those who have followed nontraditional pathways (i.e., passing a certified Navajo Bar Training Course or having served as a legal apprentice).

The irony of the Navajo justice system is that, with the emerging acceptance of the philosophy of restorative justice, American courts are trying to operate more like

Native American courts. This is especially true of juvenile justice systems in some states, where the guiding philosophy is now restorative justice. We now turn to a discussion of the historical overview of race and the courts.

Historical Overview of Race and the Courts in America

NATIVE AMERICANS

It is widely accepted that America had its beginnings long before Europeans began arriving in colonial America, and the record shows that the Native Americans who were here had already created their own norms and ways of handling deviants. As Friedman (1993) noted, once Europeans began to come en masse, there was a "clash of legal cultures" (p. 20). In many instances, this clash resulted in the conqueror Europeans imparting their system onto Native Americans and African Americans, who also began to arrive en masse after the slave trade began. Due to the smallness of 142 communities in colonial America, these early court procedures were more informal, which, as noted before, owed much to the English system. In fact, the courts were also places where community members looked for social drama (Friedman, 1993).

Early on, both Native Americans and African Americans had similarly low status in American society. As such, other than being a means of social control, the courts were generally not concerned about their lives. In Massachusetts, for example, the courts treated Native Americans with indifference (Higginbotham, 1978). For example, according to Higginbotham (1978), "The general court of the colony enforced the English right to take the Indians' 'unimproved' land in 1633" (p. 69). This action was followed by the courts treating Native Americans more harshly (Higginbotham, 1978). During this period, it was not unusual for the Massachusetts courts to use the punishment of either enslavement or banishment for Native Americans (Higginbotham, 1978). On occasion, however, magistrates were concerned about uprisings that could occur if Native Americans felt they were being mistreated by the courts. Because of this concern, on occasion, "The magistrates involved other Indians, either as viewers of the trial and punishment or ideally as witnesses of the accused" (Chapin, 1983, p. 117).

As noted by J. Smith (2009), formal court systems in Native American communities have been around since the 1880s. At that time, the Court of Indian Offenses was established by the Department of the Interior and the Bureau of Indian Affairs. These courts were created to handle minor offenses (J. Smith, 2009). In the 1930s, tribal courts were enacted in concert with the Indian Reorganization Act. Such courts were established with the intention of having Native Americans "enact their own laws and establish their own court system to reflect those laws" (J. Smith, 2009, p. 2). Notably, more than 50% of tribes have such courts today.

Whether the courts were a place where justice prevailed for Native Americans is subject to debate. However, in Chapter 6, we look a bit more closely at early sentencing patterns, shedding some light on this topic. For now, we turn our attention to a brief historical overview of African Americans and the courts.

AFRICAN AMERICANS

Much of the early race-related legislation, which guided the courts, was directed at Blacks (Higginbotham, 1978). However, some legislation did target White ethnic group members. An example of such early legislation was ACT VI, which was passed by the Virginia legislature in the 1600s and "required Irish servants arriving in the colony without indentures to serve longer terms than their English counterparts" (Higginbotham, 1978, p. 33). As one reviews the early literature, it becomes apparent that legislation contributed to many of the inequities observed in the courts. Given these connections, it is also apparent that politics played a key role in how the courts ruled. Therefore, even after the Revolutionary War and the subsequent ratification of the Constitution, southern states still had a vested interest in controlling African American slave labor. Legal scholar Kennedy (1997) wrote,

> In terms of substantive criminal law and in terms of punishment, slaves were . . . prohibited from testifying or contradicting Whites in court. Moreover, in some jurisdictions, such as Virginia, South Carolina, and Louisiana, slaves were tried before special tribunals—slave courts—designed to render quick, rough justice. (p. 77)

For those who did not go before slave courts, "plantation justice" prevailed solely under the discretion of the master or overseer (Friedman, 1993). To refer to plantation justice as being brutal is an understatement. Even minor events resulted in serious punishments. Friedman (1993) provided an example of how one slave, Eugene, was disciplined:

> William Byrd of Westover, Virginia, recorded the following in a dry, matter-of-fact tone in his diary: on November 30, 1709, Eugene, a house hand, "was whipped for pissing in bed." On December 3, Eugene repeated this offense, "for which I made him drink a pint of piss." On December 16, "Eugene was whipped for doing nothing yesterday." Three years later, on December 18, 1712, "found Eugene asleep instead of being at work, for which I beat him severely." (p. 53)

Byrd's actions were sanctioned by slave codes, which, several decades earlier, had made it a minor offense for a master to kill a slave in colonial Virginia (see Highlight Box 5.2).

With the abolition of slavery in northern states and the arrival of emancipation in 1863, it was anticipated that justice would prevail more often for African Americans. At least in the South, such conventional wisdom did not prevail. As is further discussed in Chapter 6, the courts became a cog in the operation of the convict-lease system, which, as was noted in Chapter 1, allowed states to lease out convicts to southern landowners.

> ### Highlight Box 5.2
>
> ### An Act About the Casual Killings of Slaves (1669)
>
> Whereas the only law in force for the punishment of refractory servants resisting their master, mistress, or overseer, cannot be inflicted on negroes [because the punishment was extension of time], Nor the obstinacy of many of them by other than violent meanes supprest. *Be it enacted and declared by this grand assembly*, if any slave resist his master . . . and by the extremity of the correction should chance to die, that his death shall not be accompted Felony, but the master (or that person appointed by the master to punish him) be acquit from molestation, since it cannot be presumed that propensed malice (which alone makes murther Felony) should induce any man to destroy his own estate.
>
> SOURCE: Higginbotham (1978, p. 36).

Using familiar tactics, legislators enacted laws to snare African Americans. The courts were eager participants and ensured that an ample labor supply flowed to southern landowners. When these "legal" measures failed to produce the desired outcome, southerners resorted to trickery and violence (Friedman, 1993). Such actions led many African Americans to lose faith in both the political and justice systems (Du Bois, 1901/2002).

Well into the 20th century, little changed regarding the treatment of African Americans and the courts. Higginbotham (1996) provided a classic review of how courts were bastions of racism. By the very nature of the segregated practices that prevailed in courtrooms, one could presume that the courts were places where White supremacy prevailed and African Americans received little justice. Providing reviews of cases where there were segregated courtroom seating, cafeterias, and restrooms, Higginbotham noted how such "Apartheid" stood in the way of African Americans receiving equal standing in court. Of equal concern was the practice of referring to African Americans by their first names while addressing Whites by their last names and using their appropriate titles (e.g., Mr. or Mrs.). Again, such forms of disrespect produced an air of inequality, which translated into African Americans being at a disadvantage in courts.

During the 1960s and 1970s, there were national inquiries into civil disorders (riots) and also the state of racism in the criminal justice system. One such inquiry looking at the courts was prepared by the National Minority Advisory Council on Criminal Justice (1979), which was created in 1976. The report, "Racism in the Criminal Courts," surveyed the literature to determine the prevalence of racism in the courts. Drawing on the literature and public hearings in 13 cities, the report noted particular concerns with the paucity of minority judges at both the federal and state levels (National Minority Advisory Council on Criminal Justice, 1979). Many of the other concerns expressed with the courts remain with us today (see section below on "Contemporary Issues in Race and the Courts"). For example, the council expressed concerns about unfairness in the bail process, the potentially discriminatory use of peremptory challenges in jury selection, the quality of public defense, and prosecutorial misconduct in the plea-bargaining process (National Minority Advisory Council on Criminal Justice, 1979).

LATINOS

The early literature on Latinos and the courts is quite sparse. Much of the early literature discussed Mexicans and immigration concerns. Another central focus of the literature is Mexicans' experience with the criminal justice system in the western and midwestern United States. The 1931 Wickersham Commission report, *Crime and the Foreign Born*, noted a common problem when dealing with persons, such as Mexicans, who did not speak English:

> With the best intentions in the world, an investigating officer, a prosecutor, or a court will have the greatest difficulty in getting at the exact truth and all the facts in the case when the accused is a non-English-speaking immigrant. (National Commission on Law Observance and Enforcement, 1931a, p. 173)

The report also noted that in California, Mexican defendants were abused by bail bondsmen, loan sharks, and "shyster lawyers" (p. 175). The report also mentioned that Los Angeles had an excellent public defender system in the 1920s, which "very frequently appear[ed] on behalf of the foreign born, especially the penniless Mexican" (p. 175).

Paul Warnshuis, as part of the Wickersham Commission report, conducted a study of how Mexicans were faring with criminal justice agencies in Illinois. Besides the notion that Mexicans were often arrested on flimsy charges, an overriding theme in the report is that Mexicans were at a serious disadvantage in the court process because of language concerns (Warnshuis, 1931). In many instances, interpreters, although needed, were not used, which resulted in the swift conviction of Mexicans. In some instances, judges simply ruled on the evidence and, if convinced of the defendants' guilt, proceeded without hearing testimony (Warnshuis, 1931). As one can imagine, the courts were a place where, for Mexicans, justice was not always received.

During the 1940s, the "Sleepy Lagoon Case" became the high-profile case that brought the issue of Latinos and the courts to the fore. The case revolved around the murder of Jose Diaz. Following the murder, 22 Mexicans were arrested, whereas another 17 were convicted (the convictions were overturned 2 years later). According to most reports, the trial was unfair. The media played a large role in this by creating a moral panic. Friedman (1993) wrote that the media spewed headlines calling for concern about the "zoot-suit gangsters" and "pachuco killers." One representative from law enforcement stated the following to the grand jury:

> Mexicans [have] a "biological" tendency to violence. They were the descendants of "tribes of Indians" who were given over to "human sacrifice," in

which bodies were "opened by stone knives and their hearts torn out while still beating." Mexicans had "total disregard for human life"; the Mexican, in a fight, will always use a knife; he feels "a desire . . . to kill." (Friedman, 1993, p. 382)

Outrageous statements such as these colored the way in which society viewed Mexicans. So, a year after the Sleepy Lagoon case, the city of Los Angeles exploded with the "zoot suit riots." The riots ignited after rumors spread that a Mexican had killed a serviceman, and the city broke out into 4 days of rioting. During this time, "servicemen and off-duty policemen chased, beat, and stripped 'zoot-suiters' " (Friedman, 1993, p. 382). As for the courts, little is ever mentioned of them punishing those who attacked the Mexicans.

In later years, the National Minority Advisory Council on Criminal Justice (1979) would also express concern about the treatment of Hispanics in the courts. As in the case of African Americans, they expressed concern about the treatment of Hispanics throughout the court system. Particularly disturbing to the council was the lack of Hispanics in significant judiciary roles. The report also noted that nearly 50 years after the Wickersham report, language remained a problem, commenting,

> Non-English speaking minorities have special problems of access to courts, to legal counsel and representation because of language and the failure of the courts and priorities of resources to have trained, impartial translators at every stage of courts and trial related proceedings. (National Minority Advisory Council on Criminal Justice, 1979, p. 63)

ASIAN AMERICANS

As was mentioned in Chapter 1, when Asians began arriving in the United States, they were primarily located on the West Coast. Like other immigrants, they were despised by many. Friedman (1993) described some of the more gruesome early incidents:

> In 1871, a mob in Los Angeles killed nineteen Chinese. In the 1880s, riots in Rock Springs, Wyoming, left twenty-eight Chinese dead; Whites in Tacoma, Washington, put the torch to the Chinatown in that city; and there were outrages in Oregon, Colorado, and Nevada. (p. 98)

Eventually, as in the case of other minorities, legislation in concert with the courts was used to deal with Asian Americans. But in the case of Asian Americans, the aim of such actions was to remove them from the country. Friedman (1993) put it best, writing, "The goal of anti-Chinese policy was not suppression but expulsion. . . . The keystone of Asian policy was immigration law, exclusion, and deportation" (p. 100).

Although our focus now turns to contemporary issues in race and the courts, the unfortunate reality is that many of the same issues from our historical overview are still with us. One would have anticipated that problems from centuries ago would have been resolved, but, from the literature we review next, this is not always the case.

Contemporary Issues in Race and the Courts

Many of the contemporary issues pertaining to race and the courts are related to whether there is either race and/or gender discrimination in the various court processes (Free, 2002a, 2002b; Walker, Spohn, & DeLone, 2007). In a span of about 14 years, drug courts have also emerged as an institution that has provided some hope in stemming the tide of drug addiction and its residual effects. Given the heavy presence of racial minorities circulating through drug courts, we examine the structure, philosophy, and effectiveness of these specialized courts. We begin with a review of the literature that examines whether race or gender has been found to have an influence on each respective court process. Our first review examines the role of race and gender during the pretrial processes.

BAIL AND THE PRETRIAL PROCESS

Although bail is not guaranteed by the Constitution, the Eighth Amendment does state that when given, it should not be "excessive." Since the creation of this amendment, various types of pretrial release options have been used. During the early 1950s and 1960s, the Vera Institute of New York conducted studies on bail practices in Philadelphia and New York. The early study in Philadelphia found that "18% of persons jailed pending trial because they could not afford bail were acquitted, whereas 48% of persons released on bail were acquitted" (Anderson & Newman, 1998, p. 215). In the 1960s, the Vera Institute created the Manhattan Bail Project, which was meant to see whether those released on their own recognizance (ROR) absconded any more often than those released on monetary bail. The project convincingly showed that "the rate of return for ROR releases was consistently equal to or better than the rate for those on monetary bail" (Anderson & Newman, 1998, p. 215). Mann (1993) has noted that many of the defendants in the pioneering Manhattan Bail Project were minorities, the majority of whom did return for trial. Even with these findings, during the Reagan presidency, the Bail Reform Act of 1984 provided judges more discretion as to who can be given pretrial release. In general, judges can hold defendants in jail if they consider them either a risk for flight or if they pose a danger to the community (Robinson, 2002). Because of risk concerns, the courts consider a variety of factors when making the decision as to whether to grant pretrial release on bail. In fact, some states, such as New York, have bail statutes that, when deciding whether to grant bail, require judges to consider risk factors such as the nature of the offense, prior record, previous record in appearing when required in court, employment status, community ties, and the weight of the evidence against a defendant (Inciardi, 1999).

There are generally three broad categories of pretrial release: nonfinancial release, financial release, and emergency release (Reaves & Perez, 1994). Nonfinancial release is when someone is released without having to provide any monetary collateral. This typically comes in the form of ROR, citation releases, which are typically administered by law enforcement. Other such releases include some kind of conditional release that involves having to either contact or report to some official to ensure compliance with the conditions of release (e.g., drug treatment). Under this type of release, a third party can also be entrusted to ensure the return of someone (Anderson & Newman, 1998). According to Reaves and Perez (1994),

> About 2 in 5 defendants released before case dispositions received that release through financial terms involving a surety, full cash, deposit, or property bond. Deposit, full cash, and property bonds are posted directly with the court, while surety bonds involve the services of a bail bond company. (p. 3)

The final release is referred to as an "emergency release" and occurs when, due to jail crowding, the defendant is given pretrial release under generally nonstringent release conditions.

RACE AND PRETRIAL RELEASE

Analyzing pretrial release data on 28,000 felony defendants from 75 of the largest American counties, Reaves and Perez (1994) provided some basic data on race and the pretrial process. A review of the racial characteristics of the various types of releases showed that Blacks, Whites, and Hispanics appeared to have access to financial, nonfinancial, and emergency bail. Undoubtedly, some of these figures were tied to the nature of the offenses committed by each defendant. Nevertheless, Reaves and Perez's national data provided some useful information on the race of those who failed to appear and those who were rearrested.

Data on felony defendants who failed to appear, by race, showed that 72% of Blacks, 81% of Whites, 86% of Other, and 70% of Hispanics made all court appearances. Of those who failed to appear, 19% of Blacks, 13% of Whites, 9% of Other, and 17% of Hispanic defendants eventually returned to court. Only 8% of Blacks, 6% of Whites, and 5% of Hispanics remained fugitives. In the case of Hispanics, 13% remained fugitives (Reaves & Perez, 1994).

Of those who were released prior to trial on any type of release, 85% of Blacks, 91% of Whites, 94% of Other, and 84% of Hispanics were not rearrested (Reaves & Perez, 1994). Hispanics (12%) and Blacks (11%) were rearrested for felonies more often than Whites (7%) and Other (6%). A similar pattern emerged when the authors reviewed the racial characteristics of defendants who were charged with some sort of misconduct (e.g., new charge, failure to appear, technical violation) when they were on pretrial release: Hispanics (38%) and Blacks (35%) had the highest rates, with Whites (25%) and Other (19%) having lower rates.

Although nearly 95% of criminal defendants are charged in state courts (Wolf Harlow, 2000), the remaining 5%, or 57,000 other defendants, are annually charged with a federal offense (Scalia, 1999). As with states, federal officials adhere to the Bail Reform Act of 1984 to decide who will receive pretrial release and under what circumstances. In general, persons charged with the following characteristics were more likely to be detained prior to trial: had committed violent offenses, extensive criminal histories, history of pretrial misconduct, and no established community ties (Scalia, 1999).

More than one third (34%) of federal defendants were ordered detained prior to trial. Hispanic defendants were detained at the highest rate (46.7%), with Blacks (35.9%), Other (32.8%), and Whites (19.3%) following behind (Scalia, 1999). According to Scalia (1999), the high rate of detention for Hispanics was caused because they were identified as noncitizens, which contributed in part to them not being able to show established community ties in America. In addition, they were often charged with the more severe drug-trafficking offenses (Scalia, 1999). Like Hispanics, Blacks were also charged with the more serious drug-trafficking offenses. They also had more serious criminal histories, with 75% of them having been arrested on a prior occasion (as opposed to 61% of Whites and 57% of Hispanics).

Moreover, "38.9% of Black defendants had been arrested at least five times compared to 23.1% of Hispanics, 24% of Whites, and 15% of other non-White defendants" (Scalia, 1999, p. 10). Nearly half (46%) of the Black defendants had been previously convicted of a felony, with half of them being a violent felony and another 33% being a drug offense. Given these figures, scholars have sought to determine whether discrimination (race or gender) plays any role in the bail and pretrial process. In the next section, we review some of the literature in this area.

SCHOLARSHIP ON BAIL AND PRETRIAL RELEASE

In a comprehensive review of scholarship on race and its role in the pretrial process, Free (2002b) examined numerous methodologically sophisticated studies (those using multivariate techniques) published since 1970 (after the application of the 1966 Bail Reform Act). Using the no discrimination thesis (NDT) and the discrimination thesis (DT) as his contextual framework (see Chap. 3), he identified 12 studies that supported the DT in the bail and pretrial process and 5 studies that supported the NDT (Free, 2002b).

Some of the DT literature found both race and gender differences. In one of these studies,

> Non-Whites were less likely than Whites to receive low bail (i.e., bail below the guidelines). Further, the gender differences prevalent among White defendants, in which females were more likely than males to receive low bail, were not replicated among non-White defendants. (Free, 2002b, p. 203)

Free also aptly pointed to bail severity and pretrial release as areas of interest. In both processes, his review of studies revealed that African Americans or non-Whites are at a disadvantage. Specifically, in the case of bail severity, "being lower class poses a greater disadvantage for African Americans than Whites" (Free, 2002b, p. 203). In a few of the studies he reviewed, in order to be released, African Americans and Hispanics were required more often than Whites to post cash or surety bonds (Free, 2002b). In studies related to pretrial release, non-Whites fared no better, with Whites receiving more favorable recommendations for release. When age and employment were considered, one study found that "African Americans were 1.6 times more likely than White defendants to be detained" (Free, 2002b, p. 203).

A more recent study by Demuth and Steffensmeier (2004) examining 75 of the most populous counties in the United States (from 1990 to 1996) found that, "in general, female defendants receive more favorable decisions and outcomes than males across all racial-ethnic groups" (p. 234). In addition, the research found that Hispanic and African American males were less likely than White males to be granted pretrial release. Overall, however, Hispanics received the least favorable pretrial decisions. This finding has also been confirmed by additional research. Turner and Johnson (2005), for example, found that, among a sample of more than 800 felony defendants in Nebraska, "Hispanic defendants charged with bailable offenses were found to receive higher bail amounts than other similarly situated African American and White defendants" (p. 49). Schlesinger (2005), using a national sample of 36,709 felony defendants, found that "both Blacks and Latinos receive less beneficial pretrial release decisions and outcomes than Whites" (pp. 185–186).

In contrast to the studies supporting the DT, as noted earlier, Free (2002b) found five studies that were supportive of the NDT. He noted that such studies generally suffer from several shortcomings, including the use of an additive model in assessing the impact of race and bail and pretrial decisions, inattention to the role the race of the victim might have on these decisions, use of single-stage analyses, and failure to control for evidentiary strength.

With the use of additive models that simply take into account a race effect and not the interactive role that social class and prior felony might play in the findings, Free (2002b) asserted that studies that find support for the NDT might unintentionally mask such effects. Furthermore, the race of the victim in the bail and pretrial process could also play a role in who is released prior to trial and how. In line with the work of scholars such as Russell (1998), Free also noted that single-stage research might support the NDT, but when cases are followed through other stages of the system, discrimination might be uncovered. Finally, Free noted that most of the NDT studies do not control for evidentiary strength, which might also have an impact on who gets released prior to trial. That is, if non-Whites are often brought to court on cases based on less evidence than Whites, it is likely that they would be able to secure pretrial release at a higher rate than Whites (Free, 2002b).

The decision whether to prosecute, reject, or dismiss charges represents another area of investigation related to the courts. Again, researchers have suggested that race or gender might play a role in these highly discretionary decisions. In Free's (2002b) review of the empirical literature, he found that "African Americans receive more severe dispositions than Whites for less serious offenses" (p. 206). Thus, in the few studies that have examined the topic, Whites were more likely to have their cases dismissed than African Americans. Moreover, Free also found "moderate support for the contention that prosecutors are more likely to seek the death penalty in cases involving African American defendants than White defendants" (p. 206). These findings are in stark contrast to the previously discussed work of Wilbanks (1987), who argued that such charges were largely anecdotal.

When considering these decisions and interjecting the role of gender in the court process, some researchers have noted the potential influence of stereotypes (Huey & Lynch, 1996; Young, 1986). Manatu-Rupert (2001) has argued that the media representation of Black women leads prosecutors to use stereotypes in the decision-making process. Drawing on research in several disciplines, she suggested,

> How these [Black] women are perceived in the culture, particularly by criminal justice officials, is directly linked largely to how Black women are sexually positioned in film, which may well result in their becoming vulnerable to sexual assaults and less likely to be believed when victimized. (Manatu-Rupert, 2001, p. 184)

Using Spike Lee's movie *She's Gotta Have It* and the first *Lethal Weapon* movie as case studies, Manatu-Rupert (2001) highlighted how the myth of Black female promiscuity was perpetuated. To explore her ideas, she showed a scene from the movie *She's Gotta Have It* to her students, many of whom were headed into criminal justice occupations, including the legal profession, to see their interpretation of the interaction between a male and female. Her findings supported the notion that Black women were viewed as oversexed, which, as she noted earlier, could impact their treatment in the court system, particularly if they were ever sexually assaulted.

LEGAL COUNSEL

As a result of the blatant racism in the previously discussed Scottsboro case, since 1932, the U.S. Supreme Court has mandated that indigent defendants in capital cases have the right to adequate counsel (see *Powell v. Alabama*, 1932). Four decades later, in *Gideon v. Wainwright* (1963), the Court ruled that all felony defendants have a right to counsel, a ruling that was expanded in 1972 when the Court mandated that counsel be provided for defendants in misdemeanor cases where there was the possibility of incarceration (see *Argersinger v. Hamlin*, 1972). To meet these Supreme Court mandates, states created several legal defense systems. Many states have adopted a public defender system where, as with a district attorney, the local or state government hires a full-time attorney to provide legal counsel for indigent defendants. In other cases, defendants are assigned counsel from a list of private attorneys who are selected on a case-by-case basis by a judge. Another option is the use of contract attorneys. These people are also private attorneys who are contracted to provide legal representation to an indigent defendant (DeFrances & Litras, 2000). In the following, we review some recent national governmental studies that examine the operation of these systems.

DEFENSE COUNSEL

Over the past few years, the federal government has conducted studies to determine the state of counsel in criminal cases (DeFrances, 2001; DeFrances & Litras, 2000; Wolf Harlow, 2000). According to one of these studies, the 100 most populous counties (encompassing 42% of the U.S. population) spent more than $1.2 billion on indigent defense (DeFrances & Litras, 2000). DeFrances and Litras found that of the three previously described indigent defense systems, 82% were handled by public defenders, 15% were handled by assigned counsel attorneys, and 3% were handled by contract attorneys. At the federal level, 66% of felony defendants were represented by public defenders (Wolf Harlow, 2000).

Most observers of indigent defense wonder whether there are different outcomes depending on whether you are represented by public or private counsel. Looking at the outcomes of cases at both the state and federal levels, Wolf Harlow (2000) wrote,

> In both Federal and large State courts, conviction rates were the same for defendants represented by publicly financed and private attorneys. Approximately 9 in 10 Federal defendants and 3 in 4 State defendants in the 75 largest counties were found guilty, regardless of type of attorney. (p. 1)

One difference noted in the same study was that at both court levels, those represented by public defense were more likely to be incarcerated. At the federal court level, the difference was 11% (88% public vs. 77% private). On the state level, however, the differences were more dramatic: "In large State courts 71% with public counsel and 54% with private attorneys were sentenced to incarceration" (Wolf Harlow, 2000, p. 1). It is also interesting to note that more than 90% of federal defendants were found guilty irrespective of type of counsel (Wolf Harlow, 2000).

Turning to the representation of those who were incarcerated, minorities used public defense systems more than Whites (Wolf Harlow, 2000). Approximately 75% of Black and Hispanic state inmates and 69% of Whites had public defenders or assigned counsel (Wolf Harlow, 2000). In the case of federal inmates, 65% of Black inmates had public defenders, and Hispanics and Whites used public defense at a similar level (56% vs. 57%). To provide a more detailed examination of public defense systems, we review the findings from a statewide study that examined in detail the indigent defense system in Pennsylvania.

STATEWIDE STUDY OF PENNSYLVANIA'S PUBLIC DEFENSE SYSTEM

Using Pennsylvania as an example, we examine how the public defense system is working in the sixth most populous state. In Pennsylvania, 80% of all criminal defendants rely on public defense (Pennsylvania Supreme Court, 2003). A survey sent to Pennsylvania counties provided a snapshot of the functioning of the system. The survey revealed some troubling findings. First, only the Philadelphia office provided any formalized training to their new public defenders. Even with this training, the public defenders were then thrust into positions where they were overworked and underpaid (Pennsylvania Supreme Court, 2003). They generally also had fewer resources than the district attorney's office. These findings were repeated when comparing them with the data from the other Pennsylvania jurisdictions. Most troubling was that "most court-appointed lawyers and many public defenders do not make use of investigators, and therefore do not conduct independent investigations of cases" (Pennsylvania Supreme Court, 2003, p. 185). Because of the paucity of resources, one jurisdiction pointed out "that a case that might require a psychologist or forensic expert might exhaust the whole budget" (p. 185).

There was also a serious imbalance in technology and salaries between district attorneys and public defenders. In many counties, technology in the form of computers was passed down from district attorneys' offices to public defenders' offices. In terms of salaries, researchers found that public defenders' salaries were considerably lower than those for district attorneys. Even student loan forgiveness programs attached more importance to the role of the district attorney, considering that public defenders were not eligible for the programs (Pennsylvania Supreme Court, 2003). Such findings suggest that, in Pennsylvania, defendants who are poor and people of color are being represented by inadequate counsel. Given these facts, it is no wonder that plea bargaining pervades court systems. We review plea bargaining in the next section.

PLEA BARGAINING

Before actually proceeding to the trial phase, in any given year, an estimated 96% of convictions are reconciled through a guilty plea in the plea-bargaining process (Reaves, 2001). Early in American history, however, plea bargaining was frowned on by justice system officials and, as a result, comprised a small percentage of how cases were resolved. Only since the 20th century have courts accepted it as an important part of the criminal justice process (Shelden & Brown, 2003). Predictably, a process involving no real oversight and such broad discretion has come under scrutiny because of concerns related to race and class. The intersection of race and class is well articulated by M. Robinson (2002), who wrote,

> Plea bargaining results in a bias against poor clients, who are typically minorities, as well as the uneducated, who may not even know what is being done to them in the criminal justice process. (p. 247)

An example of how racial bias can influence the plea-bargaining process can be seen by a comprehensive study of California criminal cases conducted by the *San Jose Mercury News* in 1991. Using a computer analysis of nearly 700,000 criminal cases from 1981 to 1990, the study found,

> At virtually every stage of pretrial plea bargaining Whites were more successful than minorities. All else being equal, Whites did better than African-Americans and Hispanics at getting charges dropped, getting cases dismissed, avoiding harsher punishment, avoiding extra charges, and having their records wiped clean. (Donzinger, 1996, p. 112)

In addition, the study found that, although one third of Whites who started out with felony charges had their charges reduced to misdemeanors, African Americans and Hispanics received this benefit only 25% of the time (Donzinger, 1996). Seeking to explain some of these results, one California judge stated that there was no conspiracy among judges to produce such outcomes. When put to a public defender, he explained the disparities this way:

> If a White person can put together a halfway plausible excuse, people will bend over backward to accommodate that person. It's a feeling, "You've got a nice person screwing up," as opposed to the feeling that "this minority person is on track and eventually they're going to end up in state prison." It's an unfortunate racial stereotype that pervades the system. It's an unconscious thing. (Donzinger, 1996, p. 113)

One can only imagine the compounded impact of such attitudes if they are pervasive nationwide.

In the rare event that a case is not resolved during the plea-bargaining process, the next stage of the process requires the preparation for a jury trial. Jury selection begins this phase of the process.

JURY SELECTION

Once it is decided that a case is going to trial, the case is decided by either a jury or a judge in a bench trial (where a judge is responsible for the determination of guilt or innocence). In the event of a jury trial, the jury selection process begins with a *venire*, or the selection of a jury pool. This provides the court with a list of persons from which to select the final jury members. The Constitution requires that citizens are entitled to a jury of their peers. Those in the legal profession generally agree that this should translate into juries being representative of the communities in which the defendant resides. In the jury selection process, however, race and gender concerns have continued to pervade the process. The Scottsboro case discussed in Chapter 1 was an example of how racial bias in jury selection resulted in all-White juries trying African Americans because of long-standing discrimination in the courts. Although times have changed, in some jurisdictions, when minorities are underrepresented on juries in their own communities, questions have been raised and radical action has been taken (see Highlight Box 5.3). But in other instances, the courts have refused to intervene. Recent results from a study commissioned by the Pennsylvania Supreme Court on racial and gender bias have provided some examples of how issues related to race and gender influence the makeup of jury pools (Pennsylvania Supreme Court, 2003).

Highlight Box 5.3

Judge's Decision Stirs Debate on Jury Makeup

Pittsburgh: A Judge's decision that a black murderer defendant can postpone his trial until blacks make up at least 10% of the jury pool is spurring debate about how minorities are picked for juries and whether the judge's remedy is practical.

Judge Lester Nauhaus issued Friday's ruling after a public defender argued that less than 5% of the people on jury duty in Allegheny County during one five-month period were black, even though the county's black population is about 12%.

While some analysts say the judge's ruling contradicts the ideal that justice should be colorblind, others believe the judge's decision was likely needed to spur reforms.

"If a logjam is the kind of thing needed to get the attention of the county jury commissioner, so be it," Duquesne University law professor Kenneth Hirsch said. "Once it's been out in the newspapers that this has been a ruling in one case, it won't take the other defense attorneys long to start raising the issue on a regular basis."

County officials said they'll announce various jury selection reforms at a news conference today, including tracking the race of jurors and improving efforts to contact people who don't respond to jury screening questionnaires and summonses.

Under state law, potential jurors are drawn at random from voter registration lists.

Counties may also use driver's license and telephone records—and citizens can simply sign up for jury duty.

The public defender who raised the issue, Christopher Patarini, said any system that routinely results in too few blacks on jurors is unfair.

"We're not saying it's intentional or malicious, but if the end result indicates that the system itself does not reasonably reflect a cross section of the community, it has to change," Patarini said.

The U.S. Supreme Court has ruled that defendants can't pick the racial makeup of their specific juries, but systems that don't produce jury pools that reflect a cross section of the community are illegal.

A *Pittsburgh Tribune-Review* study last year concluded that blacks were underrepresented on county juries, prompting the state to study the matter.

The Joint State Government Commission concluded that only 8.1% of county jurors were black, compared to the 12.4% black population counted in the 2000 U.S. Census.

Chris Mount, the director of court research for the New York State Unified Court System, said Allegheny County could probably improve its system—but he disagrees with Nauhaus' decision.

Mount said Chief Judge Judith Kaye began reforming New York's system in 1993, pushing to improve jury pay, remove the state's numerous exemptions from jury service—and most importantly—expand the list of potential jurors. New York picks its jurors from voter, driver, state, tax, welfare, and unemployment lists, Mount said.

The Pennsylvania Legislature is considering a bill that would add welfare rolls, private mass-marketing lists, and local tax records to the list of potential juror sources.

Hirsch advocates using census information to choose jurors from predominantly black ZIP codes.

Mount said New York considered and rejected such a plan 10 years ago.

"The policy is here that the law says jurors should be picked at random, and [using Zip codes] gets you away from that—and it also crosses a line into who decides what ZIP codes, and how many and what's an acceptable percentage," Mount said.

Hirsch said jury disparity is a serious problem that needs action now.

"Should we live in a better world than we do? Yeah," Hirsch said. "But what are we going to do about it by next Tuesday? In the real world people notice the difference between a person's color or race and it makes a difference in how they think, whether it should or not."

SOURCE: Mandak (2003, p. B7).

To investigate ethnic and gender bias in jury selection in Pennsylvania, researchers sent surveys to every county in the state to assess the various practices. More than 80% of the 67 counties returned the surveys (Pennsylvania Supreme Court, 2003). Once the surveys were tabulated, it was revealed that there were a variety of barriers to minorities and women participating in jury service. For minorities, these barriers included receiving fewer summonses for jury duty, transportation issues, child-care issues, and employer issues (Pennsylvania Supreme Court, 2003). In addition, minorities more so than others have transportation difficulties that result in them being unable to serve on juries. Often, they are also impeded from serving on juries because they are unable to secure appropriate child care. For hourly wage employees, in some instances, employers are unsupportive of jury service. The second part of the study, which looked in depth at four of Pennsylvania's largest counties (Allegheny, Lehigh, Montgomery, and Philadelphia), found that "African Americans were under-represented in juror yield in all four counties; Latinos were under-represented in juror yield in three of the counties; and Asian Americans were under-represented in juror yield in Philadelphia County" (Pennsylvania Supreme Court, 2003, p. 70).

In addition to surveys, the committee charged with completing the study also heard public testimony. When they queried one Pittsburgh attorney on the state of affairs concerning the racial composition of juries, he responded,

> In all of the cases which I have tried on behalf of African American plaintiffs in the past five years, a grand total of one African American was involved in the deliberations that determined the outcome of the case. Indeed, in most of the cases, the only African American in the courtroom was my client. (Pennsylvania Supreme Court, 2003, p. 74)

Turning to the underrepresentation of women on juries, the report revealed some of the interconnections between race and gender. Women were generally the ones responsible for child care; therefore, they were often unable to serve on juries. The committee also noted that women had difficulty reaching the courthouse, a concern shared by non-Whites. In certain situations, women, like men, were unable to serve on juries due to economic hardships. A final gender-related concern was "[the finding of] evidence that the interpersonal dynamics within the jury room can operate to the detriment of female jurors" (Pennsylvania Supreme Court, 2003, p. 106). More specifically, "Women in Pennsylvania were less likely than men to be chosen as presiding jurors" (p. 106).

Once the jury pool is formed, the opportunity for race and gender bias does not end there. There have been concerns with the subsequent *voir dire* process, where jurors are screened for their fitness to serve on a particular case.

VOIR DIRE

Considering the difficulty in locating non-White jurors, it would seem that once the final juror selection process began, non-White jurors would not be the targets for removal. To the surprise of some, this is not the case. Both the defense and prosecution often use their peremptory challenges to remove jurors based, in large part, on their race. In general, peremptory challenges can be used to remove jurors without cause. This was challenged in the 1965 case of *Swain v. Alabama*. In the case, Robert Swain, a

Black teenager, was accused and convicted of raping a White teenager (Cole, 1999). During the case, the prosecution "struck all six prospective Black jurors," and after investigation, it was revealed "that no Black had ever served on a trial jury in Talladega County, Alabama," despite the fact that Blacks made up 25% of the county population (Cole, 1999, p. 119). Based on this information, the conviction was challenged, and the Supreme Court decided to hear the case. However, the Supreme Court found no problem with striking the prospective Black jurors, indicating among other things that, "in the quest for an impartial and qualified jury, Negro and White, Protestants and Catholic, are alike subject to being challenged without cause" (cited in Cole, 1999, p. 119). But according to the 1986 Supreme Court decision in *Batson v. Kentucky*, race or gender must not be the deciding factor. It is notable that, six years later, in *Hernandez v. New York* (1991), the Supreme Court ruled

> that a criminal defendant's Fourteenth Amendment rights to equal protection were not violated when a prosecutor exercised a peremptory challenge excluding potential Latino/a jurors who understood Spanish on the basis that they might not accept the court interpreter's version as the final arbiter in the case. The defendant had argued that the elimination of potential Latino/a jurors violated his right to a trial by his peers in violation of his constitutional rights. (cited in Morin, 2005, p. 78)

In the eyes of the Supreme Court, the prosecutor's reason for the removal of Latino/a jurors was "race neutral" (Morin, 2005, p. 78).

Prior to the *Batson* decision, studies of jurisdictions in Texas and Georgia found that peremptory challenges were used to strike 90% of Black jurors (Cole, 1999). In fact, in large cities such as Philadelphia, district attorneys were given clear instructions to strike non-White jurors. A now infamous training tape by one Philadelphia assistant district attorney declared,

> Young Black women are very bad. There's an antagonism. I guess maybe they're downtrodden in two respects. They are women and they're Black . . . so they somehow want to take it out on somebody, and you don't want it to be you. (Cole, 1999, p. 118)

Scholars have also noted that, in certain instances, defense attorneys follow similar practices (Cole, 1999).

Since the *Batson* decision, prosecutors have turned to deception to strike non-White jurors. According to Cole (1999), prosecutors are masking their race-based actions by using neutral explanations to remove jurors. In some instances, for example,

> Courts have accepted explanations that the juror was too old, too young, was employed as a teacher or unemployed, or practiced a certain religion. They have accepted unverifiable explanations based on demeanor: the juror did not make eye contact or made too much eye contact, appeared inattentive or headstrong, nervous or too casual, grimaced or smiled. And they have accepted explanations that might often be correlated to race: the juror lacked education, was single or poor, lived or worked in the same neighborhood as the defendant or a witness, or had previously been involved with the criminal justice system. (Cole, 1999, pp. 120–121)

Gabbidon, Kowal, Jordan, Roberts, and Vincenzi (2008) confirmed these findings after reviewing 5 years worth of litigation in which plaintiffs sued because they felt alleged race neutral peremptory challenges were race-based. Based on an analysis of 283 cases from the U.S. Court of Appeals, the researchers found that Blacks were the ones most likely to be removed from juries. As for the explanations used to justify their removal, some include questionable body language, questionable mannerisms, medical issues, child-care issues, limited life experiences, and unemployed. The courts accepted these explanations considering that 79% of the appellants lost their cases (Gabbidon et al., 2008). For those who actually won their appeal, there was an extremely high burden of proof. However, at the heart of winning was being able to show some inconsistency on the part of the attorney who made the removals. As an example, Gabbidon et al. noted that in one instance, the prosecutor struck a juror "because he was concerned that she did not seem to understand or respond appropriately to certain dire questions" (p. 64). But after further review, "[t]he courts ruled against the prosecutor because they found that the prosecutor declined to strike white jurors whose answers were far more inappropriate or unresponsive" (Gabbidon et al., 2008, p. 64).

Sommers and Norton (2007) conducted an experiment using 90 undergraduate college students, 81 advanced law students (second and third year), and 28 practicing attorneys to determine several things including (a) to what extent does race affect jury selection judgments, and (b) if decision makers fail to report the influence of race on jury selection judgments, how do they justify their decisions? On the first point, the authors found that, "Across three samples, this investigation provides clear empirical evidence that a prospective juror's race can influence peremptory challenge use and that self-report justifications are unlikely to be useful for identifying this influence" (Sommers & Norton, 2007, p. 269). Furthermore, the researchers reported that when justifying their reason for removing Black jurors, the respondents used a race-neutral explanation. Thus, as Sommers and Norton (2007) aptly note, "The practical implications of these findings are clear: even when attorneys consider race during jury selection, there is little reason to believe that judicial questioning will produce information useful for identifying the bias" (p. 269).

Considering the frequent removal of racial and ethnic minorities from jury pools, one wonders how jurors perceive the jury selection process. Recently, McGuffee, Garland, and Eigenberg (2007) explored this question based on a survey sent to Tennessee residents summoned for jury service. In general, most of 138 respondents felt that most jury verdicts are correct (92%) and "that juries try hard to do the right thing" (p. 456). Fewer respondents reported that they felt the system was fair (83%). As for questions related to race and the jury system, the authors found that:

> About half the respondents (52%) agreed that it is important for African-Americans to have African-Americans on the jury in order to get a fair trial. Similarly one-half of the respondents (49%) reported that if you are African American you want other African-Americans on the jury to get a fair trial. (McGuffee et al., 2007, p. 457)

Other findings showed that the respondents were less likely to feel that having Whites on the jury was critical for Whites to receive a fair trial. Also, a little more than 25% of the respondents felt that African Americans were "more likely to be dismissed from jury duty than are whites. Furthermore, about one-fifth (18%) of the respondents believed that whites are more apt to convict than minorities." In terms of fairness in decision making, "Only 7% of the respondents believe that whites are fairer than minorities in jury decisions" (McGuffee et al., 2007, p. 457). It is also important to note that most of the respondents (57%) felt that race had nothing to do with jury selection or jury verdicts (68%).

Even with these generally positive sentiments from admittedly a limited sample of jurors, prosecutors and defense attorneys still resort to peremptory challenges to remove racial and ethnic minorities from jury pools. Why? Concerns regarding jury nullification likely lie at the heart of the practice. We review jury nullification and the fears surrounding it in the next section.

JURY NULLIFICATION

Given the numerous measures being taken in some jurisdictions to diversify jury pools, one would think that keeping non-Whites on the jury would be a high priority. On the contrary, as discussed in the previous section on peremptory challenges, prosecutors and defense attorneys alike often eliminate jurors based on race if they believe it will help them secure a victory. The underlying premise behind such thinking is surely a lack of trust, particularly of non-White jurors. Why else would someone not want non-Whites on juries? The concern relates to the practice of jury nullification. According to Walker et al. (2007),

> Jury nullification . . . occurs when a juror believes that the evidence presented at trial establishes the defendant's guilt, but nonetheless votes to acquit. The juror's decision may be motivated either by a belief that the law under which the defendant is being prosecuted is unfair or by an application of the law to a particular defendant. (p. 223)

In the past decade or so, there has been increasing debate on this practice. In 1995, Paul Butler, a professor of law at George Washington University, published a seminal and controversial article on the utility of African Americans engaging in jury nullification.

Since Professor Butler published his controversial article, there has been continuing discussion about the practice. During the beginning of his tenure as a federal prosecutor, Butler (1995) wrote that "we would lose many of our cases, despite having persuaded a jury beyond a reasonable doubt that the defendant was guilty. We would lose because some Black jurors would refuse to convict Black defendants who they knew were guilty" (p. 678). After considering the concept, Butler used two case studies to show how the practice can send a message to society regarding injustices. The first case study reviewed the case of *United States v. Barry* (1991). After a long and expensive federal investigation, Marion Barry, the former mayor of Washington, D.C., was observed on videotape in a sting operation smoking crack cocaine. During his trial, two prominent and controversial African American leaders, Reverend George Stallings and Minister Louis Farrakhan, were provided with passes by Barry to attend the trial. However, in both cases, they were denied access because the presiding judge felt as if they might influence juror sentiment. The American Civil Liberties Union (ACLU) took up the cases and argued that they both had a right to attend the trial. In the end, the trial court conceded and let the two attend the trial; however, they were given "special rules" that had to be followed (Butler, 1995). Although Barry was clearly guilty, at the end of the trial, he was found guilty of only 1 of 14 charges.

The second case study involved John T. Harvey III, a prominent Washington, D.C., attorney. During a trial in the early 1990s, Harvey's attire became the subject of controversy while he was representing a client. Along with a traditional suit, Harvey wore a stole made of *Kente* cloth, which has roots in African culture. During the pretrial process, a judge warned him about wearing the cloth. Apparently, the judge felt that wearing the cloth would "send a hidden message to jurors" (Butler, 1995, p. 685). Given his convictions on the matter, the judge offered Harvey three options: "He could refrain from wearing the *Kente* cloth; he could withdraw from the case; or he could agree to try the case before the judge, without a jury" (p. 685). After considerable wrangling, Harvey was allowed to wear the *Kente* cloth. The judge even suggested that Harvey be charged for the time devoted to the *Kente* cloth issue. Eventually, "Harvey's client is tried before an all-Black jury and is acquitted" (p. 686).

On the whole, Butler (1995) advocated the use of jury nullification to fight against unfair practices, which are outlined in his two case studies, as well as unjust laws, such as the controversial crack cocaine laws, which annually send a disproportionate number of people of color to jail and prison. In Butler's (1995) words,

> The Black community is better off when some nonviolent lawbreakers remain in the community rather than go to prison. The decision as to what kind of conduct by African-Americans ought to be punished is better made by African-Americans themselves, based on the costs and benefits to their community, than by the traditional criminal justice process, which is controlled by White lawmakers and White law enforcers. Legally, the doctrine of jury nullification gives the power to make this decision to [African-American jurors who sit in judgment of] African-American defendants. (p. 679)

Although acknowledging some merits of Butler's arguments, Krauss and Schulman (1997) see the general concern of Black juror nullification as being based on anecdotal evidence. In their analysis, the discussion comes down to the following equation: "Black defendant + Black jurors + non-conviction = miscarriage of justice" (Krauss & Schulman, 1997, p. 2). As they see it, those who believe in this equation ignore the fact that, in some of these cases, "Black jurors are being condemned for doing exactly what jurors are supposed to do: demanding that the prosecution prove its case beyond a reasonable doubt" (Krauss & Schulman, 1997, p. 2). Furthermore, they view the continuing outrage over jury nullification as a response to an article published in the *Wall Street Journal* shortly after the conclusion of the O. J. Simpson trial in 1995. The article provided figures showing that nationwide there was an overall acquittal rate of 17%, whereas in jurisdictions such as the Bronx, New York, and Washington, D.C., the acquittal rates were 47.6% and 28.7%, respectively. The authors countered these figures, showing that the acquittal rate nationally was closer to 28% (Krauss &

Schulman, 1997). Subsequently, contrary to the notion of jury nullification, the authors noted that the Bronx rate was likely a result of "(1) jurors doing their jobs well, [and] (2) prosecutors who are not doing their jobs well" (Krauss & Schulman, 1997, p. 3).

Krauss and Schulman (1997) also noted that there is the belief that only White jurors can be color-blind. They responded that it is impossible to have a color-blind jury because race is the first thing that people see in others. Another important point discussed by the authors relates to police testimony. Although in many instances the testimony of police officers is believed without challenge, the authors pointed out that, because of their historically negative experiences with police officers, persons of color give equal weight to police testimony and that of other witnesses even when they differ (Krauss & Schulman, 1997).

Drug Courts

Drug courts represent a fairly recent initiative. Dade County, Florida, is credited with starting the first one in 1989 (Goldkamp & Weiland, 1993). Since then, partly as a result of Title V of the Violent Crime Control and Law Enforcement Act of 1994, which awarded federal monies to drug court programs (U.S. General Accounting Office, 1997), the number of drug courts has risen to more than 2000 in 2007 (www.ndci.org; see Figure 5.1). These courts surfaced when drug-related cases began to overwhelm traditional courts, which was largely due to the wide-scale use of mandatory minimum drug sentences (Fox & Huddleston, 2003). This sentencing approach had a disparate impact on minorities (see Chap. 6), and, consequently, a large share of those diverted to drug courts have been minorities (Fielding, Tye, Ogawa, Imam, & Long, 2002). Before we discuss the effectiveness of drug courts, we review the structure and philosophy of the courts.

Figure 5.1   The Number of Drug Courts Continues to Increase Nationwide (1989–2007)

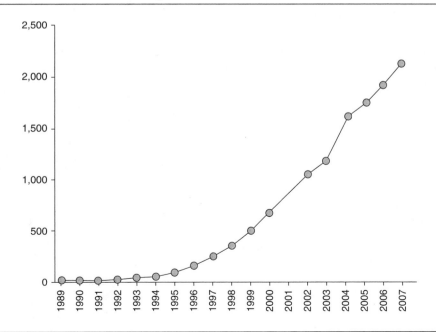

SOURCE: National Drug Court Institute (January 2008)

STRUCTURE AND PHILOSOPHY OF DRUG COURTS

Court Structure

Drug courts (also referred to as "drug treatment courts") attempt to provide "a bridge between criminal justice and health services" (Wenzel, Longshore, Turner, & Ridgely, 2001, p. 241). This bridge results in a combination of court oversight and therapeutic services (Fox & Huddleston, 2003). This results in the use of "an intense regimen of drug treatment, case management, drug testing, and supervision, while reporting to regularly scheduled status hearings before a judge" (Fox & Huddleston, 2003, p. 13). In addition to the judge, the prosecutor, defense attorney, treatment provider, law enforcement officer, probation officer, case manager, and program coordinator are part of the "drug court team" (Fox & Huddleston, 2003, p. 13).

Based on a report commissioned by the U.S. Department of Justice and the National Association of Drug Court Professionals, there are 10 key components of drug courts (Fox & Huddleston, 2003). To receive federal funding, drug courts must adhere to the following components. First, they must use a multidisciplinary process to address the needs of persons who are diverted to the court. Second, the courts are also seen as nonadversarial. Going against the traditional philosophy of courts, all parties work together, again, to ensure that the participant has the best chance at success. Early identification is the third component of drug courts. Here, the courts try to get to prospective participants as early in the criminal justice process as possible. The fourth component of the court is to offer a "continuum of treatment and rehabilitation services" (Fox & Huddleston, 2003, p. 16). This acknowledges the diverse needs of those who come under the purview of the court. Regular alcohol and drug testing is another key component. Such random testing, which should be observed by a program official, is essential to ensure participants have not had relapses.

The sixth key component of the court is to have regular meetings to discuss the progress of participants. Philosophically, these meetings should not take on a punitive tone (Fox & Huddleston, 2003). In line with the sixth component, the seventh requires that there should be ongoing interaction with the judge. Evaluation is the eighth essential component of a drug court. Using goals and objectives as a starting point, each court should periodically measure its effectiveness. The ninth component requires that team members seek out additional education or training to stay abreast of current drug court practices. The final key component of drug courts is developing community partnerships. Drug courts are encouraged to make links to "enhance program effectiveness and general local support" (Fox & Huddleston, 2003, p. 17).

Court Philosophy

In short, the underlying philosophy of the court is that persons with drug addictions need treatment vis-à-vis prison sentences. That is, because some of the problems that result from drug abuse are related to criminal justice, drug courts seek to address the addiction problem through treatment services in the hopes of producing the following positive outcomes: a reduction in drug use, less criminal activity, lower recidivism, better health, better social and family functioning, better educational/vocational status, and residential stability (Fischer, 2003; Wenzel et al., 2001). Taking into consideration the unique approach and philosophy of drug courts, the next section looks to see how effective they have been.

EFFECTIVENESS OF DRUG COURTS

Approaching nearly 20 years in existence, drug courts have been the focus of a considerable number of evaluations. In 1997, the U.S. General Accounting Office (GAO) reviewed 20 studies that evaluated the effectiveness of drug courts. The GAO noted the serious shortcomings in many of the evaluations that had been done up to that point. In general, the report found mixed results. Some programs recorded relapse rates from 7% to 80%, with some programs reporting recidivism rates ranging from 0% to 58% (U.S. General Accounting Office, 1997). In addition, completion rates were mixed, ranging from 1% to 70%, with an average of 43% (U.S. General Accounting Office, 1997). A review of studies conducted for drug courts in California, Florida, and Delaware found that Hispanics and African Americans were less likely to complete the programs. In Nevada, however, researchers reported "no difference in the termination rates between ethnic groups and between male and female participants" (U.S. General Accounting Office, 1997, p. 77).

Los Angeles County started its first drug court in 1994. By June 2001, there were "11 pre-plea drug court programs . . . operating in parts of 11 Los Angeles County's 24 judicial districts" (Fielding et al., 2002, p. 218). Like others around the country, the court "provide[s] a treatment alternative to prosecution for non-violent felony drug offenders" (p. 218). To measure the effectiveness of the drug court, the researchers used a quasi-experimental design with three sample groups. The first group included those who participated in the drug court program. The second group participated in another diversion program, whereas the third group comprised felony defendants who went to trial. The study aimed to investigate the program completion and recidivism rates of the groups. Most of the persons in the three groups were male and minorities, with an average age in the low 30s. When Fielding et al. (2002) examined recidivism among the three groups, they found that,

> Drug court participants were less likely to be re-arrested than drug diversion participants or felony defendants. However, results differ by risk strata. Low risk study participants' re-arrest rates did not differ significantly from those of the diversion sample. However, for those classified at medium or high risk, the re-arrest rate for drug court participants was significantly below those of the diversion and/or felony defendant groups. (pp. 221–222)

Moreover, the time to rearrest was longer for those who participated in drug courts than those who were in the other groups. This held true for time to new drug arrests. On the whole, drug court graduates fared better than those in the other groups. A final area of interest to the researchers was cost. The cost for drug court ranged from nearly $4,000 to nearly $9,000. Yet the alternatives to drug courts—prison ($16,500) and residential treatment centers ($13,000)—were considerably more expensive.

Other studies have also found the courts to be effective in reducing recidivism (Gottfredson, Najaka, & Kearley, 2003; Johnson Listwan, Sundt, Holsinger, & Latessa, 2003), improving the participants' job prospects and self-image (Creswell & Deschenes, 2001), and reducing costs (Carey, Finigan, Crumpton, & Waller, 2006). More recent research has examined how effective drug courts are outside urban settings.

Thus, although most of the evaluations have been of urban-based drug courts, little has been done to determine whether the success of the courts is transferrable to non-metropolitan areas. Galloway and Drapella (2006) examined the effectiveness of drug courts in such an area. Their results suggest that the success of the courts is not limited to urban areas. Their study of a drug court in a small nonmetropolitan area revealed lower recidivism among drug court participants even after controlling for age, race, gender, and other key variables.

Still other research has examined the effectiveness of family treatment drug courts (FTDCs), which are designed to get drug-addicted parents the treatment they need "to be reunited with their children" (Green, Furrer, Worcel, Burrus, & Finighan, 2007, p. 44). In such instances, the courts often work in conjunction with the child welfare system. Green et al. (2007) conducted an assessment of FTDCs at multiple sites using a quasi-experimental approach. The research revealed that,

> Participants in FTDCs entered treatment more quickly, stayed in treatment longer, and were more likely to successfully complete treatment, even when controlling for a host of risk factors. Moreover, FTDC participants were more likely to be reunited with their children, and children were placed in permanent living situations more quickly compared to those in the comparison groups. (Green et al., 2007, p. 56)

Although some of the earlier research on drug courts found mixed results (J. Anderson, 2001; Bavon, 2001; Cooper, 2003; Fischer, 2003; Harrell, 2003; Johnson Listwan et al., 2003), the vast majority of the more recent studies on drug courts have yielded positive results. That is, in most studies, drug court participants are faring better than those in comparison groups. Even when the outcomes are quite similar, the cost of the drug court is far lower than the alternative. As a result, in the 2008 National Drug Control Strategy, several pages are devoted to touting the promise of drug courts.

Conclusion

This chapter examined race and the American court system. From the earliest period of American history, race/ethnicity and class have played a role in the operation of American courts. More recent research has shown race effects during every aspect of the court process. However, unlike in the past where the discrimination was blatant, some of the bias is now masked, such as in the jury selection process and the use of peremptory challenges to remove racial/ethnic groups from jury pools. It is clear from the research in this area that both defense and prosecutors do not trust racial/ethnic minorities to make reasoned decisions when serving on juries. This is particularly disturbing because this false perception leads to actions that deny minority defendants the opportunity to have minorities (their peers) serve on their juries.

Finally, as a way to reduce the prison population, while also giving drug-addicted offenders the treatment they need, drug courts were focused on. We reviewed their structure, philosophy, and promise for dealing with some of the addiction problems. Although the results of early evaluative studies were mixed, more recently, the initiative has shown considerable promise for diverting racial minorities and other drug-addicted offenders out of regular courts. Intimately tied to the courts is the sentencing process. Chapter 6 looks at how racial minorities fare in the sentencing phase of the court process.

Discussion Questions

1. Explain how race, class, gender, or celebrity status can play a role in the American court system.

2. How do Native American courts differ in philosophy than typical American courts?

3. Discuss why the DT and the NDT are critical considerations in the American court process.

4. Explain how peremptory challenges can be misused in the American court system.

5. What program characteristics make drug courts so promising?

Internet Exercise

Go to the following Web site, http://members.tripod.com/~jctMac/jurynull.html, and read the statements from significant American figures concerning jury nullification. Do any of their thoughts sway your feelings on jury nullification? Take a look at the other links and see whether you agree with the suggestions for improving the jury system.

Internet Sites

National Center for State Courts: www.ncsconline.org/

National Criminal Justice Reference Service: www.ncjrs.org

U.S. Supreme Court: http://www.supremecourtus.gov/

National Drug Court Institute: www.ndci.org

Sentencing 6

The social responsibility for crime is so widely recognized that when the criminal is arrested, the first desire of decent modern society is to reform him, and not to avenge itself on him. Penal servitude is being recognized only as it protects society and improves the criminal, and not because it makes him suffer as his victim suffered.

—Du Bois (1920, p. 173)

Sentencing represents the stage of the criminal justice process where, following conviction, a defendant is given a sanction for committing his or her offense. Sanctions that are regularly used because they are not in violation of the Eighth Amendment's cruel and unusual punishment clause include fines (which cannot be excessive); being placed on probation, given an intermediate sanction (something in between probation and incarceration), and/or incarceration; and the death penalty. Such a process has considerable potential for discrimination and errors. It is important to consider, however, that, like the laws that send the criminal justice process in motion, state and federal legislatures are primarily responsible for determining the appropriate sentence for each offense. Although the legislature has a significant role to play in the creation of sentences, in many instances, judges also have considerable discretion in the sentencing process. Therefore, a variety of legislative enactments (e.g., sentencing guidelines) have been created to "equalize" or reduce the disparities in the sentencing process, yet judicial discretion remains a place where sentences can be influenced appreciably. Because of this, over the years, the actions of the judiciary have been scrutinized more closely by citizens and social scientists alike.

In this chapter, we examine the following three areas related to sentencing: (a) sentencing philosophies, (b) historical overview of race and sentencing, and (c) contemporary issues in race and sentencing. The first section reviews the various sentencing philosophies (also called justifications) that influence the direction of sentencing policies. Such philosophies generally shift based on political leanings, which are also discussed. Our historical overview provides some early examples of patterns in race and sentencing. Finally, the last segment of the chapter explores contemporary race and sentencing issues.

Sentencing Philosophies

The sentencing phase of the criminal justice process is where punishment is meted out. According to van den Haag (1975), "'punishment' is a depravation, or suffering, imposed by law" (p. 8). Drawing on the works of English philosopher H. L. A. Hart, Spohn (2002), a leading scholar in the area of sentencing, wrote that there are five necessary elements of punishment: It must (a) involve pain or other consequences normally considered unpleasant, (b) be enacted for an offense against legal rules, (c) be imposed on an actual or supposed offender for his offense, (d) be intentionally administered by human beings other than the offender, and (e) be imposed and administered by an authority constituted by a legal system against which the offense was committed.

Spohn (2002) added that some scholars feel that, in addition to the five elements outlined by Hart, there must also be an appropriate justification for the punishment. These justifications amount to what are generally referred to as "philosophies of punishment" or "sentencing philosophies." Generally speaking, there are four different sentencing philosophies: (a) justice/retribution, (b) incapacitation, (c) deterrence, and (d) treatment/rehabilitation. According to van den Haag (1975), "Justice [or retribution] is done by distributing punishments to offenders according to what is deserved by their offenses as specified by law" (p. 25). Incapacitation suggests that as long as offenders are detained, society is protected from them. Deterrence operates under the premise that the public will be affected by seeing others punished (van den Haag, 1975). There are two forms of deterrence: specific and general. Specific deterrence operates under the notion that when you punish someone, he or she will be deterred from further criminal activity. General deterrence suggests that when you punish someone, the larger society is deterred from committing crimes (van den Haag, 1975). Treatment or rehabilitation refers to a philosophy where making the offender better is the overriding aim of the sentence. In some instances, each of these might be present in a sentence; however, one will typically predominate (Anderson & Newman, 1998). Predictably, each of these philosophies is tied to politics. We discuss this connection next.

SENTENCING AND POLITICS

Sentencing philosophies are germane to any discussion of punishment because such considerations tend to drive the nature of what happens at the "back end" of the system (i.e., corrections). Thus, depending on the prevailing philosophy, sentences could be longer and harsher, or, conversely, sentences could be lighter and more treatment oriented. Each of these two possibilities has consequences for society as a whole. For example, when sentences are longer, jails and prisons often reach or exceed their capacity; government officials justify such expensive policies by pointing to cases where people who were released early went out and committed new crimes. However, when sentences are lighter and more treatment oriented, government officials argue that a shorter sentence under the rehabilitation approach is the cheaper and more effective approach. Therefore, more often than not, politics at the local, state, and federal levels dictate the prevailing philosophy (Beckett & Sasson, 2000; Scheingold, 1984).

A good example of this is the formulation of the crime control and due process models of criminal justice decision making constructed by Herbert Packer in the late 1960s, which 20 years later was refined by Samuel Walker (Packer, 1968; Walker, 1989). In both formulations, there are those who adhere to a more conservative approach when handling crime control, and, on the other side, there are those who adhere to a more liberal approach. In short, those who adhere to the conservative approach take a more hard-line stance on crime and are generally supportive of sentencing philosophies that are both punitive and deterrent in orientation. Adherents to the liberal philosophy lean more toward the rehabilitation/treatment approach. They tend to believe that people can be "cured" or, at the least, made better through a variety of treatments. Therefore, they generally recommend shorter jail or prison sentences combined with appropriate treatment. Such thinking reigned during the 1960s, when some literature actually pointed to the counterproductiveness of punishment (see, e.g., Menninger, 1966). However, during the punitive decades of the 1980s and 1990s, some liberals, particularly those in inner cities, were influenced by the conservatism that swept the country. This saw some people calling for "personal responsibility" and more punitive sanctions for those engaging in urban violence.

The liberal and conservative approaches have predominated for the last 40 years. As is widely acknowledged, the rehabilitation approach was dealt a near fatal blow in the 1970s by the increasing crime rate and the controversial Martinson (1974) report, which reviewed numerous treatment-oriented approaches and suggested that few were having their desired outcome. In the end, Martinson's study was interpreted as saying, when it comes to rehabilitation, "nothing works." Such a dismal forecast set in motion a movement toward the conservative philosophy, which most would agree remains the predominating approach.

Historical Overview of Race and Sentencing

PUNISHMENT IN COLONIAL AMERICA

A good place to begin a historical overview of race and sentencing is colonial America. This era provides us with a sense of the punishments that were rendered in the early years of the colony. Moreover, a review of court records and other documents also provides us with some indication as to racial/ethnic disparities in sentencing.

As has been previously noted, much of the law in colonial America was based on English practices. The remainder of the laws that were created originated from the colonists or the Bible (Chapin, 1983). The colonists took biblical influence one step further by having some cases actually handled within some religious congregations. In general, it is thought that summary proceedings administered by judges were the primary way many cases were handled. For those who desired a trial by jury, Chapin

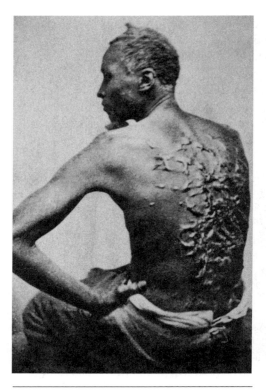

Photo 6.1   Gordon, a free slave in Baton Rouge, Louisiana, displays his whip-scarred back on April 2, 1863. He later became a corporal in the Union Army.

(1983) wrote, "A defendant could have a jury in almost any kind of case if he asked and paid for it" (p. 41). It is obvious from this practice that class bias was present from the earliest colonial times and was also present when administering punishments.

One of the first things the colonists did when they began to construct their justice system was to do away with the excessive use of the death penalty, which prevailed in England. Many of the more brutal sentences were reserved for servants and "lesser men." One such servant was Richard Barnes, whose offense, aggravated contempt, resulted in him having "his arms broken and his tongue bored through and [he] was forced to crawl through a guard of forty men who kicked him" (Chapin, 1983, p. 50). Certainly, the colonists did execute people and were fond of corporal punishment and other public displays. Some people were whipped in public (see Photo 6.1), but class lines again prevailed because those who were considered "gentlemen" could substitute some other satisfaction (likely money) to fulfill their sentences. According to Chapin (1983), however, "The most common punishment was a fine payable in money or tobacco" (p. 51). As with today, the punishments were incremental. For example, "A convicted burglar or robber in Connecticut and Massachusetts was branded on the forehead with the letter 'B' for a first offense, branded and whipped for a second, and executed as incorrigible for a third conviction" (p. 49). This example also suggests that, although the moniker might be new, three-strikes laws are not.

EARLY COLONIAL CASES

The colonists were clearly sensitive to economic issues, which, along with their xenophobia, translated into unfairness in the justice system. One way to examine this unfairness is to review early cases and their sentencing patterns. To get a glimpse of the justice that prevailed for African Americans in early America, some scholars have turned to Helen Tunnicliff Catterall's (1926/1968) four-volume work, *Judicial Cases Concerning American Slavery and the Negro*, which provides brief summaries of early relevant cases (see also Higginbotham, 1978; Russell, 1998). Four cases decided in colonial Virginia that have received considerable attention by legal scholars seeking to see how race impacted on sentences in early America are *Re Davis, Re Sweat, Re Negro John Punch,* and *Re Negro Emmanuel.*

The summary from the 1630 trial of *Re Davis* reads,

> Hugh Davis to be soundly whipt before an assembly of negroes and others for abusing himself to the dishonor of God and shame of Christianity be defiling his body in lying with a negro, which fault he is to acknowledge. Next sabbath day. (Catterall, 1926/1968, p. 77)

Here, one could interpret the transcript as saying that Davis was White and he had violated the law by sleeping with a Black woman. His obvious sentence was penance on the next Sabbath day. But the mere suggestion that he defiled his body by sleeping with the woman suggests notions of superiority and inferiority. As noted earlier by Higginbotham, religion also played a key role in maintaining African Americans and other undesirables in their "place." A decade later, *Re Sweat* (1640) also spoke to the criminal liaisons between White men and Black females, but in this case, where a child resulted, the penalties were distinctly different. The case summary reads,

> Whereas Robert Sweat hath begotten with child a negro woman servant belonging unto Lieutenant Sheppard, the court hath therefore ordered that the said negro woman shall be whipt at the whipping post and the said Sweat shall tomorrow in the forenoon do public penance for his offense at James city church in the time of divine service according to the laws of England in that case provided. (Catterall, 1926/1968, p. 78)

Here, Sweat was given part of the punishment as Davis in the prior case; however, the whipping was reserved for the Black woman. Although the punishments were obviously disparate, the sentence was likely even more stinging for the Black woman considering that, given the time period, it is likely that the relationship was not consensual. Thus, there is the possibility that a convicted rapist did penance while the victim was whipped for being raped. Such contradictions or inequities were also present in other trials.

In *Re Negro Punch,* the case record stated the following:

> Whereas Hugh Gwyn hath . . . Brought back from Maryland three servants formerly run away . . . the court doth therefore order that the said three servants shall receive the punishment of whipping and to have thirty stripes apiece one called Victor, a Dutchman, and other a Scotchman called James Gregory, shall first serve out their times with their master according to their Indentures, and one whole year apiece after the time of their service is Expired . . . and after that service . . . to serve the colony for three whole years apiece, and that the third being a negro named John Punch shall serve his said master or his assigns for the time of his natural Life here or elsewhere. (Catterall, 1926/1968, p. 78)

In this case, the punishments rendered suggest that the two White servants received less punishment than the Black servant. The two Whites received extended indentures and service to the colony, with the Dutchman also receiving a whipping. Yet the Black servant received *lifetime* servitude. Another case that shows the early disparities in sentencing is *Re Negro Emmanuel.*

In *Re Negro Emmanuel,* a plot to run away was foiled. Again, the summary of the case is illuminating:

> Complaint . . . by Capt. Wm. Pierce, Esqr. That six of his servants and a negro of Mr. Reginolds has plotted to run away unto the Dutch plantation . . . and did assay to put the same in Execution. They "had . . . taken the skiff of . . . Pierce . . . and corn, powder and shot and guns . . . which said persons sailed down . . . to Elizabeth river where they were taken . . . order that . . . Emmanuel the Negro to receive thirty [lashes] and to be burnt in the cheek with the letter R, and to work in shakle one year or more as his master shall see cause." (Catterall, 1926/1968, p. 77)

Except for Emmanuel, each of the conspirators received additional service to the colony, with one White conspirator also being branded and whipped. Although Emmanuel received no additional service, which some have suggested was an indication that he was already in lifetime servitude (Higginbotham, 1978), he was whipped, branded with "R" (presumably so everyone would know he had previously run away), and made to wear shackles for a year.

Scores of related cases existed in the colonial era. Based on an analysis of these cases, Catterall noted that a hierarchy of social precedence developed (see Figure 6.1). According to her analysis, at the top of the hierarchy were White indentured servants, whereas at the bottom of the hierarchy were Black slaves (Catterall, 1926/1968). Based on the case studies presented, one can confidently assume that justice in the courts was distributed in line with this hierarchy.

To further investigate these early disparities in case outcomes, we turn to one of the more diverse colonies, New York. From the late 1600s through the 1700s, New York was the gateway for an influx of racial and ethnic groups. Therefore, looking at data from this colony provides a more expansive picture than the cases presented by Catterall.

Figure 6.1   Catterall's Hierarchy of Social Precedence

1. White indentured servants

2. White servants without indentures

3. Christian Black servants

4. Indian servants

5. Mulatto servants (Black or Indian)

6. Indian slaves

7. Black slaves

SOURCE: Catterall (1926/1968, pp. 53–54).

CRIME AND JUSTICE IN COLONIAL NEW YORK

Greenberg (1976) provided data on more than 5,000 cases that were prosecuted in New York from 1691 to 1776. Of these cases, nearly 75% involved persons of English background, most of whom were males. As seen in Table 6.1, the remaining persons were of a variety of racial and ethnic backgrounds. What stands out most in the figures is that both the Dutch and slaves, based on their numbers in the population, were underrepresented in the criminal cases.

In the case of the Dutch, Greenberg (1976) surmised that, given that in some estimates they were nearly 50% of the population of the colony, they were likely misidentified and undercounted in the case figures. The discrepancy related to slaves was explained differently. Because they were constantly under the threat of being brutally punished, they were less likely to engage in criminal activity (Greenberg, 1976). Furthermore, as noted previously, virtually every aspect of slave life was controlled, which naturally resulted in fewer offenses. Greenberg (1976) also pointed out,

> [Another] factor that accounts for the apparently small volume of slave crime is that prosecutions for what was probably the most common and, from the master's point of view, the most costly of slave crimes—running away—appear very infrequently in the records, since slave runaways were seldom captured. (p. 45)

Before looking at the case outcomes, it is important to review the distribution of crimes by ethnic group (see Table 6.2). The review resulted in no unusual patterns except that the English and Blacks predominated in thefts. Explaining the large number of Blacks involved in theft, Greenberg (1976) wrote,

Table 6.1   Ethnic Distribution of Criminal Cases in Colonial New York, 1691–1776

Ethnic Group	Number Accused	Percentage of All Accusations
English	3,889	73.4
Dutch	693	13.1
Jewish	44	0.8
Other Whites	252	4.8
Slaves	353	6.7
Free Blacks	35	0.7
Indians	31	0.6
Total	5,297	100

SOURCE: Greenberg (1976, p. 41).

Table 6.2   *Ethnic Distribution of Major Categories of Crime in New York, 1691–1776*

Crime	English	Dutch	Jews	Whites	Blacks	Indians	N
Crimes of violence*	871	156	27	65	18	3	1,140
Thefts	585	30	4	19	86	3	727
Contempt of authority	221	79	1	10	1	0	312
Crimes by public officials	137	47	0	15	0	0	199
Disorderly houses	163	17	0	10	3	0	193
Violations of public order	795	135	6	48	7	6	997
Crimes against masters**	3	0	0	0	217	8	228

SOURCE: Greenberg (1976, p. 58).

*Against persons, not resulting in death.

** By servants or slaves.

> Prosecutions for theft were almost twice as frequent among Blacks as in the defendant population at large. This figure is a function both of the "real" position of Blacks in the eighteenth century and prevailing ideas *about* Blacks. On the one hand, slaves stole because they were deprived, and on the other, they were deprived because people believed them inherently sinful. (p. 61; italics original)

Conviction patterns also speak to how ethnic and racial groups were treated in this early period. Here, as noted in Table 6.3, although slaves represented 6.7% of the accusations, they represented nearly 10% of all convictions (Greenberg, 1976). Tables 6.3 and 6.4 illustrates that slaves had higher conviction rates than other ethnic or racial groups. With an overall conviction rate of 68.6%, American Indians had the next highest conviction rate (51.6%). Given their status within colonial America, for obvious reasons, slaves would not have been expected to receive the same consideration as other racial or ethnic groups. Their higher conviction rates speak to their low status in American society.

Closely related to conviction trends are sentencing dispositions. Although Greenberg's (1976) analysis does not provide sentences by ethnic or racial group, he does note that the nature of punishments changed in the mid-1750s. More specifically, he observed that whereas before 1750, most thefts resulted in whippings and few executions, after 1750, the number of whippings for theft dropped, whereas the use of

Table 6.3   Patterns of Judgment for Ethnic Groups in New York, 1691–1776

Ethnic Group	Number Accused	Percentage of All Accusations	Number Convicted	Percentage Convicted of Those Accused	Percentage of All Convictions	Number Acquitted	Percentage Acquitted of Those Accused	Percentage of All Acquittals
English	3,889	73.4	1,809	46.5	71.3	542	13.9	68.0
Dutch	693	13.1	329	47.5	13.0	118	17.0	14.0
Jews	44	0.8	14	31.8	0.6	16	36.4	2.0
Other Whites	252	4.8	115	45.6	4.5	48	19.0	6.0
Slaves	353	6.7	242	68.6	9.5	59	16.7	7.4
Free Blacks	35	0.7	13	37.0	0.5	6	17.1	0.8
Indians	31	0.6	16	51.6	0.6	8	25.8	1.0

SOURCE: Greenberg (1976, p. 73).

Table 6.4   Conviction Patterns for Slaves in Colonial America, 1691–1776

Crimes by Slaves	Number of Accusations	Number of Convictions	Percentage Convicted of Those Accused	Percent of All Convictions Against Slaves
Thefts	69	38	55.1	15.7
Crimes against masters	214	149	69.6	61.6
Crimes against morality (all Sabbath breach)	32	32	100.0	13.2
Crimes of violence not resulting in death	16	13	81.3	5.4
Total (these crimes)	331	232	71.0	95.9
Total (all cases involving slaves)	353	242	68.6	100.0

SOURCE: Greenberg (1976, p. 74).

branding and the number of executions increased dramatically (Greenberg, 1976). Although Greenberg did not make the connection, this change in sentencing practices could have been a response to the increasing number of racial or ethnic minorities who were engaging in theft to survive. Although the first two sections of our historical overview presented information on early cases from two distinct colonies, we now review some early national figures on race and sentencing.

EARLY NATIONAL SENTENCING STATISTICS

Following the Revolutionary War, in terms of treatment, little changed for racial or ethnic minorities. Although it was a major development when the new government created federal courts, including the Supreme Court, criminal justice remained a place where race and class bias persisted (Shelden, 2001). The development of these courts only provided new mechanisms to reinforce the race and class bias in state judiciaries (Shelden, 2001). Accurate sentencing trends from the 18th and 19th centuries are difficult to ascertain because there were no national statistics until the 1880 census (Cahalan & Parsons, 1986, p. 31). Taken from penitentiaries, the data in that first report revealed that "99% [of the offenders] were reported to be under sentence at the time of the survey . . . and 88% had sentences listed over 1 year" (Cahalan & Parsons, 1986, p. 31). In jails, "about 55% of the inmates were under sentence and of these only 8% had sentences of 1 year or longer" (p. 31).

The 1890 census provided even more detailed information of the characteristics of offenders, noting the ethnicity of those who were sentenced. At that time, the most severe average sentence was for crimes against the person at 7.8 years (Cahalan & Parsons, 1986) (see Table 6.5). Table 6.6 shows the average sentences by race, nativity, and gender. The figures show that Chinese received the longest average sentences (6.58 years), followed by Indians (5.64 years). When looking for the female group with the longest average sentence, Negroes were highest (2.8 years), with persons of Chinese origin closely behind (2.54 years) (Cahalan & Parsons, 1986). Although the data do not speak to offense trends by specific racial group, they do show that, from the earliest records available, disparities by race were present.

How far have we come in race and sentencing? Before answering this question, we examine how politics and the "get tough" philosophy have impacted on sentencing over the last thirty years. The last section of the chapter takes a deeper look at contemporary issues related to race and sentencing.

Table 6.5   Average Sentence, by Gender and Offense, 1890

Offense Against	Male	Female
Government	2.76	1.75
Society	.79	.67
Person	7.8	6.99
Property	3.9	2.29
On high seas	2.75[a]	
Other	4.67	.8

SOURCE: Cahalan and Parsons (1986, p. 39).

[a]Not separately enumerated.

Table 6.6 Average Sentence in Years, by Gender and Nativity and Race, 1890

Color, Nativity, Race	Male	Female
White	3.66	1.12
Native	3.79	1.51
Parents native	4.25	1.76
One parent foreign	3.66	1.08
Parents foreign	3.10	1.08
Unknown	2.47	1.28
Colored	5.04	2.79
Negroes	5.01	2.80
Chinese	6.58	2.54
Indians	5.64	.2

SOURCE: Cahalan and Parsons (1986, p. 39).

THE 1980s TO 2000s: THE CHANGING NATURE OF SENTENCING PRACTICES

Coinciding with the 1980 election of President Ronald Reagan was the largest number of homicides in U.S. history. That year, according to the FBI's UCRs, there were 23,040 homicides. Pointing to this unprecedented level of violent crime, to reduce these figures, President Reagan championed a conservative and more punitive approach. Following the lead of the federal government, most states moved away from sentencing policies that were generally more indeterminate in nature, which had allowed for minimum and maximum sentences and gave discretion to release offenders when they were thought to be rehabilitated. Observers from all sides felt such a sentencing approach was unfair and in too many instances resulted in bias against poor people and racial minorities. Nevertheless, over time, the public also bought into the punitive approach.

A 1995 national opinion survey on crime and justice of more than 1,000 randomly selected U.S. citizens revealed that when asked whether the government should focus on rehabilitation or punishing and putting away violent offenders, nearly 60% responded "punish," whereas only 27% said "rehabilitate" (Gerber & Engelhardt-Greer, 1996). There were, however, significant differences by race. Nearly half of the Black respondents (46%) indicated that rehabilitation should be the focus, with Hispanics (38%) also being more receptive to rehabilitation than Whites (23%) (Gerber & Engelhardt-Greer, 1996).

In line with national sentiment, states created "get tough" approaches such as determinate sentencing, sentencing guidelines that included mandatory minimum statutes, three-strikes-and-you're-out legislation, and truth-in-sentencing laws (Spohn, 2002). In addition to being more punitive, proponents of these approaches suggested that such policies also reduced judicial discretion, which some felt contributed to leniency in some instances and racial disparities in other instances. One of the first alternative sentencing approaches enacted by states was determinate sentencing.

In contrast to indeterminate sentencing, this approach required states to create "a presumptive range of confinement for various categories of offenses. The judge imposed a fixed number of years from within this range, and the offender would serve this term minus time off for good behavior" (Spohn, 2002, p. 225). In the few states that did enact this form of sentencing, the results were mixed; therefore, most states lost interest in the approach (Spohn, 2002).

One of the most popular sentencing approaches adapted during the 1970s was presumptive sentencing guidelines. Such guidelines based sentencing decisions on the various aspects of the crime and the offender history. According to Spohn (2002), there are several common aspects of this approach. First, there is generally some sentencing commission or committee who makes the recommendations for the guidelines. Second, the guidelines take into consideration "the severity of the offense and the seriousness of the offender's prior criminal record" (Spohn, 2002, p. 228). Finally, although this approach allows for some judicial discretion, judges who deviate from the specified sentence must justify their rationale. An often-cited example of this approach is the Minnesota Sentencing Guidelines (see Figure 6.2).

Following the lead of states, the federal government created its own guidelines. Once the Federal Sentencing Guidelines were enacted in 1984, federal parole was eliminated (Hickey, 1998). Moreover, during this process, in 1987, the U.S. Sentencing Commission was created. Spohn (2002) noted that the commission was created "to develop and implement presumptive sentencing guidelines designed to achieve 'honesty,' 'uniformity,' and 'proportionality' in sentencing" (p. 232).

Mandatory minimum sentences were created in the 1970s and 1980s in response to concerns about drug dealing and violent crimes involving guns. Such sentences stated a minimum penalty for a particular offense, and in some instances there were mandatory penalties for using a gun in the commission of a crime. Generally, the proscribed minimums were fairly harsh and provided little, if any, room for judicial discretion in sentencing. Prosecutors generally have the leeway to charge the offender with an offense that carries a particular sentence. As a result, under such an approach, in reality, prosecutors actually determine the nature of one's sentence (Spohn, 2002).

Popularized in the 1990s by Washington and California, three-strikes-and-you're-out legislation caught on as evidenced by the fact that, "by 1997, about half of the states and the federal government had adopted some variation" (Spohn, 2002, p. 251). Although from state to state there were different versions of the legislation, the common theme was that like a batter in a baseball game, when an offender committed their third "strike" (felony offense), they were "out" (sentenced to 25 years to life or some other draconian sentence).

Another sentencing innovation was the Truth in Sentencing Act of 1997. Clearly falling under the incapacitation philosophy, the act "awarded funds to states that kept their offenders—particularly those serving time for violent offenses—in prison for at least 85% of their sentences" (Blumstein, 2002, p. 461). It was anticipated that such a policy would have serious implications for preventing serious crimes and also for the size of correctional populations. It is notable that although crime did go down during the 1990s, whether this sentencing strategy produced the reductions has been subject

Figure 6.2 The Minnesota Sentencing Guidelines Grid

Severity Level of Conviction Offense (Common offenses listed in italics)		Criminal History Score						
		0	1	2	3	4	5	6 or more
Murder, 2nd degree (international murder; drive-by shootings)	X	306 299–313	326 319–333	346 339–353	366 359–373	386 379–393	406 399–413	426 419–433
Murder, 3rd degree; murder, 2nd degree (unintentional murder)	IX	150 144–156	165 159–171	180 174–186	195 189–201	210 204–216	225 219–231	240 231–246
Criminal sexual conduct, 1st degree; assault, 1st degree	VIII	86 81–91	98 93–105	110 105–115	122 117–127	134 129–139	146 141–151	158 153–163
Aggravated robbery, 1st degree	VII	48 44–52	58 54–62	68 64–72	78 74–82	88 84–92	98 94–102	108 104–112
Criminal sexual conduct, 2nd degree (a) & (b)	VI	21	26	30	34 33–35	44 42–46	54 50–58	65 60–70
Residential burglary; simple robbery	V	18	23	27	30 29–31	38 36–40	46 43–49	54 50–58
Nonresidential burglary	IV	12[1]	15	18	21	25 24–26	32 30–34	41 37–45
Theft crimes (over $2,500)	III	12[1]	13	15	17	19 18–20	22 21–23	25 24–26
Theft crimes ($2,500 or less); check forgery ($200–$2,500)	II	12[1]	12[1]	13	15	17	19	21 20–22
Sale of simulated controlled substance	I	12[1]	12[1]	12[1]	13	15	17	19 18–20

SOURCE: Minnesota Sentencing Guidelines Commission: Sentencing Practices, Annual Summary Statistics for Felony Offenders Sentenced in 2006 (2007).

to debate (Blumstein & Wallman, 2005). Each of the aforementioned sentencing approaches, in varying magnitudes, contributed to an increasing number of persons being incarcerated (see Chap. 8). In fact, many of those who bore the brunt of such policies were poor persons and racial minorities. Another set of legislative enactments from the 1980s through the 2000s, which influenced the nature of sentencing, targeted the "War on Drugs."

THE "WAR ON DRUGS"

Declaring a "War on Drugs" in the late 1980s was also a major contributor to the increase in punitive sentencing policies during the last 20 years. Beginning with Ronald Reagan and taking shape under George H. W. Bush, the White House Office of National Drug Control Policy, which was officially established under the Anti-Drug Abuse Act of 1988 and extended under the Violent Crime Control and Law Enforcement Act of 1994, was created to "establish policies, priorities, and objectives for the Nation's drug control program" (see http://www.whitehousedrugpolicy.gov). In line with its purpose, "The goals of the program are to reduce illicit drug use, manufacturing, and trafficking, drug-related crime and violence, and drug-related health consequences" (see http://www.whitehousedrugpolicy.gov). In some ways, the role of the director (also referred to as the drug czar) is symbolic in that that person is the face of drug control for each president. Although the director drafts and promotes the annual National Drug Control Strategy (2008) report, he or she does not actually carry out the activities proposed to meet its goals. Nevertheless, because the director is a component of the Executive Office of the President, he or she does "[advise] the President regarding changes in the organization, management, budgeting, and personnel of Federal Agencies that could affect the Nation's anti-drug efforts; and regarding Federal agency compliance with their obligations under the Strategy" (see http://www.whitehousedrugpolicy.gov).

With such an influential role, the director can push either punitive or treatment-oriented drug-control strategies. However, given that the position is an appointed one, it is likely that the director will simply follow the party line on most issues. Again, as in the sentencing strategies discussed in the previous section, the party line on such issues in the 1980s and early 1990s was punitive in orientation. Even with the election of President Bill Clinton (a Democrat) and the appointment of Lee P. Brown as director, who championed rehabilitative and education-oriented initiatives, such punitive approaches had become so institutionalized that during Clinton's administration one of the most punitive crime bills in history was enacted. During the 2000s, the collateral consequences of two decades of the punitive ideology had actually brought the system back to rehabilitation. Why? Because states finally realized that they could not afford to maintain the correctional building programs of the 1980s and 1990s. As such, states were increasingly looking for ways to divert or release nonviolent drug offenders—an option that was not on the table during the height of the "War on Drugs." We return to this issue later in the chapter.

Contemporary Issues in Race and Sentencing

We begin our discussion of contemporary issues with a review of current sentencing patterns. Following this initial discussion, we examine race and sentencing scholarship, as well as the role of judges in sentencing disparities.

In 2004, nearly 1.1 million adults were convicted of a felony; 70% of them were sentenced to either jail (30%) or prison (40%), with the remainder being sentenced to probation (Durose & Langan, 2007). In 2004, the average sentence was 4.9 years. For those who committed murder and non-negligent manslaughter, the average was 20 years. In general, most persons who committed homicide received a sentence of incarceration, with only 2% being sentenced to death. In line with the arrest statistics presented in Chapter 2, most persons convicted for felony offenses were Whites; however, Blacks were overrepresented in nearly every category (see Table 6.7). In 2004, for example, Blacks, who make up 12% of the population, represented 38% of those persons convicted of a felony and 39% of convicted felons who committed a violent crime (Durose, 2007). Other races (American Indians, Alaska Natives, Asian and Pacific Islanders) represented 3% of all felony convictions and the same percentage of convictions for violent crime (Durose, 2007). Given the large numbers of Blacks convicted of felony offenses, the next question is whether they are convicted and sentenced fairly. In recent years, there has been an explosion of literature investigating this question. We review some of these studies next.

SCHOLARSHIP ON RACE AND SENTENCING

In the past two decades, there have been numerous studies examining the relationship between race and sentencing. To provide context for these studies, we review several comprehensive studies that provide baseline summaries on the topic up to 2002 (Chircos & Crawford, 1995; O. Mitchell, 2005; Pratt, 1998; Spohn, 2000; Zatz, 1987). Following a review of these studies, more recent studies are examined.

Separating sentencing research into four waves, Zatz (1987) provided a summary of the diverse research on race and sentencing. During Wave I (1930s to mid-1960s), the literature "showed clear and consistent bias against non-Whites in sentencing" (Zatz, 1987, p. 71). Most of the early research suggested that Blacks and foreigners received biased sentences in the courts. A noted concern of this early research was its lack of methodological sophistication. On this point, Zatz wrote, "These studies were flawed by a number of serious methodological problems. The most damning flaw was the lack of controls for legally relevant factors, especially prior record" (p. 72).

Wave II (late 1960s to 1970s) followed on the heels of the successful civil rights movement. During this period, studies began to emerge that indicated that discrimination in sentencing had subsided. Moreover, more sophisticated studies began to suggest "that minorities were overrepresented in the criminal justice system and prisons because of their greater proportional involvement in crime, and

Table 6.7　　Percentage of Persons Convicted of Felonies, by Offense, 2004[a]

Most serious conviction offense	Sex			Race		
	Total	Male	Female	White	Black	Other
All offenses	100%	82%	18%	59%	38%	3%
Violent offenses	100	90	10	57	39	3
Murder, non-negligent manslaughter	100	91	9	44	53	4
Sexual assault, rape	100	98	2	72	25	3
Rape	100	98	2	68	29	3
Other sexual assault	100	97	3	75	23	3
Robbery	100	92	8	40	58	2
Aggravated assault	100	85	15	59	38	4
Other violent[b]	100	90	10	70	26	5
Property offenses	100	74	26	64	34	2
Burglary	100	90	10	65	33	3
Larceny, motor vehicle theft	100	74	26	62	36	2
Motor vehicle theft	100	87	13	71	26	3
Fraud, forgery, embezzlement	100	58	42	64	33	2
Drug offenses	100	82	18	52	46	2
Possession	100	81	19	54	44	2
Trafficking	100	83	17	51	47	2
Weapons offenses	100	96	4	43	55	2
Other offenses[c]	100	87	13	69	27	4

SOURCE: Durose and Langan (2004, p. 6).

NOTE: [a]Detail may not sum to total due to rounding. Racial categories include Hispanics.

[b]Negligent manslaughter and kidnapping.

[c]Composed of nonviolent offenses such as receiving stolen property and vandalism.

not because of any bias in the system" (Zatz, 1987, p. 73). Studies were also beginning to show that race might have a "cumulative effect" by indirectly impacting on minority offenders through other variables. In addition, "extralegal attributes of the offender could *interact* with other factors to influence decision making" (Zatz, 1987, p. 73; italics original).

In Wave III (studies conducted in 1970s and 1980s), the Wave II studies that suggested discrimination had subsided were reanalyzed and produced different results. Previously unaccounted for in earlier studies, Zatz (1987) noted that, "depending on the degree of victimization and the relative social harm perceived, minority members were treated more harshly in some situations and Whites in others" (p. 74). In addition, Zatz pointed to selection bias and specification error as two other common problems with earlier research. Here, she was concerned that some people were dropped out of study samples too early in the process, which can invalidate the results of studies. With specification error, in some instances, researchers were not considering interaction effects. Such effects highlight when

> Race/ethnicity can operate *indirectly* through its effect on other factors. It can interact with other variables to affect sentencing. That is, the impact of other variables (e.g., prior record, type of offense) could differ, depending on the defendant's race/ethnicity. (Zatz, 1987, p. 75; italics original)

Another important type of indirect effect is referred to as cumulative disadvantage. Here, race/ethnicity has a small effect on the initial stages of the process; however, "as the person moves through the system, these add up to substantial, and often statistically significant, disparities in processing and outcomes for different social groups" (p. 76).

At the time of the publication of her article, Zatz (1987) indicated that Wave IV (data from late 1970s to 1980s, conducted in 1980s) was underway and substantive issues were emerging. Determinate sentencing was beginning to garner significant attention. Researchers were beginning to investigate the outcomes being produced by this sentencing strategy.

Eight years after the publication of Zatz's article, Chircos and Crawford (1995) provided an updated summary of the literature on race and sentencing. Their review centered on whether African Americans were "more often sentenced to prison upon conviction" (Chircos & Crawford, 1995, p. 281). Moreover, they also wanted to see whether region (particularly the South), the percentage of Black population as a whole or percentage of Black population that is urban, and unemployment were influential in sentencing outcomes. To investigate the influence of these factors, Chircos and Crawford examined 38 studies published over a 16-year period (1975–1991). In terms of the effect of region, most studies, even after controlling for prior record and crime seriousness, showed that Blacks were more likely to be incarcerated than Whites (Chircos & Crawford, 1995). Furthermore, Chircos and Crawford found that "in no instance is the incarceration of White defendants significantly more likely in the south" (p. 297).

When reviewing the results of studies that examined the percentage of Blacks in the population, Chircos and Crawford (1995) noted that "it is clear that Black defendants are disadvantaged much more often when the Black population exceeds the national average than when it does not" (p. 298). Relatedly, when they reviewed the influence of percentage of Black defendants who are urban, their results paralleled those of their regional comparison. As Chircos and Crawford noted, "The disadvantage to Black defendants, as measured by significant positive results, is approximately twice as great in places where they are *less* concentrated in urban areas than in places where they are *more* concentrated" (p. 299; italics original). In their final area of focus, unemployment, Chircos and Crawford found that "the disadvantage of Black defendants is more consistently demonstrated for places with higher levels of unemployment" (p. 299).

Picking up where Chircos and Crawford (1995) left off, Pratt (1998) conducted a meta-analysis of the race and sentencing research. His analysis included race and sentencing studies published in social science journals from 1974 to 1996. In contrast to the earlier summaries of the extant race and sentencing literature, he found that "the severity of the offense is the only significant variable related to length of sentence. . . . Neither race nor prior criminal record variables were found to be significant" (Pratt, 1998, p. 517). Pratt did note that one explanation for this finding was the way in which race was operationalized in some studies. As an example, he notes that it matters whether studies use the Black/White or White/Non-White categories. More specifically, he found that "When the race variable takes the form of a White/non-White classification scheme, the likelihood of finding a significant race effect on sentence length is greater than for either of the two other racial classification schemes" (Pratt, 1998, p. 518).

Two years after Pratt's work, Spohn (2000) examined 40 methodologically sophisticated studies on race and sentencing (32 on state sentencing and 8 on federal sentencing). In her review, she noted two considerable limitations. First, most of the studies reviewed centered on Black and White defendants; few of them explored beyond this racial dichotomy. In addition, although she reviewed 32 state-level studies, only 13 states were represented. Concluding her review of these studies, Spohn (2000) wrote,

> Considered together . . . race and ethnicity do play an important role in contemporary sentencing decisions. Black and Hispanic offenders sentenced in State and Federal courts face significantly greater odds of incarceration than similarly situated White offenders. In some jurisdictions, they also may receive longer sentences or differential benefits from guideline departures than their White counterparts. (p. 458)

More specifically, the studies revealed interaction effects between race/ethnicity and several legally irrelevant offender characteristics. Studies reviewed by Spohn (2000) showed that "Racial minorities are sentenced more harshly than Whites if they (1) are young and male, (2) are unemployed, (3) are male and unemployed,

(4) are young, male, and unemployed, (5) have lower incomes, and (6) have less education" (p. 462).

Spohn (2000) also found that, in several studies, when looking at the interaction between the offenders' race and process-related factors, "Racial minorities are sentenced more harshly than Whites if they (1) are detained in jail prior to trial, (2) are represented by a public defender rather than a private attorney, (3) are convicted at trial rather than by plea, and (4) have more serious prior criminal records" (p. 462). Other studies looking at the interaction between offender race and victim race revealed that "racial minorities who victimize Whites are sentenced more harshly than other race of offender/race of victim combination" (p. 463). Finally, Spohn's (2000) review found studies that showed interaction effects between offender race and type of crime: "Racial minorities are sentenced more harshly than Whites if they are (1) convicted of less serious crimes, or (2) convicted of drug offenses or more serious drug offenses" (p. 463).

Following Spohn's (2000) comprehensive review, researchers have continued to examine the question of race and sentencing (Engen, Gainey, Crutchfield, & Weis, 2003; B. Johnson, 2003; Steffensmeier & Britt, 2001; Steffensmeier & Demuth, 2000, 2001; Ulmer & Johnson, 2004), and, unfortunately, they have continued to find that race and ethnicity matters in the sentencing process. For example, an increasing finding in studies of both state (Steffensmeier & Demuth, 2001) and federal (Steffensmeier & Demuth, 2000) courts is that Hispanics are receiving the harshest sentences. In addition, studies are beginning to look at more contextual factors, noting that African Americans and Hispanics tend to be sentenced more harshly in areas where they are the majority populations (Ulmer & Johnson, 2004).

In 2005, a meta-analysis by Ojmarrh Mitchell again sought to determine whether race matters in a large selection of empirical studies. Mitchell's analysis included both published and unpublished studies (e.g., dissertations). In all, he whittled down 331 studies to 71 that met his rigorous criteria for inclusion in his analysis. In contrast to Pratt's earlier work, O. Mitchell (2005) found that "even after taking into account offense seriousness and prior criminal history, African-Americans were punished more harshly than whites" (p. 456). He did note, however, that:

> The observed differences between whites and African-Americans generally were small, suggesting that discrimination in the sentencing stage is not the primary cause of overrepresentation of African-Americans in U.S. correctional facilities. The size of unwarranted sentencing disparities grows considerably, however, when contrasts examined drug offenses, imprisonment decisions, discretionary sentencing decisions, and recently collected Federal data. (O. Mitchell, 2005, p. 462)

More recent research has, in line with previous research (Crawford, 2000, Crawford, Chircos, & Kleck, 1998), continued to find that Blacks and Hispanics are more likely to be prosecuted under habitual offender statutes than Whites (Crow & Johnson, 2008). Given the breadth and contradictions of the race and sentencing literature, we now look at a few other important areas. Our specific areas of foci include sentencing disparities related to the "War on Drugs," the impact of minority judges in reducing sentencing disparities, and race and misdemeanor sentencing. To determine the extent of racial disparities in sentencing, we turn to studies that look at race and misdemeanor sentencing.

RACE AND MISDEMEANOR SENTENCING

Although much of the research reviewed earlier focused on felony defendants, two recent studies have examined race and misdemeanor sentencing. In the first study, Munoz and McMorris (2002) reviewed more than 8,000 misdemeanor cases in Nebraska to determine whether racial/ethnic bias was present. Their analysis found that, in line with studies focusing on felony sentencing,

> Non-Whites in comparison to Whites were charged with more serious misdemeanors and more total number of offenses. Latinos experienced unfavorable outcomes in comparison to Whites, but not to the extent Native Americans did. In addition to having the highest average total number of offenses, Native Americans were most likely to find themselves charged with alcohol/drug, property, and assault related offenses. (Munoz & McMorris, 2002, p. 254)

Mirroring felony studies, legal variables such as prior record and seriousness of the offense were the best predictors of sentence severity. Extralegal factors such as race and gender were also contributors to the observed disparities (see also Munoz, McMorris, & DeLisi, 2005).

Leiber and Blowers (2003) tested whether, in the case of misdemeanor offenses, African Americans were punished more severely than Whites. With a sample of more than 1,700 defendants, they examined the influence of extralegal variables (race, age) and legal variables (crime type, weapons use, property loss, and prior arrest). The findings of Leiber and Blowers revealed that "race does not have a direct effect on the conviction and incarceration decisions" (p. 477). They contextualized their findings by noting that the race effect was indirect. That is, in their study, the race effect was masked through the procedural variables of case status and continuance. On this finding they wrote,

> The prioritizing of a case and not having a continuance increase the chances of being convicted and incarcerated. Because African Americans are more likely to have their case classified as a priority and not have a continuance, they have a greater probability than Whites to be convicted and incarcerated. (Leiber & Blowers, 2003, p. 477)

Each of these studies provides evidence that racial disparities in sentencing are not limited to felony cases. The "War on Drugs" has also been suggested as a producer of racial disparities in criminal justice. This is discussed next.

SENTENCING DISPARITIES
AND THE "WAR ON DRUGS"

As noted earlier in the chapter, the "War on Drugs" rang in a new era of punitive sanctions. Unfortunately, as has been noted by several researchers, minorities have been impacted most by the war (Donzinger, 1996; J. Miller, 1996; Tonry, 1995). On the federal level alone, Tonry (1995) noted that, in 1980, drug offenders represented 22% of the admissions to institutions, but by 1989, this had risen to 39%, and by 1990, this had risen to 42%. Strikingly, by 1992, 58% of federal inmates were drug offenders (Tonry, 1995). At the state level, Shelden and Brown (2003) noted that, "between 1980 and 1992, sentences on drug charges increased by more than 1,000%" (p. 252). They continued by noting that when taking race into consideration, "The number of African Americans sentenced to prison on drug charges increased by over 90%, almost three times greater than White offenders" (p. 252). Shelden and Brown also pointed to figures that show that, during the 10-year period of 1985 to 1995, "the number of African American inmates sentenced for drug offenses increased by 700%" (p. 252).

During this same period (1985–1995), crack cocaine became a significant concern for the government, which led to the formulation of differential mandatory minimum penalties. For example, someone with 500 grams of powder cocaine would receive a mandatory minimum sentence of 5 years, whereas someone with 5 grams of crack cocaine would receive the same 5-year mandatory minimum penalty. For those caught with 50 grams of crack cocaine, the mandatory minimum penalty was 10 years, whereas those caught with 5,000 grams of powder cocaine also received a 10-year mandatory minimum sentence. Unfortunately, although minorities are overrepresented in the use of crack cocaine, the majority of those who use crack cocaine are White, yet most of those serving time under these federal policies are African Americans and Hispanics (Russell-Brown, 2004).

Tracing the origin of this much maligned 100-to-1 crack cocaine-to-powder cocaine sentencing differential, Russell (1998) suggested that the sudden death of University of Maryland basketball star Len Bias contributed to this "moral panic" or unsupported fear about crack cocaine (Brownstein, 1996; Jenkins, 1994). Because Bias' death was linked to crack cocaine, a cheaper form of cocaine, initially, many public officials felt that this was a sign that crack would devastate the Black community.

It is important to note that, in the ensuing legislative debates, according to Kennedy (1997), Black legislators such as Charles Rangel and Major Owens argued vociferously for their legislative colleagues to move quickly in drafting legislation to head off a potential epidemic in the African American community. Although Kennedy (1997) noted that the legislators didn't call for the differential penalties that ensued, "Eleven of the twenty-one Blacks who were then members of the House of Representatives voted in favor of the law which created the 100–1 crack-powder differential" (p. 370).

Many observers would agree the early efforts to stave off an impending epidemic were noble, but others, however, would argue that, because of these actions, an "incarceration epidemic" ensued (Mauer, 1999; Radosh, 2008; Tonry, 1995). Moreover, once it became clear that the sentencing differential was leading to massive incarceration disparities, in 1995 and 1997, the government had two opportunities to rectify the situation. But according to Russell (1998), Congress overwhelmingly voted against taking such action, although it was recommended by the U.S. Sentencing Commission. To the surprise of many in the African American community, President Clinton did not follow the recommendations of the Commission. In response to this inaction, prisoners at several federal correctional facilities rioted (Russell, 1998). Congress returned to the matter in 2002, when the Senate held hearings on the federal cocaine policy (see U.S. Senate, 2002). In preparation for this congressional hearing, in 2002, the U.S. Sentencing Commission produced the report, *Cocaine and Federal Sentencing Policy.*

The Commission based its report on empirical analyses on federal cocaine offenders sentenced in 1995 and 2000, a survey of state cocaine sentencing policies, public opinion, and public hearings. From these sources, the Commission reported the following four major findings: (a) current penalties exaggerate the relative harmlessness of crack cocaine, (b) current penalties sweep too broadly and apply most often to lower level offenders, (c) current quantity-based penalties overstate the seriousness of most crack cocaine offenses and fail to provide adequate proportionality, and (d) current penalties' severity impact mostly minorities (U.S. Sentencing Commission, 2002, pp. v–viii).

With the first finding, the Commission recognized that the form of cocaine is irrelevant; either crack or powder cocaine "produces the same physiological and psychotropic effects" (U.S. Sentencing Commission, 2002, p. v). In addition, speaking to the short-lived, unsuccessful, and, some believe, racism-based attempt of some states (e.g., South Carolina and Kentucky) to prosecute mothers (who happened to be overwhelmingly Black) who used drugs during pregnancy (Kennedy, 1997; Russell-Brown, 2004), the Commission noted that "the negative effects of prenatal crack cocaine exposure are identical to the negative effects of prenatal powder cocaine exposure and are significantly less severe than previously believed" (U.S. Sentencing Commission, 2002, pp. v–vi). Equally important, the Commission added that "the negative effects from prenatal cocaine exposure are similar to those associated with prenatal tobacco exposure and less severe than the negative effects of prenatal alcohol exposure" (U.S. Sentencing Commission, 2002, p. vi).

Under the second finding, the Commission noted that more than a quarter of the federal crack cocaine offenses involved small quantities (less than 25 grams). In contrast, "Only 2.7% of federal powder cocaine offenses involved less than 25 grams of the drug, perhaps because the statutory minimum penalties would not apply to such a small quantity of powder cocaine" (U.S. Sentencing Commission, 2002, p. vi). If what the Commission surmised is true, then those dealing in powder cocaine actually have the law on their side. To further illustrate the nature of the sentencing disparity in practice, the commission provided the following figures:

> Defendants convicted of trafficking less than 25 grams of powder cocaine received an average sentence of 13.6 months, just over one year. In contrast, defendants convicted of trafficking an equivalent amount of crack received an average sentence of 64.8 months, over five years. (U.S. Sentencing Commission, 2002, p. vi)

The third finding by the Commission argued that the crack cocaine–violence link was unsupported by the empirical evidence. As they reported, "In 2000 . . . three-quarters of federal crack cocaine offenders had no personal weapon involvement, and only 2.3% discharged a weapon" (U.S. Sentencing Commission, 2002, p. vii). In the Commission's view, the penalty "sweeps too broadly" and does not acknowledge that, although some crack cocaine offenders do engage in violence, the majority do not (U.S. Sentencing Commission, 2002). The final finding discussed by the Commission was that minorities were hit hardest by the current cocaine penalties. They noted that, in 2000, 85% of those impacted by the crack cocaine penalties were Black, whereas only 31% of those impacted by the powder cocaine laws were Black (U.S. Sentencing Commission, 2002).

Taking their findings into consideration, the Commission provided three recommendations. First, they recommended "increas[ing] the five-year mandatory minimum threshold quantity for crack cocaine offenses to at least 25 grams and the ten-year threshold quantity to at least 250 grams (and repeal the mandatory minimum for simple possession of crack cocaine)" (U.S. Sentencing Commission, 2002, p. viii). Second, rather than having a blanket mandatory minimum sentence, they recommended providing sentencing enhancements based on the nature of the offense. Such enhancements would take into consideration whether a weapon was used, bodily injury occurred, the person was a repeat offender, and so on. Finally, the Commission recommended maintaining the current powder cocaine thresholds, but also incorporating the said sentencing enhancements (U.S. Sentencing Commission, 2002).

Five years later, the Commission returned to the topic with another report, *Cocaine and Federal Sentencing Policy* (2007). The Commission returned to similar themes from the 2002 report, noting the continuing "universal criticism from representatives of the Judiciary, criminal justice practitioners, academics, and community interest groups" (U.S. Sentencing Commission, 2007, p. 2). In addition, the report noted that the major conclusions from the 2002 report remained valid. Figures 6.3 and 6.4 highlight the most recent data trends related to crack cocaine and powder cocaine offenders and prison sentences. Also, Table 6.8 provides a look at the demographics of cocaine offenders during three periods. These data, again, revealed the stark disparities related to the two forms of cocaine.

During 2007, however, the Commission had the benefit of two Supreme Court cases that provided additional impetus to change the disparity. The cases of *Blakely v. Washington* (2004) and *United States v. Booker* (2005), both involved issues related to sentencing. In the *Blakely* case, the Court ruled that judges could not enhance penalties based on facts outside of those noted by the jury and the offender. In doing so, they violated one's Sixth Amendment rights to a jury trial. In the *Booker* case, the

Figure 6.3   Trend in Number of Powder Cocaine and Crack Cocaine Offenders

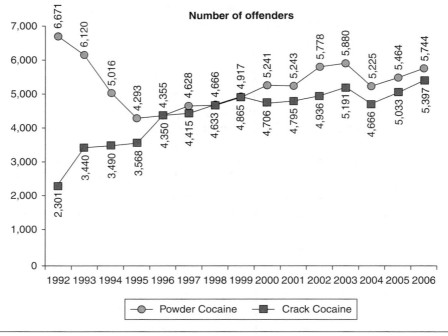

SOURCE: U.S. Sentencing Commission (2007).

Figure 6.4   Trend in Prison Sentences for Powder Cocaine and Crack Cocaine Offenders

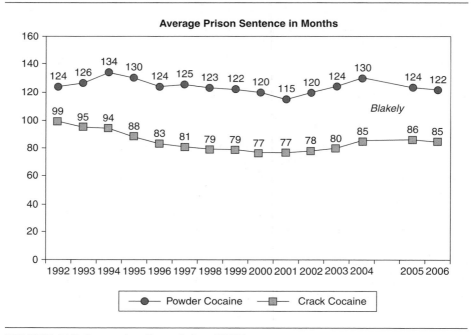

SOURCE: U.S. Sentencing Commission (2007).

Table 6.8 Demographic Characteristics of Federal Cocaine Offenders: Fiscal Years 1992, 2000, and 2006

	Powder Cocaine						Crack Cocaine					
	1992		2000		2006		1992		2000		2006	
	N	%	N	%	N	%	N	%	N	%	N	%
Race/Ethnicity												
White	2,113	32.3	932	17.8	821	14.3	74	3.2	269	5.6	474	8.8
Black	1,778	27.2	1,596	30.5	1,550	27.0	2,096	91.4	4,069	84.7	4,411	81.8
Hispanic	2,601	39.8	2,662	50.8	3,296	57.5	121	5.3	434	9.0	452	8.4
Other	44	0.7	49	0.9	66	1.2	3	0.1	33	0.7	56	1.0
Total	6,536	100.0	5,239	100.0	5,733	100.0	2,294	100.0	4,805	100.0	5,393	100.0
Citizenship												
U.S. Citizen	4,499	67.7	3,327	63.9	3,463	60.6	2,092	91.3	4,482	93.4	5,195	96.4
Non-Citizen	2,147	32.3	1,881	36.1	2,256	39.4	199	8.7	318	6.6	196	3.6
Total	6,646	100.0	5,208	100.0	5,719	100.0	2,291	100.0	4,800	100.0	5,391	100.0
Gender												
Female	787	11.8	722	13.8	561	9.8	270	11.7	476	9.9	461	8.5
Male	5,886	88.2	4,518	86.2	5,179	90.2	2,032	88.3	4,330	90.1	4,936	91.5
Total	6,673	100.0	5,240	100.0	5,740	100.0	2,302	100.0	4,806	100.0	5,397	100.0
Average Age	Average = 34		Average = 34		Average = 34		Average = 28		Average = 29		Average = 31	

This table excludes cases missing information for the variables required for analysis.

SOURCE: U.S. Sentencing Commission (2007).

Court affirmed this decision. Thus, when mandatory sentences are in place and they go against a jury's findings, the judge can view such mandatory penalties as only advisory. Thus, the draconian mandatory crack cocaine sentences can be seen as advisory, and, with these decisions, judicial discretion was permissible. In December 2007, the Supreme Court ruled in the case of *Kimbrough v. United States* that mandatory crack penalties were advisory in nature and judges could take into account the crack and powder cocaine disparities when sentencing. Given all these significant cases and the long-standing efforts of numerous entities, the Commission's recommended changes were partially adopted and made "sentences for crack offenses between two and five times longer than sentences for equal amounts of powder" (*Kimbrough v. United States*). The changes in the federal guidelines were made retroactive and took effect in early 2008, which resulted in the potential early release of nearly 20,000 offenders (see Highlight Box 6.1).

Highlight Box 6.1

Thousands of Crack Sentences Could Be Reduced

Rules taking effect this week intended to bring fairness to drug penalties

NEW YORK—Marsha Cunningham was no drug dealer. But when authorities busted her boyfriend in the 1990s for selling crack and powdered cocaine, they also arrested her on a crack possession charge.

Her sentence: Fifteen years behind bars, only two less than her boyfriend got.

But Cunningham is now one of nearly 20,000 inmates convicted of crack offenses who may see their prison terms reduced under new federal guidelines intended to bring retroactive fairness to drug sentencing.

"Marsha is a really good person," said her aunt, Ruby Jones of Houston. "She got caught up in this behind her boyfriend."

The sentencing guidelines went into effect Monday—the result of a December decision by the U.S. Sentencing Commission to ease the way the system came down far harder on crack-related crimes than on those involving powdered cocaine.

Previously, a person with one gram of crack would receive the same sentence as someone with 100 grams of the powdered form of cocaine. The disparity has been decried as racially discriminatory, since four of every five crack defendants in the U.S. are black, while most powdered-cocaine convictions involve whites.

1,600 eligible this week

"The sentences for crack cocaine have been one of the most corrosive and unjust areas of criminal law," said Michael Nachmanoff, head of the federal public defender's office for the Eastern District of Virginia. "It's really undermined respect for the criminal justice system, not only in the African-American community but throughout the country."

Nachmanoff said four clients of his office were being released under the new guidelines Monday.

About 1,600 inmates are eligible for immediate release this week, but there is no way to know exactly how many will ultimately be freed, since each prisoner has to ask for a reduction and go before a judge. The remainder of the 19,500 crack defendants will become eligible for release over the next 30 years.

"As we do with all sentencing guidelines, the department will apply the new rule as written," Justice Department spokesman Peter Carr said Monday. "We will be urging the courts not to go beyond the limited reduction that the Sentencing Commission has asked for and not to re-sentence defendants from scratch."

The Justice Department said it is more worried about crack defendants set to come up for release later, saying they include a higher share of violent offenders and potential repeat offenders than the first batch.

Violent criminals, or not?

Attorney General Michael Mukasey told a police group last week that nearly 80 percent of the crack defendants who could apply for a reduction in their sentences have some kind of criminal past.

"This tells us those who are eligible for early release are very likely to commit another crime," Mukasey told the Fraternal Order of Police. "These offenders are often violent criminals who are likely to repeat their criminal activities."

But Nachmanoff said few of the crack defendants are violent criminals.

"These are people who committed crimes and have been punished, and the sentencing commission is trying to ensure that they are not punished excessively," he said.

Cunningham, 37, was sentenced in 1998 and has a projected release date of July 24, 2011. She filed a request last week for a sentence reduction, but it has not been ruled on yet.

Her aunt said two of Cunningham's grandparents have died since she went to prison, and her father is in a nursing home.

"He's hoping that she gets out soon so that he can see her," said Jones, the aunt. "Put his arms around her."

Jones said that in phone calls to her family, Cunningham is upbeat but anxious about when she will be released.

"She's been in there for so long," Jones said. "For so long."

SOURCE: Associated Press (2008).

Throughout the last 30 years, some citizens (particularly racial and ethnic minorities) have often questioned the neutrality of judges when racial disparities are revealed. As such, a consistent theme in policy considerations related to the judiciary is that if there were more minority judges, the courts would be run more equitably. Drawing on the scholarly literature, we investigate this supposition next.

MINORITY JUDGES

More than 150 years ago, Robert Morris became the first African American to be appointed as a judge (Washington, 1994). Less than 20 years later, during the

Reconstruction period, another milestone was reached when Jonathan Jasper Wright was elected to the Supreme Court of South Carolina in 1872 (Washington, 1994). Since these early breakthroughs, few minorities have served as judges. Although this could be a product of the fact that in 2002 Blacks and Hispanics combined represented less than 8% of American lawyers (up from 3% in 1983), proactive measures have resulted in some increases in minority judicial appointments (U.S. Bureau of the Census, 2003).

Looking at judicial appointments to the U.S. district courts over the last 40 years, one does notice a considerable improvement in diversity, but only during the tenure of two presidents (Carter and Clinton) were there substantial gains (see Table 6.9). During the Carter era, nearly 14% of the appointments were African American, with nearly 7% being Hispanic (Pastore & Maguire, 2007). During this span, significant gains were also made for women. In the Johnson era, only 1.6% of the appointees were women, while during the current presidency of George W. Bush, 20% of the appointments are women. During the period selected, President George W. Bush also had the highest average rating for his appointees; nearly 70% of them were rated by the American Bar Association to be exceptionally qualified or well qualified. This figure was followed by the Clinton presidency at 59% (Pastore & Maguire, 2007). It is also notable that during Bush's presidency the percentage of Latino appointees to the U.S. district courts was higher than at any other point in history.

Turning to judicial appointments for the U.S. Court of Appeals (see Table 6.10), the Clinton presidency stands out (1993–2000). During his presidency, nearly 33% of his appointments were females, and nearly 25% were racial minorities (13.1% Black, 11.5% Hispanic) (Pastore & Maguire, 2007). Contrary to some concerns about minority judicial appointments, nearly 79% of Clinton's judicial appointments were rated as either exceptionally well qualified or well qualified by the American Bar Association, a figure that surpassed the appointment records of all other presidents over the last 40 years.

Two things stand out from these data. First, Asian Americans and Native Americans are minimally represented in the appointment figures. The absence of qualified Native American judges could be a product of many of them serving in such positions on reservations. In the case of Asian Americans, although their population figures are low, by now there are surely an ample number of potential nominees. Nevertheless, for minorities who have made it to the judiciary, are there differences between their decisions and those of White judges? We explore this question next.

Most people are familiar with the instant classic *Black Robes, White Justice* by New York State Judge Bruce Wright (1987). Judge Wright's poignant criticism of the justice system served as a much-needed case study of how, in his experience, race operated within the courts. Since the publication of his work, several observers have investigated the impact of minority judges, such as Wright, to see whether they are bringing balance to the bench. In essence, because many minority judges have experienced racism along the way, some observers believe they will be more sensitive to the plight of minorities

Table 6.9   Characteristics of Presidential Appointees to U.S. District Court Judgeships by Presidential Administration, 1963–2005

	President Johnson's appointees 1963–68 (N = 122)	President Nixon's appointees 1969–74 (N = 179)	President Ford's appointees 1974–76 (N = 52)	President Carter's appointees 1977–80 (N = 202)	President Reagan's appointees 1981–88 (N = 290)	President George H.W. Bush's appointees 1989–92 (N = 148)	President Clinton's appointees 1993–2000 (N = 305)	President George W. Bush's appointees 2001–2005 (N = 168)
Sex								
Male	98.4%	99.4%	98.1%	85.6%	91.7%	80.4%	71.5%	79.2%
Female	1.6	0.6	1.9	14.4	8.3	19.6	28.5	20.8
Race ethnicity								
White	93.4	95.5	88.5	78.2	92.4	89.2	75.1	82.1
Black	4.1	3.4	5.8	13.9	2.1	6.8	17.4	6.6
Hispanic	2.5	1.1	1.9	6.9	4.8	4.0	5.9	10.7
Asian	0	0	3.9	0.5	0.7	0	1.3	0.6
Native American	NA	NA	NA	0.5	0	0	0.3	0
Education, undergraduate								
Public-supported	38.5	41.3	48.1	55.9	37.9	46.0	44.3	47.6
Private (not Ivy League)	31.1	38.5	34.6	34.2	48.6	39.9	42.0	45.8
Ivy League	16.4	19.6	17.3	9.9	13.4	14.2	13.8	6.6
None indicated	13.9	0.6	0	0	0	0	0	0
Education, law school								
Public-supported	40.2	41.9	44.2	52.0	44.8	52.7	39.7	49.4
Private (not Ivy League)	36.9	36.9	38.5	31.2	43.4	33.1	40.7	38.1
Ivy League	21.3	21.2	17.3	16.8	11.7	14.2	19.7	12.5
Occupation at nomination or appointment								
Politics or government	21.3	10.6	21.2	5.0	13.4	10.8	11.5	8.3
Judiciary	31.1	28.5	34.6	44.6	36.9	41.9	48.2	50.6

(Continued)

(Continued)

	President Johnson's Appointees 1963–68 (N = 122)	President Nixon's appointees 1969–74 (N = 179)	President Ford's appointees 1974–76 (N = 52)	President Carter's appointees 1977–80 (N = 202)	President Reagan's appointees 1981–88 (N = 290)	President George H.W. Bush's appointees 1989–92 (N = 148)	President Clinton's appointees 1993–2000 (N = 305)	President George W. Bush's appointees 2001–2005 (N = 168)
Law firm, large	2.4	11.2	9.6	13.9	17.9	25.7	16.1	21.4
Law firm, medium	18.9	27.9	25.0	19.3	19.0	14.9	13.4	10.2
Law firm, small	23.0	19.0	9.6	13.9	10.0	4.7	8.2	6.6
Professor	3.3	2.8	0	3.0	2.1	0.7	1.6	1.2
Other	0	0	0	0.5	0.7	1.4	1.0	1.8
Occupational experience								
Judicial	34.4	35.2	42.3	54.0	46.2	46.6	52.1	56.6
Prosecutorial	45.9	41.9	50.0	38.1	44.1	39.2	41.3	44.0
Other	33.6	36.3	30.8	31.2	28.6	31.8	28.9	24.4
Political party								
Democrat	94.3	7.3	21.2	91.1	4.8	6.1	87.5	6.6
Republican	5.7	92.7	78.8	4.5	91.7	88.5	6.2	84.5
Independent or none	0	0	0	4.5	3.4	5.4	5.9	8.9
Other	NA	NA	NA	0	0	0	0.3	0
American Bar Association rating								
Exceptionally well/well qualified	48.4	45.3	46.1	51.0	53.5	57.4	59.0	70.8
Qualified	49.2	54.8	53.8	47.5	46.6	42.6	40.0	27.4
Not qualified	2.5	0	0	1.5	0	0	1.0	1.8

SOURCE: *Sourcebook of criminal justice statistics Online* http://www.albany.edu/sourcebook/pdf/11822005.pdf

Table 6.10 Characteristics of Presidential Appointees to U.S. Court of Appeals Judgeships by Presidential Administration, 1963–2005

	President Johnson's appointees 1963–68[b] (N = 40)	President Nixon's appointees 1969–74 (N = 45)	President Ford's appointees 1974–76 (N = 12)	President Carter's appointees 1977–80 (N = 56)	President Reagan's appointees 1981–88 (N = 78)	President George H.W. Bush's appointees 1989–92 (N = 37)	President Clinton's appointees 1993–2000 (N = 61)	President George W. Bush's appointees 2001–2005 (N = 34)
Sex								
Male	97.5%	100%	100%	80.4%	94.9%	81.1%	67.2%	79.4%
Female	2.5	0	0	19.6	5.1	18.9	32.8	20.6
Race, ethnicity								
White	95.0	97.8	100	78.6	97.4	89.2	738	79.4
Black	5.0	0	0	16.1	1.3	5.4	13.1	11.8
Hispanic	0	0	0	3.6	1.3	5.4	11.5	8.8
Asian	0	2.2	0	1.8	0	0	1.6	0
Education, undergraduate								
Public-supported	32.5	40.0	50.0	30.4	24.4	29.7	44.3	35.3
Private (not Ivy League)	40.0	35.6	41.7	51.8	51.3	59.5	34.4	47.1
Ivy League	17.5	20.0	8.3	17.9	24.4	10.8	21.3	17.6
None indicated	10.0	4.4	0	0	0	0	0	0
Education, law school								
Public-supported	40.0	37.8	50.0	39.3	41.0	32.4	39.3	38.2
Private (not Ivy League)	32.5	26.7	25.0	19.6	35.9	37.8	31.1	32.4
Ivy League	27.5	35.6	25.0	41.1	23.1	29.7	29.5	29.4
Occupation at nomination or appointment								
Politics or government	10.0	4.4	8.3	5.4	6.4	10.8	6.6	20.6
Judiciary	57.5	53.3	75.0	46.4	55.1	59.5	52.5	47.1

(Continued)

(Continued)

	President Johnson's appointees 1963–68[b] (N = 40)	President Nixon's appointees 1969–74 (N = 45)	President Ford's appointees 1974–76 (N = 12)	President Carter's appointees 1977–80 (N = 56)	President Reagan's appointees 1981–88 (N = 78)	President George H. W. Bush's appointees 1989–92 (N = 37)	President Clinton's appointees 1993–2000 (N = 61)	President George W. Bush's appointees 2001–2005 (N = 34)
Law firm, large	5.0	4.4	8.3	10.7	14.1	16.2	18.0	11.8
Law firm, medium	17.5	22.2	8.3	16.1	9.0	10.8	13.1	8.8
Law firm, small	7.5	6.7	0	5,4	1.3	0	1.6	2.9
Professor	2.5	2.2	0	14.3	12.8	2.7	82	5.9
Other	0	6.7	0	1.8	1.3	0	0	2.9
Occupational experience								
Judicial	65.0	57.8	75.0	53.6	60.3	62.2	59.0	61.8
Prosecutorial	47.5	46.7	25.0	30.4	28.2	29.7	37.7	35.3
Other	20.0	17.8	25.0	39.3	34.6	32.4	29.5	23.5
Political party								
Democrat	95.0	6.7	8.3	82.1	0	2.7	85.2	5.9
Republican	5.0	93.3	91.7	7.1	96.2	89.2	6.6	91.2
Independent or none	0	0	0	10.7	2.6	8.1	8.2	2.9
Other	0	0	0	0	1.3	0	0	0
American Bar Association rating								
Exceptionally well/well qualified	75.0	73.3	58.3	75.0	59.0	64.9	78.7	67.7
Qualified	20.0	26.7	33.3	25.0	41.0	35.1	21.3	32.4
Not qualified	2.5	0	8.3	0	0	0	0	0

SOURCE: *Sourcebook of criminal justice statistics Online* http://www.albany.edu/sourcebook/pdf/1822005.pdf

who come before them. Other observers believe that because of the nature of the judicial role, minority or female judges are more apt to conform to the norm so they can fit in (Spohn, 2002).

Reviewing the literature on the subject, Spohn (2002) noted that there were few differences between the sentences dispensed by Black, Hispanic, and White justices. In fact, in some instances, both Black and White judges sentenced Black defendants more harshly than White offenders. Spohn suggested that minority judges could be more sensitive to the justice given to Black *victims,* as opposed to being sensitive to the plight of Black offenders. In addition, "The fact that Black judges might see themselves as potential victims of Black-on-Black crime, could help explain the harsher sentences imposed on Black offenders by Black judges" (Spohn, 2002, p. 110). Yet others contend that at times Black judges hand down longer sentences because they work in high-crime areas (Adkins, 2002; see Highlight Box 6.2). Worse yet, minority judges might buy into the idea that minorities are more dangerous than Whites. One thing is clear from Spohn's review: The current body of macrolevel studies does not provide definitive answers to this question.

Could it be that we expect too much from minority judges? In line with this question, the late Judge A. Leon Higginbotham noted,

> No Black judge should work solely on racial matters. I think that would be a profoundly inappropriate abstention, and I would further argue that there is no special role that a Black jurist should play. (cited in Washington, 1994, p. 4)

Reviewing the decision of Supreme Court Justice Clarence Thomas in the case of *Hudson v. McMillian* (1992), where a Black prisoner was assaulted by two prison guards while he was shackled at the hands and feet, Higginbotham added,

> A Black judge, like all judges, has to decide matters on the basis of the record. But you would hope that a Black judge could never be as blind to the consequences of sanctioning violence and racism against Blacks by state officials as Clarence Thomas was in the *Hudson* case. (cited in Washington, 1994, p. 5)

So, barring exceptional cases, according to one of the leading African American jurists of the 20th century, we should not expect anything different from minority judges; irrespective of race, the facts should dictate case outcomes.

Although this debate will likely rage on in scholarly and other circles ad infinitum, the larger issue likely relates to opportunity. Whether a minority or female judge rules in accordance with Whites is, in some ways, beside the point. The more central issue is that qualified minority and female attorneys should be afforded equal opportunities to serve in these important roles. Although it could be that in modern times race discrimination continues to be a factor in not appointing more minority attorneys, such a trend could also be tied to politics. That is, which political party is in office might be an equally strong explanation as to why some minorities are selected for the judiciary while others are not.

Highlight Box 6.2

Judges' Characteristics Are a Factor

Pennsylvania's black judges were 8.9 percent more likely to sentence defendants to prison than white judges, according to a ground-breaking analysis of state criminal sentencing data for 1998 done for the York Daily Record.

The analysis was conducted for the newspaper by The Heritage Foundation, a Washington, D.C.-based research institute. Heritage analyst David Muhlhausen examined more than 53,000 criminal case sentences.

Muhlhausen's analysis found a judge's race, gender, and years of experience affected the probability of incarceration and the length of the sentence delivered when prison time was given.

Black judges, on average, handed down sentences 2.3 months longer than those given by white judges, according to the study.

Experts say the disparity stems from the location of the judge's bench.

Muhlhausen said black judges tend to dispense justice in courtrooms located in high-crime areas.

"That, along with the data, tells me that black judges are more likely to be tougher on criminals than white judges," he said. "That's based on the area that they are in." A perception exists that black judges are more sensitive to black-on-black crime and offenses that occur in primarily black neighborhoods, said Dr. Darrell Steffensmeier, professor of sociology at Penn State University.

Black judges may take a defendant's prior record more into account than their white counterparts and dispense a stricter sentence to repeat offenders, Steffensmeier said.

One reason for this behavior may originate from black judges believing they are under scrutiny by their peers.

That supposed scrutiny leads them to be more aware of criticisms in handing down lenient sentences to criminal offenders, according to a December 2001 study co-authored by Steffensmeier.

That study examined data from 1991 to 1994 to compare the sentences of 10 black male judges with those of 80 white male judges in Philadelphia, Allegheny, Dauphin, and Lawrence counties.

While black judges were more likely to send defendants to jail, female judges across the state and in York County were slightly less likely to incarcerate than their male counterparts, the study found.

Female judges in York County were 5.2 percent less likely to dispense prison time. On the state level, female judges were 4.1 percent less likely to incarcerate a guilty defendant.

Regardless, Steffensmeier said he believes judges dispense similar forms of justice.

"It's not the individual," Steffensmeier said. "It's the job. Both females and males, blacks and whites use the same guidelines." Regardless of their race or gender, judges in both York County and the state tended to assign more lenient sentences the longer they remained on the bench.

For each additional year served on the bench, incarceration sentences given by York County judges decreased by about nine days.

A York County offender's probability of going to prison decreased by 1.5 percent for each additional year the judge was on the bench.

For the state, incarceration sentences decreased by about two days for each additional year on the bench, and a guilty defendant's chance of being locked up decreased by 0.32 percent for each additional year on the bench.

As time passes, a judge may become desensitized to certain nonviolent crimes, causing similar rulings to be dispensed for some similar cases, said John Moran, an attorney in York.

"It takes a lot energy to stay mad at the defendants after 10, 20, or 30 years," he said.

"(Older) judges are more inclined to give a defendant a lot more leeway if the crime was nonviolent and victimless."

Philadelphia County

The study found that Philadelphia County handed down the longest incarceration sentences: 2.7 months more than the state average.

The probability of being incarcerated in Philadelphia County was 10.4 percent higher than in other jurisdictions, according to the study.

One assistant district attorney, however, said he has witnessed the opposite and does not agree with the study's findings.

Judges in Philadelphia County may not have reported all of their sentences to the Pennsylvania Sentencing Commission, including those handed down by the county's municipal court, said David Wasson, an assistant district attorney in Philadelphia County.

"It does seem that judges here are light on crime," he said. "Our sentences seem to be lower than the rest of the state." Philadelphia County judges may submit only sentences that are more serious in nature to the commission, Wasson said.

About the Judges

The analysis by The Heritage Foundation took into account characteristics of Pennsylvania judges who reported their sentencing decisions to the Pennsylvania Sentencing Commission in 1998.

Of the judges profiled:

280 were white

17 were black

2 were Hispanic

254 were male

45 were female

242 were white males

38 were white females

11 were black males

6 were black females

1 was a Hispanic male

1 was a Hispanic female

SOURCE: Adkins, S. (2002, December 27).

Conclusion

This chapter reviewed race-related issues in the sentencing phase of the criminal justice process. After doing a historical review of early cases that illustrated sentencing patterns, it was shown that race and class disparities have always existed in the American justice system. Following a review of the American sentencing process and its various philosophies, we noted that during the 1990s more punitive philosophies prevailed. As such, numerous "get tough" approaches were devised to deal with offenders. Most notably, mandatory minimum, three-strikes-and-you're-out, and truth-in-sentencing policies swelled prison populations with minorities and poor persons. The "War on Drugs" also contributed to the sway toward punitive policies with the passage of the 1986 and 1988 Anti-Drug Abuse Acts, which also contributed to the infamous disparities between penalties for crack cocaine and powder cocaine and further pronounced the racial disparities in sentencing outcomes. It was promising to note that, after two decades of draconian sentencing, policymakers accepted recommendations by the U.S. Sentencing Commission, which resulted in thousands of unfairly sentenced crack cocaine offenders being eligible for early release.

Next, we reviewed the extant scholarship on race and sentencing. This review showed the numerous characteristics that predict whether minorities will be sentenced more harshly than Whites. Some of these characteristics include being young, male, poor, and unemployed; being detained prior to trial; and victimizing Whites. Turning to the judiciary, we also examined the discretion wielded by judges and investigated the conventional wisdom that by having more minority judges, the scales of justice would be more balanced. Contrary to this thinking, however, it was found that the sentences administered by minority judges are in line with those of White judges. In fact, in some instances, for a variety of reasons, minority judges' sentences are actually longer than those set by Whites. We concluded that the most important concern is that qualified minority judges should be afforded equal opportunities to serve in the judiciary.

Discussion Questions

1. Discuss the sentencing disparities found in colonial America.

2. Discuss the sentencing trends from the 1980s through the 2000s.

3. What was the significance of the "War on Drugs" for racial and ethnic minorities?

4. Discuss Zatz's (1987) four waves of sentencing research.

5. Explain the trends regarding racial/ethnic minorities and judicial appointments.

Internet Exercise

Go to www.whitehousedrugpolicy.gov and examine how much of the most recent National Drug Control Strategy Report focuses on drug treatment or law enforcement.

Internet Sites

Sentencing Project: http://www.sentencingproject.org

U.S. Sentencing Commission: http://www.ussc.gov

National Drug Control Strategy: www.whitehousedrugpolicy.gov

The Death Penalty 7

The history of American Indian executions is clearly nestled within a sociopolitical context of genocidal colonialism calculated to dispossess American Indians of their Indianism by removing them from their sacred tribal territories, disrupting their traditional cultures, and continuing their marginalized status in the US society today.

—David V. Baker (2007, pp. 316–317)

O ver the years, there has been considerable debate about the merits of the death penalty as a form of punishment. Although some debates have centered on whether it is cruel and unusual punishment, a considerable portion of the debate has centered on who receives the death penalty. Racial minorities remain overrepresented as those who are sentenced to death and those who, in the end, are executed. The chapter begins with a review of several significant Supreme Court decisions (for a discussion of juveniles and the death penalty, see Chap. 9), which is followed by a historical overview of race and the death penalty. The chapter also reviews the following related areas such as current death penalty statistics; recent scholarship on race and the death penalty; public opinion and the death penalty; the Capital Jury Project; wrongful convictions; and the death penalty moratorium movement.

Significant Death Penalty Cases

The continued use of the death penalty in America can be credited to the historical and continuing strong public sentiment in favor of its use (see section on "Public Opinion and the Death Penalty") and a series of Supreme Court cases, which, over the last three decades, have reaffirmed its constitutionality. The first of these was the 1972 decision in *Furman v. Georgia*. The case centered on a 25-year-old Black man (William Henry Furman) with an IQ of 65 who was charged with killing a 30-year-old White man

(Bohm, 2007). Because of the lack of instructions given to jury members about deciding which cases warrant the death penalty and which ones do not, Furman's lawyers argued that the way the death penalty was being administered was in violation of his Fourteenth Amendment right to due process and was also a violation of the Eighth Amendment, which protects citizens against cruel and unusual punishment. In a decision where the nine justices each wrote a separate opinion, the majority agreed that the death penalty was being administered in an arbitrary and capricious manner. That is, there was little uniformity across states as to who should receive the death penalty and under what circumstances. Those in the majority also pointed out that the death penalty had been applied in a discriminatory manner. Concurring with the majority decision, Justice Douglas, for example, wrote,

> It is cruel and unusual punishment to apply the death penalty selectively to minorities whose numbers are few, who are outcasts of society, and who are unpopular, but whom society is willing to see suffer though it would not countenance general application of the same penalty across the boards. . . . (http://web.lexis-nexis.com/document, p. 1)

Given some of these considerations, in the end, Furman's sentence was commuted to life in prison (he was paroled in 1985). As Baker (2003) noted,

> The aftermath of *Furman* saw the Court vacate 120 cases immediately before it and some 645 other cases involving death row inmates. The decision rendered defective the death penalty statutes of thirty-nine states, the District of Columbia, and the federal government. (p. 180)

The *Furman* decision, however, would stand only 4 years. In the 1976 case of *Gregg v. Georgia*, the Court indicated that states that used guided discretion statutes removed concerns regarding previous procedures that were considered to be arbitrary and capricious. According to Bohm (1999), such statutes

> set standards for juries and judges when deciding whether to impose the death penalty. The guided discretion statutes struck a reasonable balance between giving the jury some discretion and allowing it to consider the defendant's background and character and circumstances of the crime. (p. 25)

The year following the *Gregg* decision, the Court provided states with further guidance on the application of the death penalty. Specifically, in *Coker v. Georgia* (1977), the Court ruled that sentencing rapists to death was cruel and unusual punishment. Considering that between 1930 and the 1970s, 405 Black men were executed in the South for the crime of rape, whereas only 48 Whites were executed for the same offense during this period (Holden-Smith, 1996), it is perplexing that the Court skirted around the racial dynamics of the historical use of executions for rapists. With about 90% of such death sentences being given to Blacks (who presumably had raped White women), it appears that this punishment was historically reserved for Blacks.

In June of the same year as the *Coker* decision, the court also ruled in *Eberheart v. Georgia* (1977) that kidnappings not resulting in death cannot be punished with the death penalty. A decade later, the 1987 decision in *McCleskey v. Kemp* represents another important Supreme Court case, which challenged the constitutionality of the death penalty based on race discrimination in its application.

Warren McCleskey, a Black man, was convicted of the 1978 shooting of a White police officer during an armed robbery. Once caught, tried, and convicted following Georgia's death penalty statute, the jury found beyond a reasonable doubt that the murder had occurred with one of their statutorily defined aggravating circumstances. In this case, however, there were two aggravating circumstances. First, McCleskey had committed the murder during an armed robbery; and second, the victim was a police officer. In the course of his appeals, McCleskey claimed that Georgia's death penalty process was being administered in a racially discriminatory manner, which violated his Eighth and Fourteenth Amendment rights. As evidence of this discrimination, McCleskey's defense showed that,

> Even after taking account of numerous non racial variables, defendants charged with killing Whites were 4.3 times more likely to receive a death sentence in Georgia as defendants charged with killing Blacks, and that Black defendants were 1.1 times as likely to receive a death sentence as other defendants. (*McCleskey v. Kemp*, 1987, p. 1)

After working its way through the courts, the case finally made its way to the Supreme Court in 1986. In an opinion written by Justice Powell and joined by four other justices (Rehnquist, White, O'Connor, and Scalia), the majority held,

> The statistical evidence was insufficient to support an inference that any of the decisionmakers in the accused's case acted with discriminatory purpose in violation of the equal protection clause of the Fourteenth Amendment, since (a) the accused offered no evidence of racial bias specific to his own case, and (b) the statistical evidence alone was not clear enough to prove discrimination in any one case; (2) the study was insufficient to prove that the state violated the equal protection clause by adopting the capital punishment statute and allowing it to remain in force despite its allegedly discriminatory application; and (3) the study was insufficient to prove that the state's capital punishment system was arbitrary and capricious in application and that therefore the accused's death sentence was excessive in violation of the Eighth Amendment. (*McCleskey v. Kemp*, 1987, pp. 1–2)

Conversely, the dissenting justices in the case (Brennan, Marshall, Stevens, and Blackmun) provided a host of contrasting points. First, as noted in previous cases, justices Brennan and Marshall reiterated that the death penalty was cruel and unusual and therefore violated the Eighth and the Fourteenth Amendments. Furthermore, in their eyes, the statistical evidence was valid and showed that there was "an intolerable risk that racial prejudice influenced his particular sentence" (*McCleskey v. Kemp*, 1987, p. 2). In his dissenting opinion, Justice Blackmun, who was joined by justices Marshall and Stevens, suggested that the statistical data showed,

> (1) The accused was a member of a group that was singled out for different treatment, (2) the difference in treatment was substantial in degree, and (3) Georgia's process for seeking the death penalty was susceptible to abuse in the form of racial discrimination. (*McCleskey v. Kemp*, 1987, p. 2)

The study that was debated by the majority has come to be known as the "Baldus study" (see Baldus, Woodworth, & Pulaski, 1990). Leading up to the *McCleskey* case, Baldus et al. (1990) had conducted two important death penalty studies: the procedural reform study and the charging and sentencing study. The first study was designed to

> compare how Georgia sentenced defendants convicted of murder at trial, before and after the statutory reforms prompted by *Furman v. Georgia*, and to assess the extent to which those reforms affected the levels of arbitrariness and discrimination observed in its sentencing decisions. (p. 42)

For the second study, Baldus and his colleagues (1990) were hired by the NAACP Legal Defense Fund to conduct the study "with the expectation that the results might be used to challenge the constitutionality of Georgia's death sentencing system as it has been applied since *Gregg v. Georgia* (1976)" (p. 44). In the end, the second study was the one considered but rejected in the *McCleskey* decision. Using a sophisticated methodological design, which controlled for hundreds of variables, the Baldus study remains the standard when examining race and the death penalty. Probably the most discussed and important finding from the study was the strong race-of-the-victim effect, which showed that Black offenders in Georgia who victimized Whites were considerably more likely to receive death sentences than White persons who victimized Blacks. In fact, looking at this dynamic nationally over a 370-year period (1608–1978), only 30 Whites in the United States have been executed for killing African Americans (Radelet, 1989).

It has been suggested that had the Supreme Court ruled in McCleskey's favor it would have opened "a Pandora's box of litigation" (Kennedy, 1997, p. 333). In short, ruling in McCleskey's favor would have, as Justice Powell stated, "throw[n] into serious question the principles that underlie our entire criminal justice system" (cited in Kennedy, 1997, p. 333). Moreover, he noted that such a decision would produce similar claims "from other members of other groups alleging bias" (Kennedy, 1997, p. 333). Nonetheless, the fact remains that those who support the death penalty outnumber those who oppose it. Even so, given the historical and contemporary racial disparities in who receives the death penalty, in 1988, Congressman John Conyers introduced the Racial Justice Act, which prohibits executions if there is a pattern of racial discrimination in death sentences at the state and federal levels. Although the bill was defeated several times, in 1998, Kentucky became the first state to enact a racial justice act.

After reviewing these significant cases, one is left to wonder whether or not race has always mattered in the application of the death penalty. We examine this question in the next section.

Historical Overview of Race and the Death Penalty

As with other sentences, the death penalty followed colonists from their homeland where English law provided for so many offenses resulting in death that it was referred to as the "bloody code." In a pioneering article on the death penalty as it relates to Native

Americans, Baker (2007) notes that besides the well-known early genocidal actions of Europeans, since their arrivals, there have been at least 450 formal executions of Native Americans. During the colonial period, he pointed to the 1639 execution of Nepauduct as the first execution of a Native American. Moreover, he noted that, during the 16th century, there were 157 executions of Native Americans (Baker, 2007).

Besides the execution of Native Americans, Banner (2002) noted that the colony had numerous death penalty statutes that were solely applied to Blacks. There were a variety of such statutes, beginning in New York in 1712. However, it is in the South where many of these statutes prevailed. Because of the greater number of slaves in the South, there was considerable concern that they might rebel; therefore, as a means of deterrence, there were wide-ranging capital statutes. For example, "In 1740 South Carolina imposed the death penalty on slaves and free Blacks for burning or destroying any grain, commodities, or manufactured goods; on slaves for enticing other slaves to run away; and on slaves maiming or bruising Whites" (Banner, 2002, pp. 8–9). Other statutes made it punishable by death for slaves to administer medicine (to guard against poisoning), "strike Whites twice, or once if a bruise resulted," or burn a house (Banner, 2002, p. 9; see Highlight Box 7.1). It is noteworthy that Black female slaves were not spared the death penalty. Thus, unlike White women who rarely received the death penalty, states did not hesitate to execute Black women. Baker (2008) noted that the first recorded execution of a Black woman was in 1681 when, Maria, a slave, was executed for arson and murder in 1681. After this initial execution, 58 more slave women were executed before 1790, with another 126 being executed prior to the Civil War (Baker, 2008).

Highlight Box 7.1

Slave Era Capital Crimes and Slave Owner Compensation in Virginia, 1774–1864

One little known fact from the slave-era relates to slave owners being compensated for capitally sentenced slaves. Thus, whenever slaves committed an executable offense, the state had a set fee that was given to slave owners. Why? Phillips (1915) asserted that the practice was put in place "to promote the suppression of crime, various colonies and states provided by law that the owners of slaves capitally sentenced should be compensated by the public at appraised valuations" (p. 336). Using Virginia as his study site, Phillips (1915) noted that the first such vouchers were distributed in the 1770s. Besides serving as a window into crimes committed by slaves, the practice also showed how critical the loss of a slave was for southern landowners. In terms of the records, though incomplete, they provide a look at 1,117 out of 1,418 capital crimes committed by slaves from the 1700s to the 1860s. An analysis of the crimes reveals that murder represented the largest share of the offenses (347). Of these, Phillips (1915) extracted a bit more context from the state records:

Murder of the master, 56; of overseer, 7; of the white man, 98; of mistress, 11; of other white woman, 13; of master's child, 2; of other white child, 7; of free negro man, 7; of slave

(Continued)

(Continued)

man, 59; of slave woman, 14; of slave child, 12 (all of which were murders by slave women of their own children); of persons not described, 60. Of the murderers 307 were men and 39 were women. (p. 337)

The remaining convictions are summarized in Table 7.1. While Phillips noted that most of the crimes were typical of the era, he did note a few unusual punishments. In one case, after one slave killed another, the offending slave "had his head cut off and stuck on a pole at the forks of the road" (p. 338). In another instance, a slave stole a silver spoon from their owner's kitchen and was put to death.

Table 7.1　　　The Capital Crimes of Slaves in Virginia: 1774–1864

Crime	Number of Offenses
Murder	346
Rape/attempted rape	105[a]
Poisoning	55[b] (40 men & 15 women)
Administering medicine to Whites	2[c]
Assault/attempted murder	111[d]
Insurrections/conspiracy	91
Arson	90
Burglary	257
Highway robberies	15
Horse thieves	20
Other theft	24
Forgery	2

SOURCE: Phillips (1915).

[a]According to records, all of these appear to have been of White women.

[b]These were mostly targeted at Whites.

[c]One of these persons was pardoned and transported away from the colony.

[d]Only two of these crimes involved Black victims.

As shown in Figure 7.1, the compensation for the lives of the capitally sentenced slaves fluctuated from the 1700s to the 1800s. The rates fluctuated from a low of less than $300 in the 1700s to a high of $4,000 in 1864. Although a host of economic factors likely accounted for these fluctuations, the reality is that the state continuously provided some compensation to slave owners for their losses attributable to slaves sentenced to death.

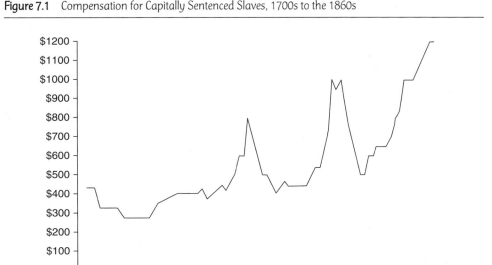

Figure 7.1 Compensation for Capitally Sentenced Slaves, 1700s to the 1860s

In general, a few critical things stand out from the Virginia data. First, the state was so wedded to the economics of the slave system that they felt a need to compensate slave owners for the loss of their "property." Second, the data revealed the wide range of offenses for which slaves were executed. It is likely that executing slaves for so many offenses was used as a tool to control the slave population. Third, it is plausible that many of the property crimes were committed by slaves in response to the deprivation inherent in the slave system, and personal crimes were likely a way for slaves to fight back against the brutal system of slavery.

SOURCE: Phillips (1915, pp. 336–340).

The obvious undercurrent with early death penalty statutes was that landowners were intent on controlling the slave population by fear and state-sanctioned brutality. As Banner (2002) noted, colonial officials "streamlined" the process by cutting out juries and using local justices of the peace to ensure the "justice" was rapid. As one can imagine, although a host of capital offenses also applied to Whites, the processes that prevailed in the slave era resulted in disparities. Looking at one state, North Carolina, the record shows that "at least one hundred slaves were executed in the quarter-century between 1748 and 1772, well more than the number of Whites executed during the colony's entire history, spanning a century" (Banner, 2002, p. 9).

Predictably, such disparities persisted well into the 19th century. Not until 1880, however, were there national statistics on capital punishment (Cahalan & Parsons, 1986). In that first report, there were 80 people listed as "present under the sentence of death" (cited in Cahalan & Parsons, 1986, p. 9).

These early reports provided information on the number of state-sanctioned executions, as well as those that were considered illegal (lynchings). Tables 7.2 and 7.3

highlight two important trends. First, Table 7.2 reveals that, from 1890 to 1984, 5,726 persons were executed, of which 54% (2,915) were non-Whites (Cahalan & Parsons, 1986). Furthermore, Table 7.3 shows that of the 4,736 illegal lynchings that occurred between the 1880s and the 1960s, nearly 73% (3,442) were of Blacks (Cahalan & Parsons, 1986). Given this early history, scholars have continued to research race and the death penalty. Our review of the death penalty continues with a review of current statistics at both the state and federal levels.

Current Statistics on the Death Penalty

STATE DEATH PENALTY STATISTICS

Using the year following the *Gregg* decision as a reference point, from 1977 to 2006, there were 1,057 inmates executed (Snell, 2007). Table 7.4 shows the racial breakdown by year of those who were executed. The figures reveal that 34% of those executed since 1977 have been Black. When reviewing data on the elapsed time from sentence to execution, for those on death row in December 2006, Hispanics averaged 121 months, Blacks averaged 136 months, and Whites averaged 137 months. Strikingly, the following five states accounted for nearly two thirds of the executions since 1977: Texas (379), Virginia (98), Oklahoma (83), Missouri (66), and Florida (64) (Snell, 2007). Table 7.5 shows that there were 3,228 persons under a death sentence at the end of 2006. In addition, 98% of these persons were male, with more than 90% of them having a high school diploma/GED or less. More than half of them had never been married (54.8%). Of those under sentence of death, 41.9% were Blacks and 55.8% were Whites. Hispanics accounted for 11.1% of those prisoners under death sentence, whereas all other races accounted for 2.3%. There were 54 females under sentence of death in 2006. Nearly two thirds (36) of them were White, 27.7% (15) were Black, and 6% (3) were categorized as "Other races" (Snell, 2007). In 2006, 53 persons were executed. Of these, 32 were White (Hispanics were included under this classification) and 21 were Black. On average, these persons had been serving a death sentence for 12 years.

FEDERAL DEATH PENALTY STATISTICS

Largely as a result of the increasing number of death penalty statutes passed during heightened concerns regarding drug-related activity, throughout the 1980s and 1990s, there was increased attention on the federal death penalty. In March 1994, the House Judiciary Subcommittee on Civil and Constitutional Rights produced a report looking at racial disparities in federal death penalty prosecutions over a 6-year period (1988–1994). Out of concern that the federal death penalty provisions in the 1988 Anti-Drug Abuse Act were contributing to this, the committee examined the existing prosecution data. What they found was troubling. When looking at data on those who

Table 7.2 Executions per Decade Under Civil Authority and Illegal Lynchings: 1890–1984

	1890s[a]	1900s	1910s	1920s	1930s	1940s	1950s	1960s	1970s	1980–1984	Total
Total under state authority	155	289	636	1,038	1,523	1,177	684	192	3	29	5,726
Race											
Number non-White	70	157	286	481	745	706	361	99	1	9	2,915
Percent non-White	55	62	47	49	52	63	56	52	33	31	54
Race unknown	(27)	(37)	(26)	(51)	(79)	(55)	(32)	(1)	(0)	(0)	(308)
Offense											
Murder	155	281	570	961	1,383	980	564	152	3	29	5,078
Rape	0	5	40	69	112	172	91	28	0	0	518
Other	0	3	26	8	28	23	19	8	0	0	115
Offense unknown	(0)	(0)	(0)	(0)	(0)	(2)	(9)	(4)	(0)	(0)	(15)
Total under local authority	1,060	901	406	131	147	110	35	0	0	0	2,790
Total under civil authority (state and local)	1,215	1,190	1,042	1,169	1,670	1,287	719	192	3	29	8,516
Illegal lynchings	1,540	895	621	315	130	33	8	1	b	b	3,543
Total per decade (legal and illegal)	2,755	1,995	1,663	1,484	1,800	1,292	721	192	3	29	12,059

SOURCES: State authority: Data for 1890–1980 tabulated from listings compiled by Negley K. Teeters and Charles J. Zibulka, 1864 to 1967, and revised by M. Watt Espy Jr. Listing published in Bowers, William; Pierce, Glen; and McDevitt, John, *Legal Homicide: Death as Punishment in America 1964–1982*. Boston: Northwestern University Press, 1984.

Local authority: Data taken from Table 2.3 in Bowers, William; Pierce, Glen; and McDevitt, John, *Legal Homicide: Death as Punishment in America 1964–1982*. Boston: Northwestern University Press, 1984. Sources for table include: Yearly Meeting of Friends, 1919, pgs. 57–58, for 1890–1917; Bedau, *The Death Penalty in America*. New York, Anchor Press, 1967, pg. 35, for 1918–1920; Barnes, J. E., and Teeters, N. K., *New Horizons in Criminology*. Englewood Cliffs, *Journal of Delinquency* 1:6, 1950, pg. 7, for 1927–1919. NPS and BJS Bulletins after 1930.

Illegal lynchings: Table published in Bureau of the Census, *Historical Statistics of the United States From Colonial Times to 1957*, and compiled by Department of Records and Research, Tuskegee Institute, Alabama, printed in Ploski, Harry, and Williams, James; *The Negro Almanac: A Reference Work on the Afro-American*, 4th ed. New York: Wiley, 1983. Updated included to Census Bureau table, 1960, pg. 218.

1980–1984: Data taken from U.S. Department of Justice, Bureau of Justice Statistics; Washington D.C. Bureau of Justice Statistics Bulletin; Capital Punishment 1984; pg. 7.

Adapted by Cahalan and Parsons (1986), p. 10.

[a] The earliest recorded execution under state authority was in 1864. Between 1864 and 1890, 57 persons were reported executed under state authority.

[b] No lynchings were reported after 1962.

Table 7.3 Illegal Lynchings by Race and Offense by Decade: 1880–1962

	1880s[a]	1890s	1900s	1910s	1920s	1930s	1940s	1950s	1960s[b]	Total
Total	1,203	1,540	895	621	315	130	33	8	1	4,736
Race										
Number Blacks	534	1,111	791	568	281	119	31	6	1	3,442
Percent Blacks	44	72	89	91	89	92	94	75	100	
Offense reportedly causing lynchings										
Homicide	537	606	372	278	100	39	5	0	0	1,937
Felonious assault	4	37	56	51	40	14	2	1	0	205
Rape	259	317	154	88	70	22	0	1	0	911
Attempted rape	9	75	99	56	22	21	6	0	0	288
Robbery and theft	58	87	33	38	6	6	4	0	0	232
Insults to White persons	4	10	11	31	17	8	2	1	0	85
All other causes	331	408	160	79	60	20	14	5	1	1,078

SOURCE: Tabulations based on data compiled by Department of Records and Research, Tuskegee Institute, Alabama. Taken from Ploski, Harry, and Williams, James; *The Negro Almanac: A Reference Work on the Afro-American*, 4th edition. New York: Wiley, 1983.

Adapted by Cahalan and Parsons (1986), p. 11.

[a] Statistics for 1880s are for 1882 through 1889.

[b] Statistics for 1960s are for 1960, 1961, and 1962; no lynchings recorded after 1962.

Table 7.4 Time Under Sentence of Death and Execution, by Race, 1977–2006

Year of execution	Number executed			Average elapsed time from sentence to execution for all inmates
	All races[a]	White[b]	Black[b]	
Total	1,057	680	362	126 mo
1977–83	11	9	2	51
1984	21	13	8	74
1985	18	11	7	71
1986	18	11	7	87
1987	25	13	12	86
1988	11	6	5	80
1989	16	8	8	95
1990	23	16	7	95
1991	14	7	7	116
1992	31	19	11	114
1993	38	23	14	113
1994	31	20	11	122
1995	56	33	22	134
1996	45	31	14	125
1997	74	45	27	133
1998	68	48	18	130
1999	98	61	33	143
2000	85	49	35	137
2001	66	48	17	142
2002	71	53	18	127
2003	65	44	20	131
2004	59	39	19	132
2005	60	41	19	147
2006	53	32	21	145

SOURCE: National Prisoner Statistics Program (NPS–8).

NOTE: Average time was calculated from the most recent sentencing date.

[a]Includes American Indians, Alaska Natives, Asians, Native Hawaiians, and other Pacific Islanders.

[b]Includes persons of Hispanic origin.

See also Methodology.

Table 7.5   Demographic Characteristics of Prisoners Under Sentence of
Death, 2006

Characteristic	Percent of prisoners under sentence of death, 2006		
	Yearend	Admissions	Removals
Total	3,228	115	132
Gender			
Male	98.3%	95.7%	98.5 %
Female	1.7	4.3	1.5
Race			
White	55.8%	62.6%	54.5%
Black	41.9	36.5	42.4
All other races*	2.3	0.9	3.0
Hispanic origin			
Hispanic	11.1%	11.5%	12.0%
Non-Hispanic	76.1	88.5	88.0
Number unknown	414	19	15
Education			
8th grade or less	13.9%	11.7%	20.5%
9th-11th grade	37.0	40.3	33.3
High school graduate/GED	40.0	40.3	35.9
Any college	9.0	7.8	10.3
Median	11th	11th	11th
Number unknown	486	38	15
Marital status			
Married	21.7%	17.7%	32.8%
Divorced/separated	20.6	22.9	17.6
Widowed	3.0	3.1	1.6
Never married	54.8	56.3	48.0
Number unknown	348	19	7

SOURCE: National Prisoner Statistics Program.

NOTE: Calculations are based on those cases for which data were reported. Detail may not
add to total due to rounding.

*At yearend 2005, inmates of "other" races consisted of 31 American Indians, 34 Asians,
and 12 self-identified Hispanics. During 2006, one Asian was admitted, and three
American Indians and one self-identified Hispanic were removed.

had been prosecuted under the so-called drug kingpin law, "Twenty-nine [78%] of the defendants have been Black and 4 have been Hispanic. All ten of the defendants approved by Attorney General Janet Reno for capital prosecution have been Black" (Subcommittee on Civil and Constitutional Rights, 1994, p. 3). The report also noted that, although Whites had traditionally been the ones executed under federal law (85% between 1930 and 1972), the new drug kingpin laws were causing a dramatic change in this trend—so dramatic, in fact, that the report called for remedial action (Subcommittee on Civil and Constitutional Rights, 1994, p. 4).

In January 1995, a policy was adopted whereby a death penalty "protocol" was instituted. This policy resulted in a process in which "United States Attorneys are required to submit for review all cases in which a defendant charged with a capital-eligible offense, regardless of whether the United States Attorney actually desires to seek the death penalty in that case" (U.S. Department of Justice, 2000, p. 2). During this process, some of the cases are withdrawn, whereas others are dismissed (some or all charges). Five years after the passage of this protocol, the Department of Justice conducted an analysis of the federal death penalty system (U.S. Department of Justice, 2000). The report began by showing that overall the use of the death penalty at the federal level pales in comparison to its use at the state level. It also showed that, between 1930 and 1999, the federal government executed 33 defendants, and up until the report was written, no one had been executed since 1963 (Timothy McVeigh and two other federal inmates have been executed since 2001). At the end of 2007, there were 55 persons with pending federal death sentences. Of these, 27 (49%) were Black, 23 (42%) were White, 4 were Latino (7%), and 1 (2%) was Native American (www.deathpenaltyinfo.org).

Even with the small numbers of persons being passed through the federal system, there were still concerns about the obvious disparities. Prior to the institution of the "protocol" procedure, between 1988 and 1994, of those being considered for the federal death penalty, 75% (39) were Black and 10% (5) were Hispanic. Once the procedures changed, out of 682 defendants reviewed between 1995 and 2000, 48% (324) were Black, 29% (195) were Hispanic, and 4% were "Other" (U.S. Department of Justice, 2000). Although 682 cases were submitted for review, only in 183 cases did the U.S. Attorneys recommend seeking the death penalty. Blacks accounted for 44% of these persons, Whites represented 26%, Hispanics represented 21%, and Other represented 8% (U.S. Department of Justice, 2000). Combined, cases involving Blacks (49%) and Hispanics (31%) accounted for 80% of the cases where U.S. Attorneys recommended against seeking the death penalty (U.S. Department of Justice, 2000).

Because most of the cases involved homicides, the U.S. Department of Justice (2000) reported that 74% (500) of the cases were intraracial (i.e., race of offender and victim were the same). U.S. Attorneys recommended the death penalty in 24% of cases where homicides were intraracial. For homicides that were interracial, U.S. Attorneys sought the death penalty in 35% of the cases (U.S. Department of Justice, 2000). In addition, in the majority of the cases where the death penalty was sought, there was only one victim (77%). Of the cases involving multiple victims (157), Blacks represented 54% (84) of these defendants, whereas Hispanics represented 27% (42) and Whites represented 17% (26) (U.S. Department of Justice, 2000).

Because the death penalty continues to be used as a punishment, scholars have sought to contextualize the application of the death penalty. The next section reviews some of the recent scholarship on race and the death penalty.

Scholarship on Race and the Death Penalty

For decades, scholars have considered a variety of issues related to the death penalty, including whether race matters when death sentences are administered. Because of the expansive nature of the literature, our emphasis here is on a brief overview of select areas of death penalty research. Our review also devotes a section to the well-known capital jury project that has enlightened scholars to the ways in which jurors make their decisions in capital cases.

More than three decades ago, Wolfgang and Riedel (1973) concluded that, after controlling for numerous factors, racial discrimination was likely the reason that Blacks received the death penalty more than Whites. Since this pioneering research, numerous studies have been done that focus on racial disparities in the administration of the death penalty (for a recent cogent review, see Baldus & Woodworth, 2003). With the development of more sophisticated methodological approaches, researchers have confirmed that discrimination permeates the initial charging and sentencing phases of capital cases (Bowers & Pierce, 1980; Paternoster, 1983). Yet as noted earlier in the chapter, it was the Baldus study that most potently brought to the fore the issue of the race-of-victim effects (Baldus, Woodworth, & Pulaski, 1990).

In recent years, researchers have continued to study the race-of-the-victim and other race/ethnicity-related aspects of the death penalty. For example, Stauffer, Smith, Cochran, Fogel, and Bjerragaard (2006) investigated whether the race-of-the-victim effect applies to situations involving not only White female victims, but females in general. Their research built on the work of M. Williams and Holcomb (2004), who found that the death penalty was more likely in cases involving White female victims. Based on data from more than 950 North Carolina capital cases, Stauffer et al. (2006) did not find any significant differences between cases involving Black and White female victims. Hence, they argued that there was a general "gender [effect] versus a racial effect in sentencing outcomes" (Stauffer et al., 2006, p. 110). Explaining this result, the authors wrote: "We may likely find that female victim cases are disproportionately selected for capital prosecution in North Carolina because they fit the profile (such as it is) of a murder that evokes a particularly severe criminal justice reaction" (Stauffer et al., 2006, p. 110). The authors also note other aggravating factors, such as females are generally perceived as being vulnerable, and their murders often involve other acts that might provoke an additional level of outrage (e.g., rape).

Although much of the past and recent death penalty research has centered on Black and White experiences with the death penalty (Keil & Vito, 2006), little research has examined how Hispanics fare regarding capital cases. To fill this gap in the literature, C. Lee (2007) analyzed death penalty data from San Joaquin County,

California. More specifically, using data from 128 death penalty cases from 1977 through 1986, she sought to determine whether Hispanics were treated more like Blacks or Whites in death penalty cases. Lee's diverse sample included defendants comprised of the following racial characteristics: 34% White, 33% Hispanic, 25% African American, and 7% Asian American. The results from her analysis were rather telling. First, cases involving White victims "netted the most death-eligible charges" (p. 21). Next, the study showed that "A death-eligible charge was never levied against Asian American defendants" (p. 21). Turning to race-of-the-victim effects, C. Lee noted the following: "In a case where the victim was Hispanic or African American, the defendant was less likely to be charged with a capital homicide than if the victim was White or Asian" (p. 21). Furthermore, as found in the study by Stauffer and her colleagues, there was a strong gender effect. Here, it was revealed that "Defendants in male victim cases were *forty-three times* less likely to face a death-eligible charge than those accused of killing a woman" (C. Lee, 2007, p. 22; italics added). To explain this effect, C. Lee turned to similar explanations postulated by Stauffer and her colleagues (e.g., vulnerability, additional offenses involved, etc.). Lee's results point to a clear need to move beyond the Black/White emphasis in death penalty research.

We turn our attention to the substitution thesis, which has, over the years, emerged as a way of contextualizing the racial disparities in the use of the death penalty in the United States.

SUBSTITUTION THESIS/ZIMRING ANALYSIS

One important area of research related to race and the death penalty has been the notion that, following the reduction in illegal lynchings, states "substituted" lynchings with state-sanctioned executions (Ogletree & Sarat, 2006; Tolnay & Beck, 1995; Vandiver, 2006). Referred to as the "substitution thesis," scholars have found support of this supposition (see J. Clarke, 1998; Tolnay & Beck, 1995; Vandiver, 2006). In recent years, Zimring (2003) has also made a strong case for this perspective. In his death penalty research, Zimring has noted that, of all the industrialized nations, America remains one of the few who use the death penalty. He noted that, within the last 50 years, European nations such as Italy (1944), West Germany (1949), Austria (1950), Britain (1969), Portugal (1976), Spain (1978), and France (1981) have all abolished the use of the death penalty (Zimring, 2003). Given this trend, he wonders why America has not abolished the death penalty. To answer the question, he turned to an analysis of executions by region, noting similarities to the current patterns in the use of the death penalty and historical lynching trends.

These data show that, from 1889 to 1918, lynchings were predominantly carried out in the South (88%). When one looks at the executions during the same period, again, the South predominated, producing 56% of the executions. During this early period, the North accounted for nearly 23%. Turning to the modern era of the death penalty (1977–2000), Zimring noted the similarities to the lynching patterns of earlier times (see Figure 7.2).

Figure 7.2   Regional Percentage Distributions of Lynchings and Executions

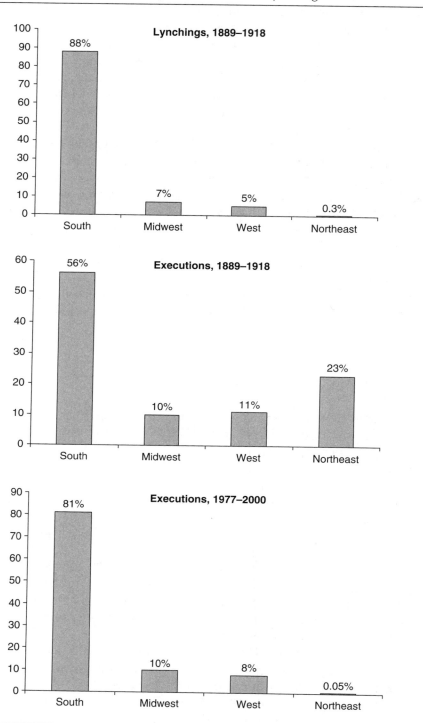

SOURCE: Zimring (2003, p. 94).

During the current period, 81% of the executions were in the South, with the Midwest (10%) and the West (8%) carrying out nearly all of the remaining executions. After considering these striking similarities, Zimring surmised that the South has a "culture of punishment." Of this culture, he wrote,

> The Southern inclination towards high levels of punishment cannot be separated from the vigilante tradition itself. Rather than two separate traditions, the values and practices associated with extensive uses of the plantation prison, high levels of corporal and capital punishment, and those associated with lynchings and vigilante nostalgia overlap extensively with each other and with racial repression. Racism, vigilantism, and high levels of punishment were concurrent conditions in the South when high levels of punishment came to characterize the region. (Zimring, 2003, p. 116)

On the racial dynamics of this culture, he opined,

> It is likely that the long tradition of viewing punishments as a community rather than state institutions and the coercive White supremacist context in which such punishments took place were defining elements of the Southern propensity to punish. . . . The larger willingness to punish was a function of the particular targets and the particular context of punishment. (Zimring, 2003, pp. 116–117)

Zimring's arguments are provocative and provide an important analysis that shows the parallels between illegal lynchings and state-sanctioned executions, which some have considered modern-day lynchings (J. Jackson & Jackson, 1996). Because of analyses such as Zimring's, there continues to be considerable debate about the application of the death penalty.

To measure public sentiment on the use of the death penalty, over the years, public opinion researchers have looked at this question. Next, we review some recent public opinion poll data on the death penalty.

Public Opinion and the Death Penalty

According to Bohm (2007), public opinion on the death penalty is important for the following five reasons. First, strong public support for the death penalty is the likely reason that it is still used as a punishment in the United States. Consequently, politicians will continue to support the death penalty as long as support from the public remains high. Second, because the public wants the death penalty, Bohm believes that prosecutors are willing to appease them by seeking the death penalty in cases where other penalties might be more suitable. Third, feeling the pressure from public sentiment, judges, like prosecutors, might feel undue pressure to impose the death sentences (Bohm, 2007). Fourth, governors are also swayed by public opinion in their decisions to commute sentences. If public support for the death penalty is strong, few governors are willing to risk political favor to go against such sentiment. Finally, because state supreme courts and the U.S. Supreme Court consider public sentiment in their decisions regarding whether the death penalty is cruel and unusual punishment, the results of public opinion polls take on an added measure of significance (Bohm, 2007).

Beginning in 1936, the Gallup poll began an annual tradition of asking Americans their opinions related to the death penalty. That early poll showed that 61% of those polled were in favor of the death penalty (Bohm, 1991). Since then, such polls have been annually conducted by several American polling organizations. Over time, support for the death penalty has ranged from a low of 42% in 1966 to a high of 80% in 1994 (Bohm, 1991). Over the 50-year period between 1936 and 1986, public support for the death penalty has averaged about 59% (Bohm, 1991). Recent polls continue to indicate strong support for the death penalty, with overall support at 69% (Gallup Organization, 2007). In some instances, even when given the option of choosing between the death sentence and life in prison without the possibility of parole, the majority of Americans (53%) remain supportive of the death penalty (J. Jones, 2003). In fact, nearly half of the American population (48%) believes that the death penalty is not imposed enough, with 26% believing that it is imposed in the right amount of cases. Less than a quarter of Americans (23%) believe that the death penalty is used too often (J. Jones, 2003). Although most Americans also believe that the death penalty is applied fairly (60%), many (73%) believe that someone who was innocent was executed within the last 5 years; even so, only if there were "quite a substantial number of innocent people" convicted of murder would a slim majority (51%) of Americans oppose the death penalty (Harris Poll, 2004).

Even with the continuing high level of support for the death penalty, the differences in levels of support for the death penalty have been attributed to a variety of characteristics, including race, political party affiliation, region, education level, occupations, religion, and gender (Longmire, 1996). Such differences have remained constant over the time period. Racial differences in support for the death penalty have also been consistent.

RACE AND SUPPORT FOR THE DEATH PENALTY

During the 1970s, there was a consistent trend in support for the death penalty; however, Black support was considerably less than Whites (Longmire, 1996). Reviewing General Social Survey (GSS) data from 1974 to 1994, Arthur (1998) found that, in 1974, only 40% of Blacks supported the death penalty. He noted that, during the 1980s, this trend began to change, with Black support for the death penalty peaking in 1989 at 59% (Arthur, 1998). During the mid-1990s, Black support had dipped a little to 57%. Characteristics such as age, income, education, marital status, and political affiliation all had an impact on which Blacks supported the death penalty.

According to Arthur (1998), "Blacks whose income exceeded $40,000 were found to be more approving of the death penalty than Blacks earning $10,000 or less" (p. 165). Thus, Blacks in the middle and upper classes were more supportive of the death penalty. Moreover, Blacks who were married and conservative showed strong support for the death penalty. Although as a whole more educated Blacks supported the death penalty (high school diploma and above), out of this educated group, those with only high school diplomas had the strongest support for the death penalty (Arthur, 1998).

As for regional considerations, "Blacks who live in large metropolitan districts and suburbs are more likely than those who live in 'other urban' and rural places to favor the death penalty for murder" (Arthur, 1998, p. 165). Another recent study looking at the geographic variation in attitudes about capital punishment also noted the significance of political affiliation, percentage of Blacks in the population, and areas with higher levels of homicides in predicting higher levels of support for the death penalty (Baumer, Messner, & Rosenfeld, 2003).

Reviewing data from the 1995 National Opinion Survey on Crime and Justice, Longmire (1996) provided a more expansive view of death penalty opinions, which included the opinions of Hispanics. The 1995 survey data revealed that Whites were most strongly in support of the death penalty (77%), followed by Hispanics (52%) and Blacks (40%) (Longmire, 1996). In line with other public opinion studies, there were also differences found by location of the respondents, with

> The strongest support [coming] from those who resided in rural areas (82%), and the strongest opposing [coming] from those residing in urban areas (25%). Urban residents also reported the lowest levels of support for the death penalty (58%) and the highest levels of uncertainty (17%). (Longmire, 1996, p. 99)

So what accounts for the racial differences in support for the death penalty? Barkan and Cohn (1994) explored the notion that racial prejudice might play a role. Using 1990 GSS data, they examined questions that might speak to prejudice, such as whether Whites strongly favor or strongly opposed "living in a neighborhood where half your neighbors were Black" and "having a close relative or family member marry a Black person" (Barkan & Cohn, 1994, p. 203). Based on questions on the survey, they constructed a racial stereotyping scale out of questions that were meant to measure "the degree to which they thought Blacks were lazy, unintelligent, desirous of living off welfare, unpatriotic, violent, and poor" (p. 203). Barkan and Cohn's results showed that "many White people are both prejudiced against Blacks and are more likely to support the death penalty" (p. 206).

Bobo and Johnson's (2004) recent research on attitudes toward the death penalty found that more than 80% of Whites and slightly more than 50% of Black respondents favored the death penalty. Such support was not mediated when the researchers introduced respondents to concerns regarding bias in the administration of criminal justice. There was, however, an effect when respondents were told that when someone murdered a White person they were more likely to receive the death penalty than Blacks. This caveat significantly reduced Black support for the death penalty, whereas White support was not significantly impacted. When Bobo and Johnson posed a hypothetical question involving executing innocent people on death row, there were only slight changes in the views of Whites and Blacks toward support for the death penalty.

Analyzing 14 years' worth of GSS data, spanning from the 1970s to 2002, related to attitudes toward the death penalty, Unnever and Cullen (2007) also reported that African Americans were significantly more likely than Whites to oppose the death penalty. Their research also found that "respondents with more years of education, who resided in the central city, and often attended church were significantly less likely to support the death penalty, and males and Americans who feared being victimized were significantly more likely to support the death penalty" (Unnever & Cullen, 2007, pp. 140–141). Income was also a significant variable in determining support for the death penalty. Here, respondents with higher incomes were more likely to support the death penalty. In general, this study, like other recent ones, found considerable support for the "racial divide" in public opinion on the death penalty (see also Cochran & Chamlin, 2006). Another consideration related to the death penalty public opinion research includes the so-called Marshall hypotheses.

THE MARSHALL HYPOTHESES

Drawing on the statements of the late Supreme Court Justice Thurgood Marshall in *Furman v. Georgia*, researchers have presented hypotheses derived from his suggestions that (a) "American citizens know almost nothing about capital punishment," and (b) "[people] fully informed as to the purposes of the penalty and its liabilities, would find the penalty shocking, unjust, and unacceptable" (http://web.lexis-nexis.com/document). Marshall qualified his second supposition by noting that this would not be the case if someone adhered to the retributive sentencing philosophy.

On the whole, scholars have found support for Marshall's first hypothesis—that the public is uninformed about the death penalty (Bohm, Clark, & Aveni, 1991; Lambert & Clarke, 2001). However, Lambert and Clarke (2001) noted that "of the five studies that directly attempt to test Marshall's second hypothesis, three found small decreases in support for the death penalty after subjects were presented with information concerning capital punishment" (p. 218). Guided by this past research, Lambert and Clarke conducted a study to determine whether reading an essay would influence college students' views on the death penalty.

To do so, Lambert and Clarke (2001) created three different essays. The first essay spoke to the major reasons that criminals are punished. The second essay presented empirical research on "the deterrence effect of the death penalty" (p. 222). The third essay discussed the "possibility and frequency of sentencing the innocent to death" (p. 222). Making use of a self-administered survey design, more than 700 students were randomly given surveys with one of these essays in it. The first section of the survey asked questions regarding the students' initial knowledge and attitudes toward the death penalty (serving as a pretest); the authors wanted to see whether their views changed depending on their exposure to one of the essays. Thus, following their review of the essay, students were again queried about their views on the death penalty. Although their research did show support for Marshall's second hypothesis, that an informed citizen would not support the death penalty, Lambert and Clarke (2001) wrote, "The type of information is critical. Although those who read the deterrence essay stated that their view of the death penalty had changed, only the innocence essay group had a statistically significant reduction in support for capital punishment" (p. 227). Although those who had read the essays on convicting innocent people showed statistically significant changes in their views, the authors did note that the change was slight.

Cochran and Chamlin (2005) used a pretest/posttest design to determine whether students' views on the death penalty would be influenced by exposure to materials presented in a course on the death penalty. The authors did find mixed support for the supposition "that death penalty attitudes and beliefs were inversely associated with student's level of knowledge" (Cochran & Chamlin, 2005, p. 582). The researchers also found that exposure to knowledge about the death penalty decreased the level of support for capital punishment and increased students' support for life without parole as an alternative to capital punishment (see also Cochran, Sanders, & Chamlin, 2006).

A. Mitchell (2006) also examined whether Marshall's suppositions could be supported using a pretest/posttest approach with a seminar course as the experimental stimulus. An interesting finding from his research was that White and Hispanic views were minimally impacted by taking the seminar. On this result, he wrote "Only black respondents showed a significant change in the level of support for the death penalty. After undergoing the seminar, blacks increased their opposition toward the death penalty" (A. Mitchell, 2006, p. 9). A. Mitchell attributed this finding to the legacy of Blacks being disproportionately sentenced to the death penalty. When comparing the experimental group to a control group that did not take the seminar, the study did find that the seminar increased the participants' knowledge of the death penalty.

Because of the importance of the Capital Jury Project, wrongful convictions, and the emerging death penalty moratorium movement, we conclude the chapter with brief discussions of these topics.

Contemporary Issues in Race and the Death Penalty

CAPITAL JURY PROJECT

Funded by the National Science Foundation, in 1991, the Capital Jury Project (CJP) was originally founded to investigate the following research objectives:

(1) to systematically describe jurors' exercise of capital sentencing discretion; (2) to assess the extent of arbitrariness in jurors' exercise of capital discretion; and (3) to evaluate the efficacy of the principal forms of capital statutes in controlling arbitrariness in capital sentencing. (www.albany.edu/scj/CJPwhat.htm)

In addition, the CJP has secured additional funds to investigate "the role played by jurors' race in making the life or death sentencing decision" (www.albany.edu/scj/CJPwhat.htm). To date, nearly 1,200 jurors from 353 capital trials in 14 states have been interviewed. Drawn from a combination of structured and unstructured questions, the interviews conducted for the project typically lasted from 3 to 4 hours with each juror.

The project has already yielded some general findings and also some related to race and capital juror decision making (Bowers, Sandys, & Steiner, 1998; Bowers, Steiner, & Sandys, 2001; Fluery-Steiner, 2004). As an example, Fluery-Steiner and Argothy (2004) considered what happens in Latino capital cases. Making use of 35 juror narratives from 14 cases involving Latino capital defendants from Texas and California, the authors illuminated the racialized perceptions that jurors had of Latinos. Such racialized views manifested themselves in narratives that included the concepts of "colorblindness," "ethnic threat," and "ethnic deceit" (p. 74). With the concept of colorblindness, it is commonly thought that there is a general denial that racism exists in American society. As such, in capital cases involving Latinos, it impacts "the ways jurors evaluate witnesses, deliberate with other jurors, and justify their decisions to impose the death penalty" (p. 75). The concept of ethnic threat centers on the notion that Latinos are prone to violence and, as a result, the "belief in 'dangerous hombres' may *itself* play a role in capital sentencing jurors' decision to impose the death penalty" (p. 77; italics original). The notion of ethnic deceit relates to the belief that Latinos will do anything to enter the United States. As such, they simply cannot be trusted. Thus, Fluery-Steiner and Argothy (2004) believe that "racial deceit may be activated at various points in the trial process, including when a Latino defendant who happens not to speak English testifies" (p. 79).

Summarizing some of the major findings from the CJP, Fluery-Steiner (2009) writes that jurors were found to "have their minds made up on punishment before the sentencing phase of the trial had begun" (p. 1). Thus, jurors are neglecting the Supreme Court mandate to consider aggravating and mitigating factors in the capital sentencing process. Jurors also "[do] not believe that a life sentence actually means the rest of the defendant's natural life" (Fluery-Steiner, 2009, p. 2). In short, because they cling to the belief that the offender might eventually be released, rather than risk the offender being released and harming someone else, they sentence the person to death. As for race, the results revealed that the possibility of being convicted was higher when there were fewer non-Whites on the jury. Even more disturbing was the finding that, in cases involving Black offenders and White victims, "a strong majority of white jurors as compared to black jurors are more likely to be predisposed to the death sentence even before the trial begins" (Fluery-Steiner, 2009, p. 2). Finally, jurors were less likely to consider mitigating circumstances in cases involving Black offenders and White victims. In an earlier study, Brewer (2004) also weighed in on this general finding and provided additional context. Using a subset of the CJP data, he found that "only when Black jurors are faced with killing an out-group member, White victim, that they become significantly more receptive to mitigation than their White colleagues on the jury" (Brewer, 2004, p. 542).

Taken together, the CJP has provided findings that align with the thoughts of some notable jurists (see Higginbotham, 1996) as well as yield new insights into capital jurors and their decision-making process. Whereas in the past researchers conducted simulations involving nonjurors to study jury decision making, the CJP has provided key insights into how *actual* jurors weigh in on capital cases. Such insights might hold the key to changing misperceptions that too often influence jury decision making. An obvious reason that this project is so critical relates to the potential for misguided verdicts and sentencing practices, which, in their worst manifestation, can result in wrongful convictions. We review various aspects of wrongful convictions in the next section.

WRONGFUL CONVICTIONS

Concerns regarding innocent people being convicted are not new. Early on, both American (Borchard, 1932) and British (Brandon & Davies, 1973) scholars investigated this issue. Generally, however, most American citizens understand that, given the nature of our justice system, at times the guilty will go free and at other times innocent people will be sent to jail or prison. This was confirmed by the previously discussed 2003 Gallup Poll, where 74% of Americans supported the death penalty, although 73% responded yes when asked,

> How often do you think that a person has been executed under the death penalty who was, in fact, innocent of the crime he or she was charged with—do you think this has happened in the past five years, or not? (J. Jones, 2003)

Therefore, when it comes to criminal justice, we generally adhere to a utilitarian philosophy. That is, because the system works for most citizens, we can tolerate it when a small number of citizens are wrongfully convicted. Essentially, this is the price we are willing to pay to maintain our adversarial "trial by combat" justice system. Although such a philosophy might be acceptable if the stakes were low, that is not the case with the American justice system. Considering that the United States still maintains the death penalty, to some, adhering to such a system leaves us open to executing innocent people. In fact, pioneering research in this area suggests that the United States has already executed hundreds of innocent people (Radelet, Bedau, & Putnam, 1992).

In the last decade, numerous authors have recently elucidated the major contributors to wrongful convictions (Castelle & Loftus, 2002; Christianson, 2004; S. Cohen, 2003; Harmon, 2001, 2004; Huff, 2004; Huff, Rattner, & Sagarin, 1996; Leo, 2002; Martin, 2002; Zimmerman, 2002). Of these, the most consistent contributors include (a) eyewitness error, (b) police misconduct, (c) prosecutorial misconduct, (d) plea bargaining, (e) community pressure for conviction, (f) inadequacy of counsel, (g) false

Photo 7.1   A gathering of exonerated death row inmates (2002)

confessions, (h) mistaken identity, (i) fabrication of evidence, (j) having a criminal record, (k) misinformation from criminal informants, and (l) race. Most of these contributors are self-explanatory; therefore, given the focus of this book, we concentrate on the role of race in wrongful-conviction cases.

Bedau and Radelet (1987) were among the first to discuss race as a factor in wrongful-conviction cases: Of the 350 capital cases they reviewed, 40% involved instances in which Blacks were wrongly accused. Huff et al. (1996) have also noted that historically there have been disproportionate numbers of Blacks and Hispanics among those wrongly convicted. He and his coauthors alluded to the fact that many of the early instances of wrongful convictions involving minorities were likely a result of being tried by racist prosecutors, who also had all-White, prejudiced juries on their sides.

According to Parker, Dewees, and Radelet (2002), the 40% figure generally holds true across studies. One exception to this is the research done by Barry Scheck and Peter Neufeld, in which their Innocence Project (housed at Cardozo Law School at Yeshiva University in New York) has found that, of those exonerated by the use of DNA, 57% have been Black (Parker et al., 2002). Parker et al. also surmised that, based on past criminal justice practices, "Among those wrongly convicted of felonies, Black defendants are significantly less likely than White defendants to be vindicated" (p. 118).

Another contribution of the Parker et al. (2002) paper is that it attempts to link theoretical explanation to the disproportionate representation of minorities among the wrongly convicted. Among their individual explanations, they pointed to racism, stereotyping, and Blacks being easy targets for a variety of reasons, including their lack of access to resources. As for structural explanations, Parker et al. relied on the conflict-oriented (see Chap. 3) power threat hypothesis; according to this perspective, "Because Blacks are perceived as a threat to Whites, they face higher conviction rates, even wrongfully as Whites respond to this perceived threat" (p. 125). The authors also noted that urban disadvantage might play a role in minorities being overrepresented in instances of wrongful convictions. As they opined,

> The racial patterns in cases of wrongful conviction, particularly the finding that Blacks are more likely to be erroneously convicted, are crystallized under these conditions. That is, these conditions—residential segregation, concentrated poverty, joblessness, and other forms of concentrated disadvantage—reinforce racial disparities in the treatment of minorities in the criminal justice system, including racial bias in convictions of the innocent. (Parker et al., 2002, p. 126)

Although there are numerous case studies that highlight the racial tenor of wrongful-conviction cases, we make note of a few. First, the "Central Park Jogger" case represents an example in which many of the contributors to wrongful convictions, especially race, played roles in the unfortunate outcomes. In the case, in April 1989, a White, 28-year-old investment banker was seriously injured when, according to the original reports, she was allegedly brutally attacked and raped by several Harlem teenagers. That night, 40 teenagers from Harlem were alleged to have engaged in "randomly molesting, robbing, and assaulting strangers who were jogging or bicycling through the Upper East Side of the park" (S. Cohen, 2003, p. 255). Following the incident, as S. Cohen (2003) noted, "The term '*wilding*' [was] introduced to the vernacular of oppression" (p. 255; italics original). Eventually, five Black youth between the ages of 15 and 17 were charged with the offense. Reviewing the facts of the case, S. Cohen (2003) wrote,

> The only physical evidence the police had were hairs on the clothing of one of the boys that were said to be consistent with the hair of the jogger. But all five youths would soon be offering detectives what passed for confessions of guilt. The videotaped statements were the backbone of the prosecutor's case. (p. 256)

All of the youth were eventually convicted of all or some of the charges. A year later, however, it was found that "semen found in the victim did not come from any of the five youths convicted of the crime" (S. Cohen, 2003, p. 257). However, not until 2002, when a convicted rapist admitted to a New York correctional officer that he had committed the offense, did anyone pay any additional attention to the fate of the youth. Although the youth were eventually released and *partially* exonerated (some officials continued to insist some of the youth participated in the crime), the racial element of the case cannot be overstated. One of the most disturbing elements of this case was that, although officials centered on the five innocent youth and looked past the obvious inconsistencies in their confessions and the overall weak evidence, the real rapist, Matias Reyes, went on "raping, assaulting, and tormenting" women (S. Cohen, 2003, p. 259). In fact, in the weeks following the arrest of the five youth, he also murdered a pregnant woman (S. Cohen, 2003). In the end, had the racial elements of the case not overtaken the fears of Whites, it is likely that the evidence in the case would have led them to conduct a more thorough investigation, which would have surely exonerated the youth shortly after their arrest.

The second case took place in March 2006. At that time, several White Duke University lacrosse players were arrested on charges of raping a Black stripper who was invited to perform at a college party. After the party, the stripper provided the sordid details of the incident, which was later proved to be a hoax. Nevertheless, during the incident, racial tensions at Duke and around the country were divided on the issue. Because of the furor over the incident, the prosecutor, Michael Niflong, vigorously pursued the case, which, in the end, revealed the innocence of the Duke students. At the conclusion of the case, Niflong was disbarred for his misconduct in the case and also faced the additional charge of criminal contempt of court for lying to a judge about the details of the case (Beard, 2007). The youth and their families have also filed a $30 million federal civil rights lawsuit against the city of Durham. In contrast to the Central Park Jogger case, some observers suggest that the economic status of the Duke students played a key role in securing their freedom (see Highlight Box 7.2).

Highlight Box 7.2

Duke Trio Learned That Justice Is Not Blind

By Connie Schultz

The lives of three privileged white men were almost destroyed by false testimony, a ruthless prosecutor and a rabid public fueled by the media. Emphasis on almost.

For Dave Evans, Collin Finnerty and Reade Seligmann, justice prevailed. The former Duke lacrosse players were cleared of rape charges. Their accuser, Crystal Gail Mangum, was ousted as a liar. And the public outcry on their behalf led to North Carolina disbarring District Attorney Mike Niflong.

These young men are a year older but decades wiser. They now know that at least 203 other wrongly accused men never had their swing at justice because they had no family members with the means to wage the costly fight for their freedom.

They know that prosecutors are seldom punished for misconduct. They know this means something is wrong with our system of justice. They could have closed the book on this horrific chapter in their lives but chose instead to meet some of the men whose stories mirror theirs in ways they could never have imagined a year ago.

There was scant coverage of this because the Duke players insisted they weren't the point. I only found out after calling Eric Ferrero at the Innocence Project, which works to free the wrongly incarcerated through post-conviction DNA testing.

"You know, they came to our benefit," he said. "We invited them, and we knew they bought tickets, but we didn't know if they would attend."

The first-ever-fund-raiser was in New York on April 24. Before the event, Ferrero stood next to the project's co-founder, lawyer Barry Scheck, who was holding a news conference outside. That's when Ferrero spotted the lacrosse players.

They started to walk toward the entrance, then darted out of view when they saw the cameras. Ferrero quietly escorted them to their table.

They didn't want to talk to reporters, Ferrero said. "We had about 20 exonerees at the dinner, and they [the former Duke students] kept saying the focus should be on those men. They kept mentioning how they would have gone to prison if they hadn't had the money to fight. We were all impressed that they could acknowledge that."

So far, the Innocence Project has helped to free 203 innocent men. Fifteen of them spent time on death row. Most were indigent upon release.

In the first 74 cases alone, 33 involved prosecutorial misconduct, Ferrero said. Current numbers aren't available because the project is trying to figure out how to codify the misconduct to avoid legal challenges.

"Even when DNA proves their innocence, we're still in a position not to antagonize district attorneys because they can decide to retry the case," he said. "Or they can drag their feet for a year or more, delaying release."

As for public outcry, it tends to be muted. "Usually, there's an initial outpouring of support for the victims," Ferrero said. "People send clothes and canned goods, offer jobs and computer training. The public feels bad, but they don't tend to go after the prosecutors because they don't want to believe these are aberrations. They don't want to alienate the sources of their beat." Any reporter more worried about contacts than coverage is a reporter in need of a new beat.

Currently, the Innocence Project has 160 active cases. Roughly 200 new requests for help come in every month. The future is grim for nearly all of them, including those who are innocent.

I never joined the bandwagon against the Duke players, and I part company with those deifying them now. They hired strippers for their party, which hardly qualifies them for sainthood.

But they didn't deserve what happened to them, and they are willing to acknowledge what most of us don't even want to think about: Yes, money talks.

And, when it comes to justice, money walks, too.

SOURCE: *Patriot-News* (July 1, 2007, p. F5).

Not long after the Duke case, the Jena 6 case reached the national spotlight. In the small town of Jena, Louisiana, a Black youth asked one of his high school teachers for permission to sit under a tree that was believed to be reserved for White youth. After receiving permission to sit under the tree, and doing so, the following day there were three nooses hanging from the tree. The details remain murky after this. It has been suggested that six Black youth (Robert Bailey, Carwin Jones, Bryant Purvis, Theodore Shaw, Jesse Beard, and Mychal Bell [the Jena 6]) got into an altercation with two White youth who were believed to have placed the nooses on the tree. During the altercation, one of the White youth, Justin Barker, was injured; as a result, the Black youths were charged with second-degree murder and conspiracy (Sims, 2009). As one of the Black youth, Mychal Bell, moved through the justice system, there was increasing outrage concerning the case. Observers wondered why such serious charges were being levied against the Black youth while nothing had happened to the White youth. This disparity in treatment led to nationwide protests, as well as a significant protest march in Jena that was attended by tens of thousands of protesters, including national civil rights leaders. In the wake of these protests, the District Attorney, Reed Walters, reduced the charges of the youth. In addition, it was later ruled that some of the youth should have been tried as juveniles; thus, their cases were moved to the juvenile court. It is likely that without national attention the Black youth would have been the victims of a rogue district attorney who overcharged the Black youth while neglecting to file any charges against the White youth. Again, such a case had the potential to lead to youth being wrongly prosecuted.

All three cases highlighted here illustrate the problems related to the "rush to judgment" that often takes place when high-profile cases involving race occur. However, this approach often leads to both mistakes and unethical coverups that can often lead to misguided prosecutions or, in the worst-case scenarios, wrongful convictions. Because of the high-profile nature of wrongful conviction cases in which persons on death row have been exonerated, scholars have continued to call for an increased emphasis on wrongful convictions in the discipline (Leo, 2005), as well as continuing to investigate the nature and scope of wrongful convictions (see Denov & Campbell, 2005; Ramsey & Frank, 2007; Zalman, 2006).

In closing, C. Ronald Huff devoted his 2001 American Society of Criminology presidential address to the topic of wrongful conviction and public policy (Huff, 2002). Although he repeated many of the previously discussed facts and figures, his address laid out several important policy considerations aimed at reducing the number of persons wrongfully convicted. First, Huff recommended that those wrongly convicted should be adequately compensated and also provided the appropriate social services to deal with the trauma and reintegration. Second, he suggested replacing the death penalty with sentences of 20 years, 30 years, or life imprisonment without parole. Third, he advocated granting prisoners access to DNA tests. In line with this policy suggestion, the Innocence Protection Act of 2003, if enacted, would require DNA tests in instances where federal inmates claim they are innocent. Fourth, Huff suggested ensuring that qualified expert witnesses be used to minimize misidentification, which is a major contributor to wrongful convictions. Fifth, legal counsel should be present when any identification procedure is used. Sixth, police interrogations of suspects should be taped. Seventh, criminal justice officials who engage in unethical, unprofessional, or illegal activity should be removed and prosecuted. Finally, Huff recommended that individual states and the federal government establish innocence commissions to handle these situations.

There can be no doubt that instituting Huff's proposed policies would go a long way toward reducing the number of wrongful convictions in general. Some states have taken heed of the concerns surrounding wrongful convictions and enacted moratoriums on the death penalty. We briefly review this movement next.

DEATH PENALTY MORATORIUM MOVEMENT

Because of the increasing recognition that errors were being made in cases where people had been convicted of capital offenses, states have begun to consider moratoriums in executions. Governor George Ryan of Illinois (a recent nominee for the Nobel Peace Prize) was the first governor to take such drastic action. When asked why he felt a need to enact such a policy, he noted,

> We have now freed more people than we have put to death under our system—
> 13 people have been exonerated and 12 have been put to death. . . . There is a
> flaw in the system, without question, and it needs to be studied. ("Illinois
> Suspends Death Penalty," 2000)

Like Illinois, other states such as Nebraska and Pennsylvania have considered such action ("Illinois Suspends Death Penalty," 2000; "Pennsylvania Panel Advises Death Penalty Moratorium," 2003), but it has not passed muster with the legislatures of these states. In 2002, former Maryland governor Parris Glendening issued a moratorium, only to have his successor, governor Robert Ehrlich, lift it shortly after he was sworn in the next year. Citizens in some states have formed groups, such as New Jerseyans for a Death Penalty Moratorium (see http://www.njmoratorium.org), that are aimed at pressuring state governments to enact moratoriums in order to study the fairness of the death penalty. Although the movement is clearly picking up momentum, few other states that currently use the death penalty have gone as far as Illinois did without turning back. A recent poll sheds some light on why more states have not followed Illinois' lead. A January 2003 ABC/*Washington Post* national poll of 1,133 citizens found that when asked, "Would you support or oppose it if the governor in your state changed the sentence of every death row inmate to life in prison instead?" only 39% supported such a policy, whereas the majority (58%) did not (ABC/*Washington Post* Poll, 2003). Such figures indicate that death penalty opponents have considerable work to do in order to increase the number of moratoriums nationwide. It is notable, however, that as a result of the efforts of the New Jerseyans for a Death Penalty Moratorium, in December 2007, the New Jersey legislature approved legislation to abolish the death penalty in the state (Hester & Feeney, 2007).

Conclusion

This chapter reviewed several aspects of race and the death penalty. Our historical overview showed that the death penalty has been applied in a discriminatory fashion since colonial times. Sadly, recent state and federal death penalty figures do not show much promise for resolving this issue in the immediate future. After reviewing several significant Supreme Court cases on the death penalty, we reviewed public opinion on the subject. Although support has remained strong, it has varied based on characteristics such as race, class, area of residence, education level, political party affiliation, and religion.

We concluded the chapter by focusing on the Capital Jury Project, wrongful convictions, and the death penalty moratorium movement. These three areas are clearly intertwined in the search for justice. Moreover, this literature has shown that the criminal justice system is not infallible. In fact, several hundred people have been wrongly convicted of capital offenses in the United States, and some of these persons have actually been executed. Scholars have suggested a variety of reasons for these errors, all of which should be the focus of policymakers to ensure that the justice system works. Because of the furor over wrongful convictions involving death sentences, several states have discussed or instituted moratoriums on executions, with states like Illinois commuting death row inmates' sentences to life without parole.

In Chapter 8, we turn our attention to corrections and the race-related issues that plague the so-called back end of the criminal justice system.

Discussion Questions

1. What was the nature of capital punishment during the colonial era?

2. Describe the significance of *Gregg v. Georgia* and *McCleskey v. Kemp*.

3. Discuss two significant findings regarding public opinion polls on the death penalty.

4. Explain the significance of the Capital Jury Project.

5. What are five of the most significant contributors to wrongful convictions?

Internet Exercise

Go to the Web site www.deathpenaltyinfo.org, and go to the link for the list of exonerees from death row since 1973. Using the data provided, summarize the overall racial trends and also the average time between conviction and exoneration.

Internet Sites

Capital Jury Project: http://www.cjp.neu.edu

Death Penalty Information Center: http://www.deathpenaltyinfo.org

The Innocence Project: http://www.innocenceproject.com

Corrections 8

In many prisons, one feels as if there is an invisible sign on the front door that reads: Only Blacks and Hispanics Need Apply.

—National Trust for the Development of African American Men
(cited in Petersilia, 2003, p. 26)

Once offenders are convicted and sentenced, the courts generally turn them over to correctional officials. Correctional departments typically oversee inmates sentenced to probation, jail, and prison. Another key component of the criminal justice system, corrections represents one of the most expensive expenditures of the system. Billions of dollars are spent each year to carry out this difficult function. To reduce such expenditures, over the last several decades, more nontraditional approaches, such as community-based initiatives, have expanded the umbrella of corrections (see Clear & Dammer, 2003). Even with the adoption of more innovative initiatives, prison overcrowding has continued to plague correctional systems throughout the United States. Moreover, corrections has also remained an arena in which race-related concerns have persisted. Considering that 58% of sentenced prisoners in state and federal facilities in the United States are mostly Black and Hispanic (Sabol, Couture, & Harrison, 2007), questions have arisen about social justice. In line with Goldkamp's belief perspectives presented in Chapter 4, one question of critical concern is whether disparities corrections are based on discrimination or whether they exist because minorities commit more serious offenses and should expect to be overrepresented in prison populations (Delisi & Regoli, 1999; Mann, 1993; Wilbanks, 1987). As noted next, this chapter aims to explore this and other related questions.

The purpose of this chapter is to provide an overview of corrections and the race-related issues that are connected to its operation. Our coverage of corrections begins with a brief overview of the structure, function, and public opinion on corrections in America. A brief historical overview of race and American corrections is presented next. We then review the current state of corrections in America. The chapter concludes by focusing on several contemporary issues in corrections. Specifically, our focus here is reviewing explanations for disparities in correctional populations, prisoner reentry into the community, felony disenfranchisement, and political prisoners.

Overview of American Corrections

Often referred to as the "back end" of the criminal justice system, corrections is generally the place where the system ends. Although there is the perception that *corrections* is synonymous with incarceration, 70% of those persons in the corrections phase of the criminal justice system are actually on probation (Glaze & Bonczar, 2007). Clear and Cole (2000) suggested that "the central purpose of corrections is to carry out the criminal sentence" (p. 7). Furthermore, describing the scope of American corrections, they wrote that "corrections . . . encompasses . . . the variety of programs, services, facilities, and organizations responsible for managing people accused or convicted of criminal offenses" (p. 8). Within this description, they are distinguishing between the various correctional options. We discuss these options in turn.

American corrections includes a continuum of sanctions that range from being incarcerated in a jail or prison facility to being sentenced to a fine or restitution (Clear & Cole, 2000). Depending on the nature of their offenses, those who are incarcerated are placed in minimum-, medium-, or maximum-security prisons. Nationwide, there are also approximately 57 "supermax" prisons in 40 states (Mears & Watson, 2006). According to Clear and Cole (2003), "These institutions are designed to hold the most disruptive, violent, and incorrigible prisoners" (p. 252). Persons who are awaiting trial and are unable to secure bail are held in jails. In addition, persons who are sentenced to a year or less are typically housed in jails. Because of overcrowding, however, some jurisdictions have turned to housing more serious offenders in jails (Allen, Simonsen, & Latessa, 2004). Other sanctions falling under the corrections umbrella include fines, community service, drug and alcohol treatment, probation, home confinement, and intensive probation supervision. Each of these sanctions provides nonincarcerative options for offenders. Such options are also attractive to corrections officials because they are considerably cheaper than incarcerating offenders.

All of these various correctional options have also resulted in considerable employment opportunities. For example, in 1983, there were 146,000 correctional officers (U.S. Bureau of the Census, 2003); however, nearly two decades later, in 2006, there were 451,000 correctional officers. Blacks (24.2%) and Hispanics (7.4%) were beneficiaries of this growth. Females also took advantage of this growth, representing more than a quarter (28.2%) of those employed as correctional officers (U.S. Bureau of the Census, 2008).

As with sentencing philosophies, public opinion and politics play important roles in the way the correctional system operates. Given the importance of public opinion research, we review public opinion on corrections for the last four decades.

PUBLIC OPINION AND CORRECTIONS

Because of the various "get tough" laws that were legislated in the 1980s and 1990s, jails and prisons have become considerably overcrowded. This overcrowding coincided with the decline in support for rehabilitation. For example, during the early 1970s, 76% of Americans were in support of rehabilitation (Flanagan, 1996). By 1980, however, only a little more than 50% of the public supported such an approach. Summarizing the public opinion literature from 1968 to 1982, Flanagan (1996) wrote, "The proportion of Americans selecting rehabilitation as the main emphasis of prisons declined 40%; the proportion selecting punishment rose 171%, and the proportion of protection of society rose 166%" (p. 79). Even with the support for this punitive approach being strong nationally, there remain considerable differences in opinion by race and gender.

In the 1995 National Opinion Survey on Crime and Justice (NOSCJ), respondents were asked whether violent offenders could be rehabilitated and what portion of them (most, some, only a few, or none of them). Responses varied by race, with 26% of Blacks feeling that most of such offenders could be rehabilitated; only 17% of Hispanics and 13% of Whites felt this way (Flanagan, 1996). Females were also less optimistic than Blacks; 17% of them felt that way (Flanagan, 1996). Turning to prison programs (e.g., prisoners learning trades, literacy programs, etc.), there was a high level of support among all racial groups for such programs (Flanagan, 1996). One area of difference was that more Whites (79%) than Blacks (52%) were opposed to giving parole violators another chance if they had previously failed on parole; Hispanics fell in the middle: 65% of them were against giving offenders another chance (Flanagan, 1996).

The final area investigated by the NOSCJ related to corrections had to do with acceptable alternatives to prison and solutions to overcrowding. Here, the focus was on determining whether the public favored or opposed various strategies to reduce prison overcrowding. Except for raising taxes to build more prisons, Blacks had stronger support for all the proposals to reduce prison overcrowding than did other racial groups. There were also significant differences by gender. A national *Parade Magazine* poll conducted in June 2001 found that, to reduce prison overcrowding, Americans still leaned heavily on increasing more community-oriented sanctions, especially for nonviolent offenders. Less than half (42%) of the respondents endorsed "building more prisons" to alleviate prison crowding (*Parade Magazine* Poll, 2002).

A 2002 national ABC News Poll also provided insights into the public's views on how to handle certain offenders. The poll found that when asked whether someone convicted of using drugs should go to jail or receive treatment, 89% responded "treatment" (ABC News Poll, 2002). Only after multiple convictions for using drugs was the majority of the public (66%) in favor of jailing drug addicts (ABC News Poll, 2002). Overall, the public still remained committed to the punitive sentencing strategies, with one poll finding that 98% supported giving repeat offenders longer sentences for committing violent crimes (*Parade Magazine* Poll, 2002). The ABC News Poll confirmed this sentiment, finding that 82% of the public supported three-strikes-and-you're-out policies, which obviously has serious implications for corrections (ABC News Poll, 2002). The majority of the public (76%), however, was not supportive of such a policy for nonviolent offenders who had committed their third offense (ABC News Poll, 2002). Because of budget shortfalls across the country, the handling of nonviolent offenders by correctional departments has become a major issue. In California, for example, Governor Schwarzenegger proposed the release of 22,000 nonviolent offenders to ease state spending on prison and correction. For a state that had previously championed the three-strikes-and-you're-out policies, a state-wide poll found that nearly half the population was in favor of this proposal. As expected, the support varied by political affiliation, with Democratic support at 55%, whereas Republican support was nearly 20 points lower at 37% (Public Policy Institute of California, 2008).

On the whole, racial minorities have been the ones disproportionately affected by what some have called America's "imprisonment binge" (J. Austin & Irwin, 2001; Baadsager, Sims, Baer, & Chambliss, 2000; Clear, 2007; Western, 2006). This was borne out by a national poll of African Americans revealing that 19% of African Americans had an immediate family member incarcerated in either a prison or juvenile detention facility (National Urban League Poll, 2001).

A recent study by Unnever (2008) examined public opinion as to why African Americans were disproportionately imprisoned. In general, African Americans and Whites had differing views. African Americans more so than Whites felt that police bias (71% v. 37%) and unfair courts (67% v. 28%) were big reasons why African Americans were disproportionately incarcerated. Another big difference in opinion was with denial of jobs. African Americans were much more likely than Whites to see this as a big reason (59% v. 37%). Both African Americans and Whites felt that poverty was a big reason (67% v. 63%) for the disproportionate incarceration of African Americans. Unnever (2008) noted that a big reason for the "racial divide" in opinions has to do with personal experience. As he writes:

> The more African Americans report that they have encountered racial discrimination the more likely they are to attribute the disproportionately high rates of imprisonment among black males to structural disadvantages such as bad schools and a lack of job opportunities and to racial discrimination within the criminal justice system. (Unnever, 2008, pp. 531–532)

Has race always mattered in corrections? The next section looks into this question by providing a historical overview of race and corrections.

Historical Overview of Race and Corrections

Because of the uneven nature of the historical scholarship and statistics on groups other than African Americans, chronicling the history of race and corrections is a considerable challenge. However, even with these limitations, we attempt to provide some insight into how race has historically intersected with American corrections. We begin our discussion with a brief overview of the history of corrections.

American corrections had its origins in the European workhouses, which were places where vagrants and other minor offenders were sent (Langbein, 1976). According to Shelden (2001), the first such institution to house offenders opened in Amsterdam in 1596. Referred to as the *Rasphaus*, the purpose of this facility was to "discipline the inmates into accepting a regimen analogous to an 'ideal factory,' in which the norms required for capitalist accumulation were ingrained in the code of discipline" (Shank, 1978; cited in Shelden, 2001). Shelden proposed that these facilities were more appropriately called "poorhouses," which were essentially modeled to support the need for industrial workers. As a result, "When released from these workhouses, the inmates would theoretically willingly adapt to the regimentation of the factory and other forms of labor under the new capitalist system" (Shelden, 2001, p. 155).

Over time, England developed three types of facilities: (a) jail, (b) house of corrections, and (c) workhouse (Collins, 1997). By the founding of colonial America in the early 17th century, some of these facilities made their way to America. Similar to today, in colonial times, jail and prisons were expensive to build and maintain (Chapin, 1983). As a result, most offenders were rarely sentenced to prison, with fines, corporal punishments (e.g., whippings), and banishment serving as the most common punishments (Chapin, 1983; Shelden, 2001). Chapin (1983) noted that "jails were used most commonly to hold persons accused of serious crimes before trial and to detain convicted persons until they could pay fines or make restitution" (p. 52).

Because of economic considerations, it was rare to find slaves, Native Americans, or indentured servants incarcerated. On this subject, Collins (1997) wrote,

> Jailing of slaves was not profitable for the slave owners, so very few slaves were ever incarcerated for an extended period of time. Instead, prior to the Civil War, Black slaves would be imprisoned in plantation built jails and punished for crimes committed (for example, running away, stealing, assaulting an overseer, or disobeying an order) by the slave master who had unlimited power, including deadly force. (p. 6)

To legitimize their actions, slave masters created "Negro courts," which meted out punishments (Sellin, 1976).

In 1790, legislation was passed to create an institution where solitary confinement and hard labor were required (Clear & Cole, 2000). In response to the legislation, the Quakers restructured the Walnut Street Jail (which began to receive prisoners in 1776) to fit the required specifications. At that moment, the Walnut Street Jail became America's first prison (Shelden, 2001). Reviewing the early prison records (1795–1826), McIntyre (1992) noted that the free African American population in Philadelphia ranged from 4.6% in 1790 to 9.4% in 1810. When she looked at prisoner statistics in the Walnut Street Jail, McIntyre (1992) found,

> Throughout the period from 1795 to 1826 in Philadelphia City and County, Blacks comprised 35% with "Mulattoes" equaling an additional 9%. This 44% reflected an inmate population for African Americans more than 13 times greater than the state's and nearly 5 times greater than the city's total African American population. (pp. 170–171)

As one might expect, soon the Walnut Street facility became overcrowded. To alleviate this overcrowding, as happens today, the legislature approved the construction of two prisons: Western Penitentiary (1826) and Eastern Penitentiary (1829). As a prelude of things to come, Clear and Cole (2000) noted that Prisoner Number 1, who arrived at the Eastern Penitentiary (near Philadelphia) on October 25, 1829, was "Charles Williams, an 18-year-old African American from Delaware County, Pennsylvania . . . serving a two-year sentence for larceny" (p. 35). Inmate records from 1829 to 1841 show that a steady stream of African Americans followed Williams into Eastern Penitentiary. An inspector's report of the records revealed that there were 1,353 prisoners from 1829 to 1841, "with a breakdown of 846 White inmates (823 males and 23 females) and 508 Black inmates (456 males and 52 females). The Black men represented 37.5% of all males, and the Black women equaled 66% of the females" (McIntyre, 1992, p. 171).

Other states had similar trends in relation to the racial composition of early correctional facilities. For example, from 1812 to 1832, the Maryland state prison in Baltimore "held 45% African American males and 68% African American females for a 51% overall Black inmate population" (McIntyre, 1992, p. 172). Over a 10-month period, from 1832 to 1833, records show that in Richmond, Virginia, on average 28% of the inmates were African American. Although African American men represented 22% of this number, African American women represented 100% of the female inmates (McIntyre, 1992).

So, according to these statistics, White women did not commit any crimes requiring incarceration during this period. A more likely explanation is that "the chivalry factor" was in effect for White women, but not Black women. That is, southerners might have done all they could to protect White women from entering prisons, but such a consideration was not given to Black women.

So how did early observers of the system explain these disparities? When Beaumont and Tocqueville (1833/1964) visited the United States in the early 1800s to examine the feasibility of applying the American penitentiary system to France, they provided some early commentary on these disparities. Examining the trends in crime across the country, Beaumont and Tocqueville wrote, "In order to establish well-founded points of comparison between the various states, it would be necessary to deduct from the population of each the foreigners, and to compare only the crimes committed by the settled population" (p. 93). Even after doing this, Beaumont and Tocqueville (1833/1964) noted that Maryland had a high crime rate, which was, as they put it,

> explained by a cause peculiar to the southern states—the colored race. In general it has been observed, that in those states in which there exists one Negro to thirty Whites, the prisons contain one Negro to four White persons. (p. 93)

In their view, "The states which have many Negroes must therefore produce more crimes" (p. 93). Beaumont and Tocqueville (1833/1964) speculated that crime was high not just in southern states, but, more specifically, those states that manumitted slaves. On this point, they opined,

> We should deceive ourselves greatly were we to believe that the crimes of the Negroes are avoided by giving them liberty; experience proves, on the contrary, that in the south the number of criminals increases with that of manumitted persons; thus, for the very reason that slavery draws nearer to its ruin, the number of freed persons will increase for a long time in the south, and with it the number of criminals. (p. 93)

Without much more said, such statements could have been interpreted as arguing for the continued enslavement of Blacks. In addition, Christianson (1998) has aptly wondered why the two men did not see slavery as being criminogenic.

The mid-1800s brought African Americans closer to their eventual emancipation. However, as noted in Chapter 1, following their emancipation in 1863 and the passage of the Thirteenth Amendment in 1865, southern landowners were devastated. Sellin (1976) noted that, following the passage of the Thirteenth Amendment,

> The penal laws of the southern states became applicable to all offenders regardless of race. This was a distressing prospect for states which had created industrial penitentiaries for offenders from the master class and now faced the rapidly growing criminality of poor, unskilled, bewildered ex-slaves cast into a freedom for which few of them were prepared. (p. 145)

Sellin added that, although Blacks were legally free, southern landowners "did not change their opinions on the status of Blacks in a society dominated by Whites" (p. 145). Because of Whites' "reluctance to labor," they created a system to maintain an able-bodied labor force (Sellin, 1976). Taking advantage of the language in the Thirteenth Amendment, which allowed for slavery and involuntary servitude, southern landowners created the convict-lease system, which

> Reintroduce[d] a species of slavery for Negro criminals and lower-class Whites. They were to be forced to do work which would more than compensate the state for their keep. The sole aim of the convict lease system was financial profit of the lessees who exploited the labor of the prisoners to the fullest, and to the government which sold the convicts to the lessees. (Sellin, 1976, p. 146)

Southern states invested in the system because it was profitable while also keeping taxes down (Oshinsky, 1996). To make the system work, the change in the racial composition of prison populations was dramatic. Oshinsky (1996) wrote,

> In Alabama and Arkansas, Texas and Virginia, Florida and Georgia, North and South Carolina, Louisiana and Mississippi, the convict populations were overwhelmingly Black. Of South Carolina's 431 state prisoners in 1880, only 25 were White; of Georgia's 1,200 state prisoners in that year, almost 1,100 were Negro. (p. 63)

Furthermore, in Georgia, "Between 1870 and 1910, the convict population grew ten times faster than the general one. Prisoners became younger and Blacker, and the length of their sentences soared" (Oshinsky, 1996, p. 63). Records from Georgia prisons in the late 1800s showed that Blacks were serving sentences twice as long as

Whites, with 50% of the inmates serving sentences of more than 10 years (Oshinsky, 1996).

Myers (1998) has shown the full extent of the racial dimension of punishment in Georgia. Looking at the Georgia system from 1870 to 1940, she chronicled the ups and downs of the convict-lease system, as well as the use of other punishments (e.g., chain gangs) to provide services to the state (e.g., public roads projects). She noted that the trends in punishments were often dictated by economic factors such as depressions and recessions (Myers, 1998).

During the 1800s, race also intersected with gender in corrections. For a long time, women prisoners were housed in the same facilities as men (Young & Reviere, 2006). Because of overcrowding issues and scandals such as women prisoners becoming impregnated, during this time, several states built reformatories for women (Collins, 1997). Although White women were initially housed in these facilities, Black women were later housed in segregated cottages. Prison officials justified their actions with ridiculous rationales, such as "a peculiar attraction has been found to exist between 'colored' and White women in confinement which intensifies much danger, always present in an institution, of homosexual involvement" (Lekkerkerker, 1931; cited in Collins, 1997, p. 10). Some officials, such as Katherine Bemet Davis, the first female corrections commissioner of New York City, refused to segregate female prisoners. More significant segregation took place in the housing of Black women in penitentiaries, whereas White women were housed in treatment-oriented reformatories. This segregation reduced the chance that Black women would be successful on their release from prison (Collins, 1997).

Early National Prison Statistics

The earliest national statistics on corrections were recorded in 1850, when the national census on prisons took a 1-day count and recorded 6,737 prisoners. By 1860, there were an additional 13,000 prisoners, with records indicating 19,086 prisoners. Dramatic increases would occur over the next several decades: 1870 (32,901), 1880 (30,659), 1890 (45,233), and 1904 (57,070) (Cahalan & Parsons, 1986). Tracking the trends by race and ethnicity, Table 8.1 shows that not only did state and federal prison populations increase, but the percentage of Blacks incarcerated also increased, from 31% in 1923 to 37% in 1960 (Cahalan & Parsons, 1986).

Figures from the 1931 Wickersham Commission Report show that in 1926, native-born Whites represented 68% of those in federal and state prisons. The racial breakdown for the remainder of the inmates was 8% foreign-born White, 21% Negro, and 3% Other (Indian, Mexican, Chinese, Japanese, and all other races) (National Commission on Law Observance and Enforcement, 1931a). During this same period, Black females were also considerably represented in the nation's prisons. Russell-Brown (2004) presented figures that show that, from 1926 to 1946, the percentage of Black female prisoners fluctuated from a low of 21% to a high of 40% of the female inmates.

Table 8.1   Characteristics of Persons in State and Federal Prisons, Institutional Population Census Data, 1910–1980

Year	Number of Prisons	Inmates Present	Percentage Female	Percentage Foreign-Born	Percentage White	Percentage Black	Percentage Other Races	Percentage Spanish Origin[a]	Percentage Juvenile (under 18)
1910	61	67,871	4	b	b	b	b	b	b
1923	64	80,935	4	12	68	31	1	b	2.0
1933[b]	117	137,997	3	(5)	(74)	(23)	(3)	b	(3.9)
1950[c]	158	178,065	4	3	65	34	1	b	2.9
1960[d]	1,072	226,344	4	1	61	37	2	b	2.3
1970[e]	633	198,831	3	b	58	41	b	7	2.2
1980[f]	2,560[g]	302,377	5	3	47	44	b	10	b

SOURCE: Cahalan and Parsons (1986, p. 65).

[a]Persons of Spanish origin may be of any race.

[b]Not available or not obtained.

[c]Except for the 3% female prison population, which is calculated on the basis of inmates present (137,997), details are calculated on the basis of prisoners received from courts (62,801) according to available data.

[d]1940 is excluded because juvenile facilities were not separated from state and federal. However, only those over 14 were enumerated. Detail data in 1950 was calculated on 3½% sample with an estimated base of 181,080 for total prison population; the complete count is 178,065.

[e]Data based on 25% sample.

[f]Data based on 20% sample.

[g]Counted each budget unit as individual facility.

Photo 8.1   Two chain gang prisoners in
Georgia (1937)

In the decades between the Reconstruction period and 1960, although much changed in terms of overall advancements for racial minorities, much stayed the same, with Blacks continuing to be over-represented in state and federal correctional institutions. However, the 1960s saw the civil rights movement, the increasing visibility of Black Muslims, and the Black power movement influencing various segments of the Black community.

Although each of the aforementioned events can be credited with bringing to the fore issues related to police brutality, they each played a role in the evolution of corrections in the 1960s and 1970s as well. Prior to this period, Black inmates essentially accepted their status without any resistance (Reasons, Conley, & Debro, 2002). Reasons et al. (2002) noted that, although Blacks were exposed to inhumane conditions, they did not protest because "(1) the courts had a hands-off policy with respect to penal conditions and issues, and (2) the institutions were located primarily in rural areas and thus functioned in relative isolation, and the guards were all White" (p. 271). Discussing the importance of the mass migration from the South to northern and western states among African Americans, Reasons et al. wrote, "This demographic shift made racial segregation more difficult and expensive to maintain in all institutions, including penal institutions" (p. 271). Another contributor to change was the fact that leaders from the civil rights movement (e.g., Martin Luther King, Jr., Ralph Abernathy, and Medgar Evers), Black Panther movement (e.g., Stokely Carmichael, H. Rap Brown, Huey Newton, and Angela Davis), and Black Muslim movement (most notably, Malcolm X) were all incarcerated at some point and spoke out about their experiences (Reasons et al., 2002). It was, however, the Black Muslim and Black Panther movements that had the greatest influence on correctional systems.

According to Conley and Debro (2002), the Black Muslim (also known as the "Nation of Islam") movement in California prisons can be traced to San Quentin in the late 1950s. On the one hand, the movement was popular among Black inmates because, along with preaching racial pride, instituting strict discipline, and pushing for economic self-sufficiency, the group preached that Whites were evil and the cause of the current plight of African Americans (Reasons et al., 2002). On the other hand, correctional officials saw the movement as both a threat and "management problem," so they took steps to suppress the movement using a variety of strategies, including dispersing the members. Even with these strategies, the Black Muslims continued to challenge correctional policies in court.

Beginning with the 1962 case of *Fulwood v. Clemmer*, courts began to recognize the Black Muslim religion. Furthermore, this decision sparked the prisoners' rights movement, which resulted in thousands of lawsuits filed by inmates. Some of these lawsuits resulted in substantive changes in the way correctional institutions were run. For example, in *Battle v. Anderson* (1974), following an earlier loss in the courts to maintain the right to read literature that was considered inflammatory, the courts reinforced inmates' First Amendment rights and provided that "prison officials had the burden of proving to the court that the publications *Elijah Muhammad Speaks* and *The Message to the Black Man in America* present a threat to security, discipline, and order within the institution" (Palmer & Palmer, 1999, p. 66). A related case, *Northern v. Nelson* (1970), ruled that Black Muslims should have access to religious literature. Other cases challenged the request of Black Muslims to be served a pork-free diet (*Young v. Robinson*, 1981), the right to correspond with their religious leaders (*Desmond v. Blackwell*, 1964), and the right to free access to ministers (*Jones v. Willingham*, 1965). Although there were victories in the latter two matters, the overall move by Black Muslims to challenge correctional officials represents a monumental contribution to prisoner's rights.

Although one could argue that the early concerns expressed by correctional officials were somewhat exaggerated, it is likely that some of the disturbances caused by Black Muslims were instigated by prison officials (Conley & Debro, 2002). In recent years, the group has continued to be involved in correctional settings through the Nation of Islam Prison Reform Ministry, which has received several awards (Gabbidon, 2004).

During the 1960s and 1970s, the Black Panther Party also impacted corrections through its nationwide efforts to steer Black youth away from the criminal justice system. But it was the party's influence on several high-profile inmates that highlighted the abuses in correctional facilities. Inmates such as George Jackson typified the inmates who adhered to the "Black power" philosophy of the Black Panther Party. Jackson was 18 years old when

> [He] was accused of stealing $70 from a gas station in Los Angeles. Though there was evidence of his innocence, his court-appointed lawyer maintained that because Jackson had a record (two previous instances of petty crime), he should plead guilty in exchange for a light sentence in the county jail. He did, and received an indeterminate sentence of one to life. Jackson spent the next ten years in Soledad Prison, seven and a half of them in solitary confinement. (G. Jackson, 1970, p. ix)

Jackson's case became a cause célèbre and was brought to even more prominence with his critically acclaimed work *Soledad Brother: The Prison Letters of George Jackson*. The year after his book was published, Jackson was shot, on August 21, 1971, allegedly trying to escape. Jackson's influence was so wide, even among prisoners in other states, that the next day at Attica prison in New York,

Inmates . . . graphically demonstrated their reaction to the shooting of Jackson. Instead of the usual banter and conversation of inmates coming out of their cells to line up for the march to breakfast, officers on many companies were greeted by somber inmates who moved silently out of their cells and lined up in rows of twos with a Black man at the head of each row; many of them wore Black armbands. (New York State Special Commission on Attica, 1972)

A few weeks later, Attica erupted in a riot that lasted 4 days, with 43 people being killed and 80 being wounded. Most of the dead (39 people) and all of the wounded were the result of the state police operation to take back the institution. Referring to that infamous assault, the preface to the report investigating the riot noted, "With the exception of the Indian massacres in the late 19th century, the State Police assault which ended the four-day prison uprising was the bloodiest one-day encounter between Americans since the Civil War" (New York State Special Commission on Attica, 1972, p. xi). In 2000, more than 25 years after the riot, former inmates and families of those inmates killed received a settlement of $12 million. However, the families of the employees of Attica never received any compensation. In an effort to remedy this, New York Governor George Pataki appointed a task force to look into the claims of former correctional employees (Public Report of the Forgotten Victims of Attica, 2003).

The Attica riot rang in an era in which more scholars began to examine correctional institutions. What they found were highly segregated institutions with increasing Black and Hispanic populations (L. Carroll, 1974; Davidson, 1974). Each of these populations was struggling for control, which increased the level of violence within institutions. Coupled with the inhumane treatment and emerging overcrowding, the prisons throughout the country were susceptible to "rage riots," which, unlike the political riots of the late 1960s and early 1970s, were "very spontaneous and expressive, as most of the violence was directed at fellow inmates, rather than prison officials or the 'system' itself. Often they were the result of racial conflicts, especially between rival gangs" (Shelden & Brown, 2003, p. 304).

Photo 8.2   Member of the Texas Aryan Circle prison gang (2001)

Prison Gangs

One of the earliest prison gangs was the "Gypsy Jokers" in Washington State prisons, where gangs have been present since the 1950s (Fleisher & Decker, 2001). Today, researchers have estimated that the population of prison gang members ranges from 15,000 (Trulson, Marquart, &

Kawuncha, 2006) to somewhere between 50,000 and 100,000 (Fleisher & Decker, 2001). Over time, the gangs have formed along racial/ethnic lines, which, with the burgeoning prison population, has perpetuated conflict within institutions. Some of the more infamous gangs include the Mexican Mafia (*La Eme*), the Black Guerilla Family, the Aryan Brotherhood, *La Nuestra Familia* ("our family"), and the Texas Syndicate (Fleisher & Decker, 2001). The Mexican Mafia was the first prison gang with nationwide ties, whereas the Black Guerilla Family had its origins in the larger Black Panther movement and was politically oriented, having Marxist-Leninist leanings (Fleisher & Decker, 2001). The Aryan Brotherhood, a White supremacist gang, was started in 1967 "by inmates who wanted to oppose the racial threat of Black and Hispanic and/or counter the organization and activities of Black and Hispanic gangs" (Fleisher & Decker, 2001, p. 4). *La Nuestra Familia* is a Hispanic prison gang that formed to provide protection against the Mexican Mafia. The Texas Syndicate consists of Mexicans, Latin Americans, and Guamanians.

Highlight Box 8.1

Ex-Gang Member Testifies of Prison Killing

Prosecutors lay out case in Aryan Brotherhood trial

Thursday, March 16, 2006; Posted: 5:19 a.m. EST (10:19 GMT) www.CNN.com

SANTA ANA, California (AP)—A former prison gang member told jurors he read Machiavelli and helped kill another inmate to impress gang leaders as testimony began Wednesday in the federal government's racketeering case against four reputed leaders of the Aryan Brotherhood.

Clifford Smith, a convicted murderer and Aryan Brotherhood member from 1978 to 1984, was the first witness in the case alleging a gang conspiracy to kill inmates who cheated on drug deals or snitched to prison authorities.

Wearing an eye patch and prison jumpsuit, Smith told the jury how his initiation included helping kill one gang enemy and stabbing another. The gang killed as a way to keep the power needed to conduct criminal activities involving drugs, extortion, fraud and identity theft, he said.

"Not everybody is willing to kill somebody," Smith said. "Some people are kind of squeamish about that stuff. I wanted to let them know I wasn't."

Authorities arrested 40 alleged Aryan Brotherhood members in 2002 after a six-year investigation. Nineteen struck plea bargains, one died and 16 others could face the death penalty in one of the largest capital punishment cases in U.S. history.

The four now on trial have been described as gang leaders: Barry "The Baron" Mills, 57; Tyler Davis "The Hulk" Bingham, 58; Edgar "The Snail" Hevle, 54; and Christopher Overton Gibson, 46. Mills and Bingham could face the death penalty; Hevle and Gibson could get life in prison.

The indictment alleges members of the white supremacist gang orchestrated a web of conspiracies, including starting a prison war against a black gang that resulted in at least two killings.

Prosecutors opened their case with a simple slide: "The Aryan Brotherhood: Blood in, Blood out."

(Continued)

(Continued)

The phrase—borrowed from the gang itself—means that inmates must kill to join the gang and can only leave when they die, Assistant U.S. Attorney Michael Emmick said in his opening statement.

Emmick said the gang even went after its own members to maintain discipline and inspire fear.

Defense attorney H. Dean Steward rejected Emmick's claims that the crimes were ordered by the gang's leadership. He said most crimes were committed by individuals who had personal conflicts.

"The murders and assaults happened," Steward told jurors. "There's no dispute. The question is 'Why'"?

Steward, who represents Mills, said nearly all of the government's case was based on 42 prison informants who had been coached and offered incentives including immunity, reduced sentences and cash payments.

During his testimony Wednesday, Smith described gang members communicating through "runners," usually female friends, who would visit them in prison and transport tiny messages, drugs and even small knives.

Sometimes they would use codes—"Lady from Bristol" meant pistol; "bottle stopper" stood for a guard or police officer; and "rough and smooth" was heroin, he said.

Smith said that when he was initiated to the gang, he was told to read a number of books, including works by Friedrich Nietzsche and Niccolo Machiavelli.

"That's the theme of most of these books: the individual, going outside the herd, being the alpha-male," he said.

In other testimony, Glen West, an Aryan Brotherhood member from 1981 to 2003, said that Mills told him he had killed one inmate and ordered a hit on another. West, now in the witness protection program, was expected to continue testifying Thursday.

SOURCE: Associated Press (2006).

Other well-known gangs such as the Crips and the Bloods have also been found in correctional institutions around the country. Each of these racial/ethnic gangs and the new ones that continue to emerge have caused their share of problems within correctional institutions. Some have argued that a considerable amount of the violence and drugs in prisons are attributable to prison gangs (G. M. Camp & Camp, 1985; Fleisher & Decker, 2001; Gaes, Wallace, Gilman, Klein-Saffran, & Suppa, 2002; Huebner, 2003).

Griffin and Hepburn (2006) recently examined the notion that prison gangs impact on violence in prisons. In their study of a diverse sample of more than 2,000 Arizona inmates, they found support for this belief, reporting that: "Inmates with no gang affiliation were significantly less likely than those with either street gang affiliation or prison gang affiliation to have been guilty of violent misconduct in the first 3 years of confinement" (p. 434). The authors also found that:

Assault misconduct was significantly more likely to occur among younger inmates and among White inmates. Compared to White inmates, African American inmates, Mexican American inmates, and inmates who were Mexican Nationals were significantly less likely to be guilty of a major misconduct for assault. Native American inmates, in contrast, were no more or less likely than White inmates to commit assault. (Griffin & Hepburn, 2006, p. 436)

Given the security issues raised by the presence of prison gangs, correctional officials have had to use intelligence and other strategies to control institutions (Carlson, 2001; Scott, 2001). One strategy that had been used previously in California was to segregate offenders by race. However, in the Supreme Court case *Johnson v. California* (2005), it was ruled that such practices were unconstitutional. Nonetheless, correctional institutions continue to struggle with these problems in the midst of an incarceration boom that has only minimally begun to subside. The next section reviews the current state of corrections.

Contemporary State of Corrections

There were nearly 2.4 million people incarcerated in 2006 in the United States (Sabol, Couture, & Harrison, 2007). This figure translates into an incarceration rate of 501 persons per 100,000 persons in the U.S. population. Table 8.2 shows the massive increase in the incarceration rate from 1980 to 2006. The figures for probation and parole are equally dramatic. In 1995, there were 3,757,282 people on probation and parole. By 2006, there were 5,035,225 on probation and parole; overall, this represented approximately a 25% increase in the number of persons on probation and parole (Glaze & Bonczar, 2007).

The breakdown of these increases by most serious offense is also illuminating. Although there were increases in incarceration for violent, property, drug, and public order offenses, the trends for drug offenses and public order offenses are most dramatic. In 1980, there were 19,000 persons incarcerated for drug offenses. Twenty-four years later, in 2004, there were 249,400 people incarcerated for drug offenses (Bureau of Justice Statistics, 2003; Sabol, Couture, & Harrison, 2007).

Reviewing the trends by race/ethnicity, Blacks remained the largest proportion of sentenced inmates (38%), and also had the highest incarceration rates at 3,042 per 100,000 in 2006. Hispanics/Latinos followed Blacks with an incarceration rate of 1,261 per 100,000, and Whites had the lowest rate at 487 per 100,000 (Sabol et al., 2007). It is important to note, however, that data from 2000 and 2006 actually reveal a declining number of Blacks among sentenced state and federal prisoners (see Table 8.3). Even so, it is equally important to note that Black men comprised the largest share of inmates in state or federal prisons or local jails (41% or 836,800 inmates). Whites were second at 718,100 inmates, and there were 426,900 Hispanic males incarcerated in 2006 (Sabol, Minton, & Harrison, 2007). Turning to the state of incarcerated women, the data also show that "White women were about one-third as likely as black women to be incarcerated and slightly more than half as likely as Hispanic women" (Sabol, Couture, & Harrison, 2007, p. 7). This trend also resulted in lowered incarceration rates for Black women from 175 per 100,000 to 148 per 100,000. Although this remained the highest rate among women, it was notable that the rates increased for Hispanic women from

Table 8.2   Incarceration Rate, 1980–2006

Number of sentenced inmates incarcerated under state and
federal jurisdiction per 100,000, 1980–2006

	Incarcerated population per 100,000
1980	139
1981	154
1982	171
1983	179
1984	188
1985	202
1986	217
1987	231
1988	247
1989	276
1990	297
1991	313
1992	332
1993	359
1994	389
1995	411
1996	427
1997	444
1998	461
1999	476
2000	478
2001	470
2002	476
2003	482
2004	486
2005	491
2006	501

SOURCE: Bureau of Justice Statistics (2003).

NOTE: Correctional populations in the United States, 1997, and prisoners in 2005

Table 8.3   Percentage of Sentenced State or Federal Prisoners, by Race, and Hispanic Origin, 2000 and 2006

| | Percent of sentenced state or federal prisoners[a] | | | |
| | Estimates[b] | | Administrative data[c] | |
	2000	2006	2000	2006
Total	100%	100%	100%	100%
White[d]	32.7	35.1	35.7	40.0
Black or African American[d]	42.4	37.5	46.2	41.6
Other[d,e]	2.4	3.7	1.7	2.7
Two or more races[d]	3.2	3.2	–	0.2
Hispanic or Latino	19.2	20.5	16.4	15.5

SOURCE: Sabol, Couture, & Harrison (2007).

NOTE: See Appendix (in original source) table 8 for age distribution by race and gender. – Not reported.

[a]Based on jurisdiction counts of inmates with a sentence of more than 1 year.

[b]Estimates for state prisoners based on inmates' self-report of race and Hispanic origin from the 2004 Survey of Inmates in State Correctional Facilities and updated from jurisdiction counts at yearend. Estimates for federal prisoners based on Federal Justice Statistics Program data. See *Methodology.*

[c]Yearend reports of race of prisoners under state or federal jurisdictions, as reported by correctional administrators in BJS NPS-1 survey. See *Methodology.*

[d]Excludes Hispanic or Latino persons; administrative data on race of prisoner may include Hispanic or Latino persons.

[e]Includes Asian, American Indian or Alaska Native, and Native Hawaiian or other Pacific Islander.

78 per 100,000 to 81 per 100,000 and White women from 33 per 100,000 to 48 per 100,000 (Sabol, Couture, & Harrison, 2007). In actual numbers, there were more White women incarcerated (95,300) than Black women (68,800) and Hispanic women (32,400) (Sabol, Minton, & Harrison, 2007).

Jails

Table 8.4 shows the significant increases in jail populations from 1990 to 2007. During the decade, the number of Whites in jails increased from 169,400 in 1990 to 338,400 in 2007, whereas the number of Blacks increased from 172,300 to 301,900. Over the same period, the number of Hispanics held in jails increased from 58,000 in 1990 to 125,600

Table 8.4   Jail Populations by Race and Ethnicity, 1990–2007

Jail populations by race and ethnicity, 1990–2007 Year	Number of jail inmates (one-day count)		
	White non-Hispanic	Black non-Hispanic	Hispanic of any race
1990	169,400	172,300	58,000
1991	175,300	185,100	60,600
1992	178,300	196,300	64,500
1993	180,700	203,200	69,400
1994	191,800	215,300	75,500
1995	206,600	224,100	75,700
1996	215,700	213,100	80,900
1997	230,300	237,900	88,900
1998	244,900	244,000	91,800
1999	249,900	251,800	93,800
2000	260,500	256,300	94,100
2001	271,700	256,200	93,000
2002	291,800	264,900	98,000
2003	301,200	271,000	106,600
2004	317,400	275,400	108,300
2005	331,000	290,500	111,900
2006	336,600	296,000	119,200

SOURCE: Correctional Populations in the United States, 1997, and Prison and Jail Inmates at Midyear Series, 1998–2006, and Jail Inmates at Midyear 2007.

in 2007 (Bureau of Justice Statistics, 2008). In 2007, the racial composition of jails was as follows: 43.3% White, 38.7% Black, and 16.1% Hispanic (Sabol & Minton, 2008).

Although some Native Americans are figured into the previously reviewed jail figures, there are 54,915 American Indians being supervised by state, federal, local, and tribal authorities (see Table 8.5). There are also separate jail facilities in Indian country. According to Minton (2006), in 2004, "a total of 68 jails in Indian country held 1,745 inmates" (p. 1). Of the inmates incarcerated in these facilities, 39% were being held for violent offenses, 14% for DWI/DUI, and 7% for drug law violations. Most of the convicted offenders in Native American jails were being held for misdemeanors (88%) (Minton, 2006). Overall, these facilities were operating at 81% of their capacity.

Table 8.5   Location of Native American Inmates

	Number of American Indians or Alaska Natives
Total	54,915
In custody, midyear 2004	23,177
Local jails*	7,500
Jails in Indian country	1,745
State prisons	11,485
Federal prisons	2,447
Under community supervision	31,738
Probation	25,844
Parole	5,894

*Estimated from the Annual Survey of Jails. 2004.

PROBATION AND PAROLE

In 2006, there were 5,035,225 men and women being supervised on probation or parole. Reviewing racial and ethnic data from 2006 for those on probation reveals that Whites comprised 55% of those on probation, with Blacks representing 29%, Hispanics 13%, and Asians and American Indians representing 1% (Glaze & Bonczar, 2007). Of those on parole in 2006, Blacks represented 39%, Whites 41%, Hispanics 18%, and 1% Asians and American Indians (Glaze & Bonczar, 2007). At the federal level, in 2004, Whites represented 66.9% of those on probation and 51.2% of those on parole, with Blacks (26.8% and 36.0%) and Hispanics (17.5% and 20.9%) being somewhat overrepresented (Bureau of Justice Statistics, 2006). American Indians represented 2.8% of those on federal probation and 1.9% of those on federal parole. Asians/Pacific Islanders also represented a small number of those persons being supervised under federal probation (3.5%) or parole (0.5%) (Bureau of Justice Statistics, 2006).

Although locating data on corrections is fairly easy, making sense of such data presents more of a challenge. One thing remains a reality, however: Blacks and Hispanics are overrepresented in nearly all areas of corrections. Whether this is a product of discrimination has been an ongoing question during the last two decades (Mann, 1993; McDonald, 2003; Morgan & Smith, 2008; Walker, Spohn, & DeLone, 2007; Wilbanks, 1987). Along with this question, race-related topics surface when considering correctional systems in America. In the next section, we begin with a review of the most salient scholarship that has been used to explain the disparities reviewed here. Furthermore, we look at several other contemporary issues related to race and corrections: prisoner reentry, felony disenfranchisement, and political prisoners.

Contemporary Issues in Race and Corrections

EXPLAINING RACIAL
DISPARITIES IN CORRECTIONS

During the early 1980s, scholars began to examine in earnest the racial disparities in corrections (see, e.g., Christianson, 1981; Petersilia, 1983). It was, however, the work of Blumstein (1982) that is most used as the benchmark study for explaining racial disproportionality in prisons. In Blumstein's state-level study, he investigated the role of discrimination in Black overrepresentation in prisons. Taking into account arrest patterns of Blacks and Whites, Blumstein found "that 80% of the actual racial disproportionality in incarceration rates is accounted for by differential involvement in arrest" (pp. 1267–1268). Although he noted that this explains the majority of the racial differences in incarceration, Blumstein also noted that if the unexplained 20% (which at the time translated into 10,500 prisoners) of overrepresentation of Blacks in prison "were attributable to discrimination, that would be a distressing level of discrimination" (p. 1268). It is important to note that, although Blumstein's work has become one of the benchmark studies in this area, his analysis is based on arrest statistics, which, as noted in Chapter 2, have serious limitations.

Pointing to other possible explanations for the disproportionality, Blumstein (1982) suggested that, although Blacks are more involved in the most serious types of offenses, they may also be involved in the more serious "versions *within* each of the offense types (e.g., *stranger-stranger* homicides, in the *armed* robberies, etc.)" (p. 1268; italics original). Furthermore, he pointed to the possibility that Black offenders might accumulate longer criminal records. Noting the complexity of various extraneous factors that could impact on the accuracy of studies like his, Blumstein pointed to research that showed that, in some instances, "Discrimination in the criminal justice system might work in the opposite direction, resulting in Black offenders receiving more favorable treatment than White offenders" (p. 1269). More specifically, using rape cases as an example, he wrote,

> Because less certain and less severe punishment results when the victim is Black, and because the victims of Black offenders more often *are* Black, this could result in Black defendants being treated less severely than White defendants. Thus, this act of discrimination against Black *victims* could result in discrimination in favor of Black *offenders*. (p. 1269; italics original)

Blumstein also suggested that regional issues, educational issues, and socioeconomic factors could all play roles in explaining the unexplained 20%. He noted, however,

> Even after taking into account all factors that are at least arguably legitimate and that could explain the racial disproportionality in prison, it would certainly not be surprising to find a residual effect that is explainable only as racial discrimination. (Blumstein, 1982, p. 1270)

Following the publication of Blumstein's work, other scholars sought to test some of his findings. Focusing on arrest and prison admissions data for 3 years in North Carolina, Hawkins (1986) found that from 1978 to 1979, 30% of the prison disproportionality was explained by arrest patterns.

For the subsequent 2 years, Hawkins (1986) noted that the figures increased to 40% and 42%. In line with Blumstein's earlier supposition for certain crimes, Hawkins's research showed that Blacks received more favorable sentences than Whites. Specifically, Hawkins found that "fewer Blacks than White assault offenders received prison sentences. Fewer Blacks than Whites also received prison terms for larceny and armed robbery" (p. 260). These findings were in line with Hawkins's earlier work noting that, in certain contexts, Black life was devalued and resulted in justice officials minimizing Black-on-Black offenses, which manifested itself in less serious punishments for such offenses (Hawkins, 1983).

Using state-level arrest and prison data for 39 states (with a 1% or greater Black population), Hawkins and Hardy (1989) found some variation across states. For example, their study revealed,

> In nine states the level of arrest explains only 40% or less of Black-White imprisonment rate differences. On the other hand, for six other states the level of arrests explains more than 80%. Thus, even allowing for some discrepancy due to differences in data sources, Blumstein's figure of 80% would not seem to be a good approximation for all states. (Hawkins & Hardy, 1989, p. 79)

The 1990s found more researchers concentrating on disparities in corrections. At the beginning of the decade, two nonprofit organizations, the Sentencing Project and the National Center on Institutions and Alternatives (NCIA), came out with reports that highlighted the control rates for African Americans. Russell-Brown (2004) noted that such rates "refer to the percentage of a population that is under the jurisdiction of the criminal justice system—on probation, parole, in jail, or in prison. It provides a snapshot of a group's overall involvement in the justice system" (pp. 123–124).

In 1990, the Sentencing Project came out with a report that showed that "almost one in four African American males in the age group of 20–29 was under some form of criminal justice supervision" (Mauer, 1990). In April 1992, J. G. Miller (1992a) of the NCIA reported,

> On an average day in 1991, 21,800 (42%) of Washington, D.C.'s 53,375 African American males ages 18 through 35 were either in jail or prison, on probation or parole, out on bond awaiting disposition of criminal charges or being sought on an arrest warrant. (p. 1)

Five months later, J. G. Miller (1992b) returned to the subject, reporting that "of the 60,715 African American males age 18–35 in Baltimore, 56% were under criminal justice supervision on any given day in 1991" (p. 1). Updating their report in 1995, the Sentencing Project found things had worsened since 1990, reporting that "nearly one in three (32.2%) of African American males in the age group 20–29—827,440—is under criminal justice supervision on any given day—in prison, or jail, on probation or parole" (Mauer & Huling, 1995). An update of the NCIA's reports also noted that things had worsened in Washington, D.C. (Lotke, 1998).

During the release of these important reports, Blumstein (1993) also returned to the task of seeking to explain some of these disparities. Blumstein noted that the situation had marginally worsened, with 24% of the disproportionality of Blacks in prison not being explained by offending patterns. More specifically, he wrote,

> The bulk of the disproportionality is a consequence of the differential involvement in the most serious kinds of crime like homicide and robbery, where the ratio of arrests is between five and ten to one. For these crimes, the race ratio in prison is still very close to that at arrest. (Blumstein, 1993, p. 6)

However, there was a disparity related to less serious offenses (i.e., drug offenses). To explain these disparities, Blumstein (1993) surmised that, because of the increased level of discretion in less serious offenses, factors such as discrimination could be contributing to the disparity.

The early 1990s also saw concerns being expressed about the increasing expenditures on prisons and the decreasing spending on education (Chambliss, 1991). Sometime during the decade, prisons became "hot commodities"; unlike in prior years, communities, especially in rural areas hit hard by economic downturns, saw them as desirable for their overall economic impact (Lotke, 1996). Some states, like Florida, created brochures to promote the economic impact of prisons for prospective communities. Residents were told that a 1,100-person rated capacity prison could produce $25 million annually in revenue and create 350 jobs (Lotke, 1996). Downtrodden rural communities bit on the carrot, which is reflected by the fact that 5% of the increase in rural populations between 1980 and 1990 was attributed to prisons (Lotke, 1996).

Huling (2002) noted that, prior to 1990, 36% of prisons were built in nonmetropolitan areas. Furthermore, according to her figures, "Between 1990 and 1999, 245 prisons were built in rural and small town communities—with a prison opening somewhere in rural America every fifteen days" (p. 198). Noting that the economic impact is often overstated, she pointed to the "hidden" costs, such as increasing cost for local court and police services and the fact that some industries might be discouraged from investing in an area where a prison was located. Another unanticipated consequence of placing prisons in rural areas is the prevalence of racism in many rural communities. Such racism is a problem for guards and inmates. Huling (2002) provided an illustration of such problems:

> In at least six states, guards have appeared in mock Klan attire in recent years. Guards have also been accused of race-based threats, beatings and shootings in ten states. Lawsuits have been filed in at least thirteen states by Black guards alleging racist harassment or violence from White colleagues. (pp. 208–209)

In Washington State, a rural institution (Clallam Bay Correctional Center) had only 4 Black officers out of 326 correctional officers. The Black officers eventually filed a lawsuit (which was settled out of court for $250,000), claiming,

> Black officers were denied promotions, subject to threats and racial epithets like "coon," and the minority prisoners were harassed and set up for beatings. Some White guards

had taken to calling Martin Luther King, Jr., Day "Happy Nigger Day" and a handful of guards openly bragged about associations with hate groups such as the Ku Klux Klan. (Huling, 2002, p. 209)

The late 1990s and early 2000s saw scholars seeking to explain disparities in corrections with radical critiques that referred to the "prison-industrial complex." Writers such as Parenti (1999) and Dyer (2000) have argued how government and citizens were profiting from the mass incarceration of principally African Americans and Hispanics. Private prison companies such as Corrections Corporation of America, which constitutes the sixth largest prison system in the nation, were in some cases found unknowingly in the retirement portfolios of many Americans (Dyer, 2000). As such, society was "investing" in prisons, which obviously relied on an ample supply of prisoners to keep the prison boom going (Dyer, 2000; Hallett, 2006; Price, 2006). Other writers of the period stressed the rebirth of the convict-lease system in the form of the modern-day use of cheap prison labor by private companies partnering with state corrections departments (Davis, 1997, 2000, 2003). Overall, scholars of this genre believed that because of the continuing need for bodies to keep the prison-industrial complex going, disparities in corrections will likely persist.

In response to the growing prison-industrial complex that has become a global problem (see Sudbury, 2005), the critical resistance movement was forged after a national conference in 1998 (see http://www.criticalresistance.org). The aim of the movement is to dismantle the prison-industrial complex "by challenging the belief that policing, surveillance, imprisonment, and similar forms of control make . . . communities [of color] safe" (see http://www.criticalresistance.org). To date, the movement has sued the California Department of Corrections to prevent it from building a new maximum-security prison. In addition, several regional chapters have been formed, and the organization has focused on educating communities about the prison-industrial complex. In September 2008, the organization held its 10th anniversary conference in Oakland, California.

In the 2000s, scholars have continued to examine disparities (Crutchfield, 2004; Fernandez & Bowman, 2004; Mauer, 2004, 2006; Sorenson, Hope, & Stemen, 2003; Western, 2006), but there has been more of an emphasis on the consequences of mass incarceration on communities in general and minority communities in particular (see Clear, 2007). Disparities in prisons create what some call "collateral consequences" or byproducts of the decision to incarcerate so many minority offenders. Two such collateral consequences are prisoner reentry concerns and felony disenfranchisement. We discuss prisoner reentry concerns first.

PRISONER REENTRY CONCERNS

Because of the mass incarceration that took place in the 1980s and 1990s, America has more offenders being released than ever before (Petersilia, 2003; Travis, 2005). Moreover, because sentences were extended and 16 states eliminated parole (Hughes, James Wilson, & Beck, 2001), more prisoners are spending extended periods of time incarcerated than ever before. Glaze and Bonczar (2007) indicated that, at the end of 2006, there were 798,202 persons on parole. In addition to those on parole, thousands of persons "max out" or complete their entire sentences and leave prison without any supervision. Considering that a large share of those persons sentenced to jails and prisons are minorities, it is only logical that the majority of those coming home will also be minorities. Consequently, Black and Hispanic males represented 57% of persons on parole in 2006 (Glaze & Bonczar, 2007). Unfortunately, as noted in Table 8.6, in 2006, "about 4 in 10 parolees exited supervision because they were returned to incarceration for a new offense or a technical violation" (Glaze & Bonczar, 2007, p. 7). Such figures speak to the need to explore prisoner reentry issues.

Petersilia (2003) defined *prisoner reentry* as "all activities and programming con-ducted to prepare ex-convicts to return safely to the community and to live as law-abiding citizens" (p. 3). According to Petersilia (2002), some of the key factors that contribute to failure on both probation and parole (which has been eliminated at the federal level and in some states) include the following:

> Conviction crime (property offenders have higher rates), prior criminal record (the more convictions the higher the recidivism), employment (unemployment is associated with higher recidivism), age (younger offenders have higher rates), family composition (persons living with spouse or children have lower rates) and drug use (heroin addicts have the highest recidivism rates). (p. 491)

Because minorities are more likely to fit some of these characteristics, they are at even higher risk for reentry problems (Pager, 2007a, 2007b). Employment and maintaining family ties are especially big concerns for minorities (Bushway, Stoll, &

Table 8.6   Outcomes of Adults on Parole, 2000 and 2006

	Percent of adults exiting parole	
Type of exit	*2000*	*2006*
Completions	43%	44%
Returned to incarceration	42	39
With new sentence	11	11
With revocation	30	26
Other/Unknown	1	2
Absconder	9	11
Other unsatisfactory	2	2
Transferred	1	1
Death	1	1
Other	2	3
Total estimated exits	459,400	519,200

Weiman, 2007; Visher, 2007). To address these issues, the federal government passed the Serious and Violent Offender Reentry Initiative in 2003 (Lattimore, 2007). This act, which allocated $100 million for grants, focuses on "employment-based programming for inmates" (Lattimore, 2007, p. 88). Moreover, the Marriage and Incarceration Act was enacted to address "family programming for adult male inmates" (Lattimore, 2007, p. 88). Even with these initiatives, because of the bias against ex-cons, some have turned to creative measures to get their foot in the door of employers (see Highlight Box 8.2).

Highlight Box 8.2

Silence Is Golden Rule for Resumes of People Who Have Broken It

By JOANN LUBLIN
Wall Street Journal Online
October 03, 2007 6:01 AM

Joanne Jester has an impressive résumé. It describes her many duties as office manager for cdm eCycling, a small Baltimore company. She also lists prior employers, including State Use Industries, from 2000 to 2003.

But there's a catch: State Use Industries was the prison-industry arm of Maryland's correctional system. Ms. Jester worked as a telemarketing office assistant there while serving a four-year sentence for theft and related charges. Her résumé reveals nothing about her imprisonment.

Ms. Jester took the right approach, according to several career coaches and legal specialists. Facing huge barriers to re-employment, former prisoners should craft résumés focused on their marketable skills, training, work and life experiences, not their troubled pasts. "You never address the fact that you were incarcerated in your résumé," says Wendy Enelow, an author and career consultant in Coleman Falls, Va. "You're giving them a reason to exclude you."

She believes this strategy could apply to anyone with gaps in employment, from cancer patients to recovering alcoholics and individuals fired for cause. "It's not just ex-offenders who have challenges returning to work. It's a problem for lots of people," says Ms. Enelow, co-author with Louise Kursmark of "Expert Résumés for People Returning to Work."

Their book offers discreet ways to account for work performed behind bars. One sample résumé calls the gigs "temporary assignments," although they lasted eight years. Another identifies the prison employer as "State of California."

Federal and state institutions released nearly 700,000 inmates in 2005, about 15% more than in 2000, the Justice Department reported. Hoping to curb recidivism, certain states such as Maryland prepare prisoners to rejoin the work force long before their release. Ms. Jester, for instance, took office-skills courses throughout her incarceration. She revised her résumé with help from Tricia Hopkins, a transition coordinator who counsels on job hunting. The revamped document identified "Maryland State Department of Education" as the training provider. It also stated that she "graduated with recommendations" from her final course.

Ms. Jester says she informed potential employers that she broke the law to survive marriage to an abusive heroin addict. You must "sell yourself and be proud of what you've become," she explains.

(Continued)

(Continued)

"You have to learn to admit your mistakes." She joined cdm as a data-entry clerk in December 2004, after Ms. Hopkins persuaded the computer-recycling company to consider hiring an ex-con for the first time.

Other former offenders avoid discussing their crimes with hiring managers, thanks to carefully worded résumés and employers' limited probes of applicants' personal records. Consider a 30-year-old computer technician whom Comcast hired soon after he completed a 13-year sentence for reckless endangerment and handgun use last spring. The charges arose from his fight with another youth in which a stray bullet seriously hurt a bystander.

The eighth-grade dropout completed community college and earned certification in several building trades during his prison stay. He prepared a "functional" résumé, which emphasizes qualifications and competencies but plays down a job seeker's work history. In such résumés, a section typically labeled "Areas of Accomplishment" can cite transferable talents, such as the ability to communicate well and meet deadlines.

His functional résumé highlights his educational achievements. The top portion states that he graduated magna cum laude from his community college. Next is his long catalog of skills, including certifications, problem-solving ability and a "strong hands-on approach to get the job done." But he doesn't mention that he attended college from prison and omits dates for his limited work experience there.

During his Comcast interview, the technician says he vaguely alluded to "something in my past." A recruiter for the nation's largest cable-TV provider replied that he had nothing to worry about if the misdeed occurred more than seven years ago.

"We conduct background checks as far back as state law permits—up to a maximum of 10 years," a Comcast spokeswoman says. If the check uncovers a conviction, she continues, officials weigh the nature of the crime, evidence of rehabilitation and other factors.

About 25% of employers in some industries refrain from asking would-be staffers about convictions more than seven years ago, estimates Kevin Lindsey, an employment lawyer at Halleland Lewis Nilan & Johnson in Minneapolis. The growing trend reflects "enlightened self interest in expanding the applicant pool rather than fear of violating a specific state law," he observes.

What about criminals without meaningful jobs behind bars? They could minimize attention to their work-force absence by describing the duration of prior positions, career consultants suggest. For instance, a résumé might read: "Gap store manager (10 years)," followed by a detailed description of the duties involved.

"Are we fooling anybody? No," Ms. Enelow concedes. "Résumés don't get you jobs. They just open doors."

SOURCE: Lublin (2007, October 3).

In his highly acclaimed work *When Work Disappears*, W. J. Wilson (1996) noted the stereotypical views of employers in the Chicago region. When he asked employers about their perceptions of inner-city workers (especially young Black males), many of the 197 participating firms referred to them as "uneducated," "uncooperative," and "dishonest" (W. J. Wilson, 1996, p. 111). Such perceptions have obvious implications for ex-cons. Building on the work of Wilson, Pager (2007b) sought to determine the impact of race and criminal histories on employment in entry-level jobs. Her experimental study of Milwaukee-area businesses found that Whites with criminal records had a better chance at employment than Blacks without criminal records. Such findings can only continue the cycle of recidivism among Blacks.

As for maintaining family ties, concerns about Black inmates and their families are not new (Swan, 1977). However, given the overrepresentation of Blacks and Hispanics among those who are incarcerated and eventually come home, it is important to note the consequences of fractured families.

Studies have shown that "imprisonment and parole affects family stability and childhood development" (Petersilia, 2002, p. 494). Incarcerated and paroled inmates have increased chances of separation and divorce, which in the long run can impact their children. Petersilia (2002) wrote, "Children of incarcerated and released parents often suffer confusion, sadness, and social stigma; and these feelings often result in school-related difficulties, low self-esteem, aggressive behavior, and general emotional dysfunction" (p. 494). More disturbingly, "Children of incarcerated parents are five times more likely to serve time in prison than are children whose parents are not incarcerated" (p. 494). Employment also intersects with family concerns in that among other things, being unemployed raises the risk that there will be violence in the home of a recently released offender (Petersilia, 2002). Recent scholarship has continued to confirm some of these problems associated with families coping with the incarceration of a loved one (see Braman, 2002; Richie, 2002).

With a change in tone at the state and federal levels regarding crime and criminals, there has been considerable effort to address the reentry problem (Travis, 2007; Visher, 2007). One way to combat reentry concerns is through the development of reentry courts. Jeremy Travis developed this concept in 1999. The idea moves the role of reentry to the judicial branch of government. Based largely on the concept of drug courts, reentry courts require that a parolee and the parole officer work with the courts to formulate a reentry plan. Each month the parolee is required to provide evidence of their progress. Travis (2005) argues that the court can be effective because:

> The judges are able to marshal community resources, and wield both "carrots" and "sticks" in their efforts to promote successful reintegration. The carrots are services, positive reinforcement, family and community support, and a forum for the acknowledgement of success. The sticks are enhanced levels of supervision (such as curfews, more intensive drug treatment, or more frequent drug testing), and ultimately short periods of incarceration, typically measured in days rather than months. (p. 59)

The reentry court model is still new, with pilot sites in the following states: Delaware, Florida, Iowa, Kentucky, New York, Ohio, and West Virginia (E. Miller, 2007). These courts, drawing on Travis's model, are based on four core components: a reentry transition plan, a range of supportive services, regular appearances for oversight of the plan, and accountability to victims or communities (E. Miller, 2007). To date, E. Miller (2007) has identified three major challenges faced by reentry courts: health, employment, and housing. As for health, E. Miller notes that inmates are typically troubled by numerous ailments, and this is yet another challenge reentry courts need to consider. As for employment, it goes without saying that this represents an essential key to the successful reentry of formerly incarcerated people. Programs such as the

Center for Employment Opportunities (CEO) in New York could serve as models for other communities struggling with this issue. The nonprofit organization "provides immediate, paid employment for people coming home with criminal records, followed by placement in a permanent job and job retention services" (Tarlow & Nelson, 2007, p. 138). Their program includes a few days of job-readiness instruction, after which, there are 2 months of paid transitional work, followed by permanent employment largely in the private sector. Because of various legislative mandates, securing housing can be quite a challenge during the reentry process. As such, it might actually require changes to current policies to open more options for ex-inmates (E. Miller, 2007).

Notably, faith-based reentry programs also have flourished and worked in conjunction with reentry courts (Herz & Walsh, 2004). As with other faith-based initiatives, faith-based reentry programs are eligible for federal funds. In general, faith-based initiatives range from "faith-saturated" to those that have no apparent faith component (Mears, 2007). Whatever pole such programs fall under, they "cannot expend funds for inherently religious activities such as worship, religious instruction or proselytization" (Mears, 2007, p. 30). It is also critical to note that, to date, little substantive research has examined the effectiveness of these programs. Consequently, there is little evidence that faith-based reentry programs are more effective than other reentry programs (Mears, 2007).

Although many reentry issues concern all inmates, as noted earlier, minorities are particularly hard hit with such concerns. Until these issues are addressed in a serious way, the cycle of recidivism will continue to pervade minority communities. A promising development occurred when President George W. Bush announced in his 2004 State of the Union address that he was planning to implement a $300 million mentoring program to help ex-felons entering the job market (Kroeger, 2004). As part of this commitment, the proposed Second Chances Act of 2007 has numerous provisions for reentry, including monies for jurisdictions to start reentry courts (E. Miller, 2007). Given the magnitude of the reentry problem, however, this can only be seen as a start.

Another collateral consequence of being convicted of a felony is the loss of the right to vote. In recent years, felon disenfranchisement has become a major focus of scholars and community activists. In the next section, we review the history, current state, and implications of this practice.

FELON DISENFRANCHISEMENT

Being able to vote is a fundamental right in a democracy. Nevertheless, beginning in ancient times, taking away a criminal offender's right to vote "was thought to offer both retribution and a deterrent to future offending" (Behrens, Uggen, & Manza, 2003, p. 562). Although such laws may seem "race neutral," recent research has challenged these notions (Behrens et al., 2003; Fellner & Mauer, 1998; Manza & Uggen, 2006; Preuhs, 2001; Shapiro, 1997; Uggen & Manza, 2002; Uggen, Manza, & Behrens, 2003). The United States' adoption of the practice dates to the colonial era. Initially, only a few offenses resulted in disenfranchisement. But according to Behrens et al. (2003), "Many states enacted felon disenfranchisement provisions in the aftermath of the Civil War. Such laws diluted the voting strength of newly enfranchised racial minority groups, particularly in the Deep South but in the North as well" (p. 563). Since these early times, the prevalence of such laws has grown. For example, "Whereas 35% of states had a broad felon disenfranchisement law in 1850, fully 96% had such a law by 2002, when only Maine and Vermont had yet to restrict felon voting rights" (Behrens et al., 2003, p. 564).

Behrens et al. (2003) sought to determine whether the composition of prison populations is associated with the enactment of state felony disenfranchisement laws from 1850 to 2002. Their results showed,

> The racial composition of state prisons is firmly associated with the adoption of state felon disenfranchisement laws. States with greater non-White prison populations have been more likely to ban convicted felons from voting than states with proportionately fewer non-Whites in the criminal justice system. (Behrens et al., 2003, p. 596)

Such findings suggest an effort to minimize the impact of the votes of felons.

A study by Uggen and Manza (2002) investigated the potential impact of felony disenfranchisement on past elections. Noting that the disenfranchised population represents 2.3% of the electorate, Uggen and Manza controlled for turnout and voter choice, which estimates the level at which felons would have participated in the election process and their party preferences. Their estimates revealed that, "On average . . . about 35% of disenfranchised felons would have turned out to vote in presidential elections, and that about 24% would have participated in Senate elections during non-presidential election years" (Uggen & Manza, 2002, p. 786). Estimates for political party preference showed,

> Democratic candidates would have received about 7 of every 10 votes cast by the felons and ex-felons in 14 of the last 15 U.S. Senate election years. By removing those Democratic preferences from the pool of eligible voters, felon disenfranchisement has provided a small but very clear advantage to Republican candidates in every presidential and senatorial election from 1972 to 2000. (Uggen & Manza, 2002, pp. 786–787)

When Uggen and Manza (2002) examined the impact of these figures on senatorial elections from 1978 to 2000, they found "7 outcomes that may have been reversed if not for the disenfranchisement of felons and ex-felons" (p. 789). Although seven outcomes might not seem substantial, they noted that this could have resulted in the Democrats controlling the Senate during the 1990s, which could have impacted on the nature of the punishment policies during the decade. Finally, when examining the 2000 presidential election, Uggen and Manza found that, without the restrictions on felons and ex-offenders, Al Gore would have won the popular election by 1 million votes, as opposed to 500,000, and, more important, "If disenfranchised felons in Florida had been permitted to vote, Democrat Gore would certainly have carried the state, and the election" (p. 792).

Given the implications of felony disenfranchisement, one wonders how the public feels about the practice. Manza, Brooks, and Uggen (2004) investigated this question in a recent national poll. More specifically, they were interested in finding out whether the public supported the enfranchisement of offenders and whether this support was contingent on "the level of supervision or the specific nature of the crime" (Manza et al., 2004). Their poll showed that the majority of the public was willing to enfranchise probationers (68%); however, with a slight change in the phrasing of the question, which included the word *prison*, the support slipped to 60%. The aversion to supporting enfranchisement for those imprisoned was also seen with a direct question on the matter, which showed only 31% of the respondents supporting such a policy. Eighty percent of the public supported enfranchising ex-felons. But when the ex-felon category was disaggregated by offense, in some cases, the support was lower, with 63% supporting enfranchisement for white-collar offenders and 66% supporting enfranchisement for violent offenders (Manza et al., 2004). The lowest level of support was found for enfranchising sex offenders (52%).

In general, it appears that there is a political dimension to felony disenfranchisement laws. Such laws disproportionately impact the voting power of African Americans and Hispanics. Given the level of national support for enfranchising some of these offenders and ex-cons, it remains perplexing as to why changes to these policies have not been forthcoming. Another contemporary issue that raises similar concern is the incarceration of "political prisoners." We briefly discuss this in the next section.

POLITICAL PRISONERS

Political prisoners are individuals who, based on their political, social, or environmental beliefs, are brought within the jurisdiction of the criminal justice system, with most of them eventually being incarcerated. Historically, the term *railroaded* was used to describe the swift and unfair manner in which political prisoners were handled. Using a more expansive definition, B. Aptheker (1971) separated political prisoners into four categories. First, there are "those who become effective political leaders in their communities, and therefore become the victims of politically-inspired frameups. They are not imprisoned for any violation of the law; but for their political beliefs" (p. 46). Second, there are those individuals who engage in civil disobedience or refuse to be drafted into the armed forces, which, although illegal, are considered political acts. Third, B. Aptheker (1971) noted that there are

> thousands of originally non-political people who are the victims of class, racial and national oppression. Arrested for an assortment of alleged crimes, and lacking adequate legal or political redress they are imprisoned for long years, in violation of fundamental civil and human rights though they are innocent. (p. 47)

Last, there are those persons who, while imprisoned, develop a political consciousness. Of these prisoners, B. Aptheker (1971) wrote that, "as soon as they give expression to their political views they become victims of politically inspired actions against them by the prison administration and the parole boards" (p. 47).

In discussing the origins of the first American jail, which was built around 1655, Takagi (1975) wrote that the colonists built the facility because they were "concerned with the discipline of laborers, servants, debtors, and *political prisoners*" (p. 19; italics added). Thus, from the beginning of the use of incarceration in America, political prisoners have been found in jails and prisons. Since these early times, political prisoners (of all races) have remained a part of the American landscape. The 20th century saw quite a few movements that resulted in the incarceration of political prisoners. Some of these movements include the repression of the Industrial Workers of the World (Wobblies), the Palmer raids against the Communist Party, World War II war resisters, the McCarthy-era witch hunts, the Puerto Rican Nationalists, the Black Liberation Movement, and the American Indian Movement (J. Lopez, 2000).

Each of these movements has had its martyrs. For example, J. Lopez (2000) noted that, during their uprising in the 1950s, the Puerto Rican Nationals were brutalized (with some being murdered), and more than 3,000 of them were imprisoned. During the 1960s, the FBI's counterintelligence program (COINTELPRO) stepped up its surveillance of Black civil rights leaders and organizations, as well as more radical organizations such as the Black Panther Party (Willmott, 2000). According to Willmott, this increase in activity by the COINTELPRO program resulted in the frameups and killing of 30 Black Panther members. During this time, Angela Davis, George Jackson, and numerous others were universally considered political prisoners because of their political views (James, 2003). In recent years, long-time political prisoner Geronimo Pratt has been released and exonerated, but others such as Native American activist Leonard Peltier (see Messerschmidt, 1983), whom many believe was framed for the shooting of two FBI agents, remains incarcerated. More recently, activists have centered on the case of Mumia Abu Jamal (see Abu-Jamal, 1995; Hanrahan, 2001; O'Connor, 2008), who, because of his writing, has become the most identifiable political prisoner in America.

One thing is clear from our review of political prisoners. Radical political movements have always garnered considerable attention from the government. Such movements have always had visible leaders, who more often than not have come under the scrutiny of the criminal justice system. Unfortunately, in some of these instances, they have been neutralized by corrections—although, in many instances, not silenced.

Conclusion

This chapter began by examining the structure of American corrections. The review showed that the correctional system in America includes both community-based and incarceration options. Furthermore, a historical overview of race and corrections showed that race, class, and gender have always had a place in American corrections. Currently, Black and Hispanic prisoners represent a large share of the incarcerated population in the United States. Furthermore, with hundreds of thousands of prisoners annually returning to their communities in the United States, prisoner reentry has become a major concern in recent years. One innovative approach that has been used to handle this issue is reentry courts, which are based on the drug courts model.

Although new, it is hoped that these courts will help stem the tide of recidivism among recently released inmates.

Our review of the extent and implications of felony disenfranchisement in the United States showed that Blacks and Hispanics have been disproportionately impacted by such laws. In fact, although public support is generally in favor of enfranchising most ex-cons, only two states have no restrictions on voting rights related to felony convictions. Scholars and activists continue to lobby for the revision of these policies. The chapter ended with a discussion of political prisoners. Such prisoners have existed since the beginning of America. Historically, movements deemed "radical" by the government have come under extra scrutiny by government agents often resulting in conflict and, in too many instances, the incarceration of political activists.

Although much of our discussion in the first seven chapters has pertained to the adult criminal justice system, in Chapter 9, we turn our attention to the juvenile justice system. Operating under a separate philosophy, the juvenile justice system has also had to deal with the overrepresentation of racial minorities. Chapter 9 reviews some of the issues related to this phenomenon.

Discussion Questions

1. Discuss two historical trends related to race and corrections.

2. What is the significance of Blumstein's work related to explaining disparities in corrections?

3. How has felon disenfranchisement impacted on American politics?

4. What are reentry courts?

5. What are political prisoners?

Internet Exercise

Look at the data provided on the publications page of the Federal Bureau of Prisons Web site: www.bop.gov. Examine the statistical trends by race and note whether any patterns have developed during the years provided. What do you believe explains the trend(s) revealed?

Internet Sites

Amnesty International links regarding the death penalty: http://www.amnesty.org

Information about Native American activist Leonard Peltier: http://www.freepeltier.org

Links related to political prisoners and the prison moratorium movement:

http://www.prisonactivist.org

Juvenile Justice $\mathbf{9}$

> *Thus, there is evidence to suggest that processing decisions in many state and local juvenile justice systems may not be racially neutral. Race effects may occur at various decision points, they may be direct or indirect, and they may accumulate as youths are processed through the system.*
>
> —Pope & Feyerherm (1990, p. 331)

The previous quotation appears in a review of 46 studies on race effects in juvenile justice processing funded by the federal Office of Juvenile Justice and Delinquency Prevention (OJJDP). The most visible outcome of race effects is *disproportionate minority confinement* (DMC), a term used to refer to the proportion of juvenile minorities in confinement exceeding their proportion of the general population (Devine, Coolbaugh, & Jenkins, 1998). This contemporary challenge for juvenile justice is discussed in more detail later in this chapter. Most textbooks on juvenile delinquency and juvenile justice provide only a cursory discussion of race (Taylor Greene, Gabbidon, & Ebersole, 2001), focusing primarily on statistics, police arrest decisions, gangs, and, more recently, the issue of DMC. The history of discrimination and segregation in juvenile facilities since their inception in the 1800s is often omitted, as is the important role played by minority communities in protecting their youth from involvement in crime.

Since the late 1800s, historically Black colleges and universities (HBCUs) were in the vanguard of efforts to improve their race and instrumental in calling attention to Black youth. Early in the 20th century, African American women who attended these institutions were prepared to be teachers, although it did not take long for them to realize that they would also need to become involved in community programs. For example, at the 1907 Hampton Conference, several women joined together to form the Virginia State Federation of Colored Women's Clubs. These women were at the forefront of the "Black child savers" discussed later in this chapter. Similar efforts were surely undertaken by Asian Americans, Hispanics, Native Americans, and White immigrants in their communities as well.

Despite the overrepresentation of other minority youth in juvenile justice, most historical and contemporary information on race and juvenile justice focuses on White and Black youth. This is due, at least in part, to how data were collected. There is considerable more research on Latino/as (Bond-Maupin, 1998; Cintron, 2005; Madriz, 1997; Males & Macallair, 2000; Maupin & Maupin, 1998; McCluskey, 2002; McCluskey, Krohn, Lizotte, & Rodriguez, 2002; McCluskey & Tovar, 2003; Vega & Gil, 1998) than Native Americans and Asian Americans. Contemporary research is more inclusive of Latino/as and Native Americans and focuses on several topics, including coping strategies (McGee, Barber, Joseph, Dudley, & Howell, 2005); family functioning and delinquency (Dillon, Robbins, Szapocznik, & Pantin, 2008), prevention (McKinney, 2003), strain theory (McCluskey, 2002), and youth development (Gallegos-Castillo & Patino, 2006). A few edited volumes focus on minority youth, delinquency, and justice (see, e.g., Hawkins & Kempf-Leonard, 2005; Leonard, Pope, & Feyerherm, 1995; Penn, Taylor Greene, & Gabbidon, 2005). Here we focus on historical and contemporary research.

The purpose of this chapter is explore race effects in juvenile justice. It begins with a brief overview of juvenile justice in the United States, presents the history of race and juvenile justice, explains the extent of juvenile crimes and victimization, and discusses several contemporary issues, including DMC, minority female delinquency, juveniles and the death penalty, and delinquency prevention.

Overview of Juvenile Justice

Many students are surprised to learn that the concepts of juvenile delinquency and juvenile justice are of recent origin. Prior to the 19th century, youth were the primary responsibility of their families and communities. Those who could not be controlled were punished like adults or sent to live with other families, usually as apprentices to learn a skill or trade. During the 1800s, separate facilities for youth in trouble were established. These early institutions included asylums, orphanages, houses of refuge, and reformatories.

Later in the century, the first juvenile court opened in Cook County (Chicago), Illinois. Today, the phrase *juvenile justice system* is used to refer to the agencies and processes responsible for the prevention and control of juvenile delinquency. In fact, according to the National Center for Juvenile Justice (2004), there are actually 51 separate juvenile justice systems in the United States. Figure 9.1 presents an overview of case flow through a contemporary juvenile justice system. Like the criminal justice system, the juvenile system includes law enforcement, courts, and corrections, and both systems have discretionary decision making at several stages. The systems intersect because some juveniles are tried in criminal courts, detained in jails, and sentenced to adult prisons. Like adults, juveniles can be detained without bail before their trials (adjudicatory hearing), have procedural safeguards, have the right to an appeal (in some states), and be placed on probation. In addition to these similarities, there are

Figure 9.1 What are the stages of delinquency case processing in the juvenile justice system?

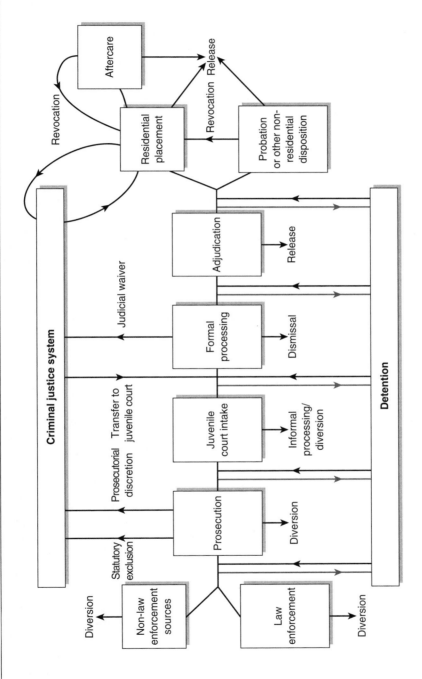

SOURCE: Snyder and Sickmund (2006).

many differences between the two systems, including more parental involvement in the juvenile system, different terminology, and the fact that juveniles do not have a constitutional right to a jury trial (Siegel, 2002). More recently, teen (youth) courts and juvenile drug courts have been established in some jurisdictions.

The most important stages of the juvenile justice process are referrals (usually by police), intake, adjudication, and disposition. Juveniles may be diverted from the process at any stage by police, probation officers, and judges. For example, a juvenile referred to the court for a minor offense can be diverted from the process by an intake/probation officer. A juvenile who commits a more serious crime, like rape, might be formally petitioned to appear in juvenile court or transferred (waived) to adult court. Another discretionary decision is then made about whether the juvenile will be detained until appearing before a judge. If the juvenile who committed the rape is adjudicated as a delinquent, several dispositions are available, including probation, restitution, community service, and secure confinement.

Another unique characteristic of the juvenile justice system is the jurisdiction of the juvenile court over delinquent youth, status offenders, and youth who are dependent, neglected, and abused. *Juvenile delinquency* is described by Greenwald (1993) as a "euphemism for behavioral problems of children and youth that reach beyond mere non-conformity and existed since earliest man" (p. 735). In the early 19th century, it encompassed any illegal behavior by a minor who fell under a statutory age limit and was labeled a serious social problem (Clement, 1993). In most states today, juveniles include youth who are under the age of 18. Delinquent acts include both criminal and status offenses. Unlike adults, juveniles can be apprehended for offenses due solely to their status as children or adolescents. Curfew violations, underage drinking, incorrigibility, and running away are examples of status offenses. In some states, these juveniles are referred to as either "children or persons in need of supervision."

The juvenile justice system today is quite different from a century ago. Initially, the goal of the early courts was to rehabilitate youth and act in the best interest of the child. In the 1960s, a fundamental shift away from the idea of rehabilitation began to occur (Urban, St. Cyr, & Decker, 2003). Several scholars have observed a gradual transition to what is now a more accountability-driven and punitive system (Benekos & Merlo, 2008; Feld, 1999; Hinton, Sims, Adams, & West, 2007; Urban et al., 2003; Ward, 2001). Today, the competing goals of rehabilitation and punishment have resulted in a system that places the personal, social, and educational needs of juvenile delinquents second to the need to punish them (Urban et al., 2003). Bernard (1992) identified the cycle of juvenile justice as being driven by three ideas: Crime is exceptionally high, the present policies make the problem worse, and changing the policies will reduce juvenile crime. According to Bernard, the cycle of juvenile justice continues because the social conditions that foster juvenile delinquency are never adequately addressed. How did we go from a system that focused primarily on the best interest of the child to one that emphasizes punishment? When and why did we return to policies that blur the distinction between juveniles and adults? The following section provides some answers to these questions.

Historical Overview of Race and Juvenile Justice

Children and youth have always misbehaved, and there have always been varying attitudes and practices toward what has come to be known as juvenile delinquency. Clement (1993) noted that, in colonial America, southerners were more tolerant of misbehavior than were colonists in Massachusetts, Pennsylvania, and New York. More specifically, southerners tolerated troublesome White males, but slave owners did not tolerate misbehavior of young (or adult) slaves. In the North, as the population in cities rapidly increased due primarily to urbanization, industrialization, and immigration, there was growing concern about youth crime. During the 1700s, the problem of youth misbehavior common to all children began to be referred to as the "crimes and conditions of poor children" (Mennel, 1973, p. xxvi). By the 1800s, concerned citizens referred to as "child savers" coalesced to protect children and youth and work on their behalf. Early in the century, these individuals and their organizations were instrumental in establishing separate facilities for youth; later in the century, they were instrumental in the creation of the first juvenile court. From the beginning of the movement to salvage youth, Black children were excluded and treated differently.

As early as 1819, a group of individuals concerned about pauperism and the plight of youth in New York City formed the Society for the Prevention of Pauperism, which later became known as the Society for the Reformation of Juvenile Delinquents. On January 1, 1825, this organization opened the New York House of Refuge; Boston and Philadelphia soon followed suit, opening "houses of refuge" prior to the 1830s. Most youth in the early houses of refuge were not committing crimes, but rather were impoverished, neglected, and homeless adolescent White males. There was a considerable amount of prejudice in these early facilities toward immigrants, especially Irish youth, girls, and Black delinquents. The New York and Boston houses of refuge admitted Whites and Blacks, although they were segregated. In 1849, Philadelphia opened a separate house of refuge for "colored juvenile delinquents" (Frey, 1981; Mennel, 1973). Interestingly, although all Black children born in Pennsylvania after 1780 were free, prejudice and separation of the races were the norm well into the 20th century. Although the houses of refuge in Philadelphia were segregated, both sought to maintain social control over poor and delinquent youth and to instill in them a basic education, the desire to work, and a moral foundation (Frey, 1981). Early houses maintained a practice of placing out youth to apprenticeships for training or to rural farms. Because White girls and boys were preferred, Black youth stayed in the houses for longer periods of time (Ward, 2001).

Unlike northern states, most southern states moved slowly to open separate facilities for youth. Young (1993) noted "that the house of refuge established in Maryland in 1840 and opened in 1855 was restricted to White children due to the slave status of Black children" (p. 557). In Kentucky, the Louisville House of Refuge established in 1854 was for White children only (Young, 1994a). Black youth were excluded from most facilities or treated quite differently. Managers believed that placing White and Black children together would be degrading to the White children (Mennel, 1973). Before and after the Civil War, in both the North and the South, Black youth were more likely to be sent to adult jails and prisons than to juvenile facilities. In Maryland, the children of free Blacks were perceived as a threat (as were their parents) and could be either sold or bound out if their parents could not provide for them (Young, 1993).

By the 1850s, houses of refuge that had opened earlier in the century were criticized and eventually replaced with reformatories. Massachusetts opened its reform school for boys in 1847 and for girls in 1856. These institutions differed from houses of refuge in several ways. First, smaller buildings that were maintained by the occupants were utilized. Second, there was a greater emphasis on education. By 1890, most states outside the South had reform schools for boys and girls (Mennel, 1973).

Like the houses of refuge, the early reformatories excluded Blacks and instead sent them to adult jails and prisons. There is little information about the treatment of Black youth in adult facilities. We do know that the convict-lease system in jails and prisons in the South was difficult for all prisoners (Du Bois, 1901/2002; Oshinsky, 1996; Woodward, 1971; Work, 1939). As noted in earlier chapters, many southern states used the convict-lease system as both a means of generating revenue to maintain their penal institutions and for profit. Punishment was excessive, death rates were high, sexual assaults and rapes were ignored, and oversight was rare (Colvin, 1997). Black youth outnumbered other juveniles in these facilities (Ward, 2001) and were subjected to the most brutal form of punishment. It was not until 1873 that one of the first separate reformatories for "colored boys" opened in Baltimore (Ward, 2001; Young, 1993). It would take several decades for other reformatories to open for Black youth. Virginia opened its first industrial home for "wayward colored girls" in 1915 and a facility for "wayward boys" in 1920 (Young, 1994a). These segregated and often inferior reformatories were due, in large part, to the efforts of the "child savers."

THE CHILD SAVERS

As previously mentioned, child savers were concerned with the plight of poor, vagrant, and neglected children (Siegel, 2002). These activists comprised middle- and upper class females, criminal justice practitioners, and various organizations that were primarily interested in saving White children, who usually came from immigrant families struggling with the transition to life in a new country. Most of the White child savers were not concerned about Negro children, who faced similar, if not worse, challenges than immigrant youth. Although Platt (1969) and Shelden and Osborne (1989) questioned the benevolence of the child savers and posited that they were motivated by self-interest, they do not address how child savers overlooked Negro youth and the role of Black child savers (Ward, 2001).

Youth facilities were originally opened to separate delinquents from adults and the poor conditions found in prisons. It also was believed that youth were malleable and could be "saved" from the harmful effects of poverty and harsh living conditions. Unlike White youth, Black youth often were placed in adult facilities, evidence of what Ward (2001) referred to as "historical racial inequality" in juvenile justice. In addition, ideas of Black childhood varied between the races: Whites viewed Black childhood as developmentally limited, whereas Blacks viewed their development as critical to the future of the race (Ward, 2001).

Black child savers emerged in response to the unfair and prejudicial handling of Black youth. As early as 1899, at the Hampton Negro Conference, the issues of the need for clubs for girls and the dangers faced by girls migrating to the North were discussed. Like White child savers, the early Black child savers were initially women of a higher social class. However, they existed in a distinctly racially segregated community where the meaning of class was quite different. Although they held status and power in their own communities, Black child savers faced a greater challenge: a racialized justice system that was unwilling to invest in the rehabilitation of Black youth. They had to both challenge racial disparities and develop resources to finance facilities on their own (Ward, 2001).

Ward (2001) described at least three different types of Black child savers. The early child savers were forced to work on the periphery of the justice system. The Federation of Colored Women's Clubs was at the forefront of the early Black child savers movement. These groups of women were active across the country, especially in the South, and were instrumental in creating the first reformatories for female and male Negro juveniles (Neverdon-Morton, 1989). The next group of Black child savers was more reform oriented and focused more directly on working as practitioners not only in juvenile justice, but in other agencies as well. They emerged early in the 20th century, when "African Americans were beginning to move from the periphery of juvenile justice, as community-based providers, toward its administrative center" (Ward, 2001, p. 164). These child savers were "African American participants in juvenile justice administration who aimed to affect outcomes in the legal processing of Black delinquents especially, though not exclusively" (Ward, 2001, p. 120). The National Association for the Advancement of Colored People (NAACP) can be viewed as the third type of Black child savers because the organization devoted a tremendous amount of time and resources to the plight of Black youth in the justice system in the early part of the 20th century. Ward (2001) did not idealize the Black child savers in light of their incredible obstacles, although he does conclude that they probably did "improve the circumstances of Black juvenile delinquents" (p. 186). In addition, Black and White child savers coalesced in some places.

JUVENILE COURTS

One of the most important developments in the history of juvenile justice was the creation of the juvenile court. The juvenile court was viewed as a progressive idea that recognized the special needs of children not only in juvenile facilities, but in court as well. As previously stated, the primary goal of juvenile courts was rehabilitation and to act in the best interest of the child. By the mid-1930s, most states had legislated some form of juvenile courts. Feld (1999) argued that, since its inception, one of the most important functions of the juvenile court was to control ethnic and racial minorities:

> The Progressives created the juvenile court to assimilate, integrate, and control the children of the eastern European immigrants pouring into cities of the East and Midwest at the turn of the century. In postindustrial American cities today, juvenile courts function to maintain social control of minority youths, predominantly young Black males. (p. 5)

Juvenile courts were initially criticized for lacking procedural safeguards, insufficient training of judges and staff, and failing to control juvenile crime. The due process revolution that occurred in the 1960s was in large part a response to the punitive function of juvenile courts (Mears, 2002). Despite procedural safeguards, during the latter part of the 20th century, juvenile courts shifted from the welfare and rehabilitation of youth to an even more punitive "second rate criminal court for young offenders" (Feld, 1999, p. 5).

Although we know quite a bit about the development of juvenile courts, little information is available about the treatment of Black and minority youth in these courts prior to the 1980s. Frazier (1939, 1949) provided some information on Negro youth appearing before the children's court in the District of Columbia; Nashville, Tennessee; New York City; and other locales in the early 20th century, although he did not address the race issue (Frazier emphasized community and family factors). In light of segregation and treatment in the early facilities, it is doubtful that Negro youth fared much better in juvenile courts. The Black child savers and other professionals may have played an important role in early juvenile courts, but there were not enough of them. Although they aimed to work in the best interest of the child, they could change neither the structural and economic conditions nor the stereotypical attitudes of other (White) justice professionals. Differential treatment of Black youth was tolerated and accepted, and practitioners were aware of the overrepresentation of Black and other minority youth well into the 20th century.

During the past two decades, the hybrid mix of rehabilitation and punishment has resulted in the implementation of numerous conflicting policies and programs to prevent and control juvenile delinquency. Boot camps, mentoring programs, family therapy, teen (youth) courts, juvenile drug courts, trying juveniles as adults, mandatory sentences, and longer sentences are common. Some of the more punitive approaches were a reaction to increases in juvenile involvement in violent crime that occurred in the 1980s and early 1990s. Although the number of juveniles arrested for violent crime is still a problem, there has been a substantial decrease since 1994. This issue is discussed in more detail next.

Juvenile Crime and Victimization

For decades, juvenile involvement in crime as perpetrators and victims has received considerable attention. Part of the concern is related to the disproportionate number of youth, especially Black and other minority youth, who are arrested and confined. Howell (2003) argued that one of the most damaging myths about juvenile crime and violence was the prediction about a new breed of juvenile offenders ostensibly labeled as *superpredators* (DiIulio, 1995, 1996). Implicit in the notion of the superpredator was the demonization of Black and Hispanic youth; these and other chronic offenders were projected as contributing to a youth violence epidemic in our country in the coming years (Blumstein, 1996). Although there was no increase in the number of juvenile arrests after these dire predictions, there was a "moral panic over juvenile delinquency" that resulted in more punitive measures and "a crisis of overload in the juvenile justice system" (Howell, 2003, p. 24).

Chapter 2 provided an overview of several sources of crime and victimization data, including the Federal Bureau of Investigation (FBI) Uniform Crime Reports (UCR) and National Incident Based Reporting System (NIBRS), as well as the Bureau of Justice Statistics (BJS) National Crime Victimization Survey (NCVS), which provide information on juveniles. There are also self-report studies, publications that present data on juvenile drug use, and data on youth in the juvenile justice system, including juvenile court statistics and data on youth in residential confinement. Chapter 2 also addressed the strengths and limitations of available data. Some of these limitations are important for understanding juvenile crime and victimization. First, Hispanics are included as "Whites" in FBI arrest data. Second, because FBI data are aggregated, we cannot examine arrests by race and gender simultaneously. Third, UCR arrest data are based on estimates, and secondary analyses of UCR juvenile arrests vary considerably from the UCR estimates (see, e.g., Snyder, 1997, 2000, 2003). Fourth, although NIBRS data provide details on crime incidents including sexual assaults and kidnapping, most states are not participating in the program.

NCVS data are also problematic. First, the NCVS excludes youth under 12 years of age. Second, NCVS data were disaggregated by race, age, and gender in the past, but not in more recent publications. Third, although the NCVS does include information on Hispanic victims, it is aggregated as well. Several self-report studies, such as the Monitoring the Future survey, the National Adolescent Survey, National Survey on Drug Use and Health (formerly the National Household Survey on Drug Abuse), and the Youth Risk Behavior Surveillance System, include useful information about juveniles. Juvenile justice data (juvenile court and confinement) are informative, but dated. For example, a recent compilation entitled *Juvenile Offenders and Victims: 2006 National Report* includes data for 2003 and earlier. Despite these limitations, the available data do help us understand juvenile participation in crime as arrestees, victims, and those who are processed in the juvenile justice system.

Juvenile arrests have fluctuated over time. During the early 1980s, juvenile arrest rates declined between 1980 and 1984, increased after 1984 until 1987, and then decreased again until 1989 (Feld, 1999). Although juvenile arrests increased in the early 1990s, they have leveled off since then. In 1995, juveniles comprised 19% (147,000) of violent crime index (VCI) arrests and 35% (737,400) of property crime index (PCI) arrests (Snyder, 1997). Stated another way, less than one half of 1% of all people ages 10 to 17 were arrested for VCI offenses. According to the UCR, in 2001, an estimated 1.5 million juveniles were arrested (Federal Bureau of Investigation, 2001). In 2002 the number of persons arrested increased 0.5% for the first time in several years to 1,624,192 (Federal Bureau of Investigation, 2002). In 2006, an estimated 1,626,523 juveniles were arrested, a slight increase over 2005 (Federal Bureau of Investigation, 2006) (see Table 9.1).

The majority of arrestees are males, although the number of female arrestees has increased in recent years. For example, in 1980, only 20% of juveniles arrested were females (Snyder & Sickmund, 2008); in 2006, 29% were females (Office of Juvenile Justice and Delinquency Prevention, 2008). Most juveniles arrested for Part I (Crime

Table 9.1   Juvenile Arrest Estimates, 2001–2006

Year	Arrests	Total Arrests (%)
2001	1,558,496	16.7
2002	1,624,192	16.5
2003	1,563,149	16.2
2004	1,578,893	15.8
2005	1,582,068	15.3
2006	1,626,523	15.5

SOURCE: Federal Bureau of Investigation (2001–2006, Table 38).

Index) violent and property crimes, regardless of age, race, and gender, are arrested for larceny/theft. In 2006, juveniles under the age of 15 were responsible for 29% of arrests, including 58% of persons arrested for arson, 47% for sex offenses (other than rape and prostitution), and 41% of juvenile arrests for vandalism. Some juveniles arrested belong to gangs, although we cannot determine this from UCR arrest data as currently reported. Youth gang members are involved in a disproportionate amount of serious and violent crime (Thornberry, 1998; Thornberry, Huizinga, & Loeber, 2004) (see Highlight Box 9.1). They are also more likely to own and use firearms in the commission of violent crimes (National Youth Gang Center, 2008).

Highlight Box 9.1

Youth Gangs

What is a youth gang?

There is no single, accepted nationwide definition of youth gangs. A youth gang is commonly thought of as a self-formed association of peers having the following characteristics: three or more members, generally ages 12 to 24; a name and some sense of identity, generally indicated by such symbols as style of clothing, graffiti, and hand signs; some degree of permanence and organization; and an elevated level of involvement in delinquent or criminal activity.

Is a new wave of gang activity developing in the United States?

Youth gang activity tends to follow a cyclical pattern with upswings followed by downturns (Klein, 1995), and these occur at multiple levels—within regions, cities, and neighborhoods (Egley, Howell, & Major, 2004). To illustrate, the largest cities reported persistent gang problems from 1996 to 2002 in the NYGS; in contrast, smaller cities and suburban and rural counties tended to report variable gang problems during this period (Egley et al., 2004; see also Egley, 2002). Some cities may experience a large and sudden flare-up in gang violence in a given year. Such upswings are typically local in nature and not indicative of a nationwide trend (Tita & Abrahamse, 2004). Further complicating matters, local officials may be reluctant to acknowledge their gang problem until it publicly surfaces in a tragic event, or they may declare they have successfully dealt with it, only to see it surface again.

Are today's youth gangs different from gangs in the past?

Some of the gangs that have emerged in the past decade are noticeably different from those that emerged before the mid-1980s (Howell, Egley, & Gleason, 2002; Howell, Moore, & Egley, 2002). These gangs are commonly described as having a "hybrid gang culture," meaning they do not follow the same rules or methods of operation, making documentation and categorization difficult (Starbuck, Howell, & Lindquist, 2001). They may have several of the following characteristics: a mixture of racial/ethnic groups, a mixture of symbols and graffiti associated with different gangs, wearing colors traditionally associated with a rival gang, less concern over turf or territory, and members who sometimes switch from one gang to another. Members of contemporary gangs often "cut and paste" bits of Hollywood images and big-city gang lore into their local versions of gangs. Small town and rural gangs also differ from urban gangs in other important respects (Howell, Egley, & Gleason, 2002), hence urban models of gang development and response do not necessarily apply in rural areas (Weisheit & Wells, 2004).

Is gang migration a common problem?

The most common reasons for migration are social considerations affecting individual gang members, including family relocation to improve the quality of life or to be near relatives and friends. Moreover, in the 2004 NYGS, a majority (60 percent) of respondents reported no or few (less than 25 percent of documented gang members) such migrants. Agencies that experienced the highest levels of gang-member migration were significantly more likely to report migration for social reasons (Egley & Ritz, 2006).

What is the racial and ethnic composition of youth gangs?

According to the 2001 NYGS respondents, nearly half (49 percent) of all gang members are Hispanic/Latino, 34 percent are African American/black, 10 percent are Caucasian/white, 6 percent are Asian, and the remainder are of some other race/ethnicity (Egley & Ritz, 2006). However, the racial composition of gangs varies considerably by locality. The newest gang-problem areas (i.e., emergence within the past decade) report, on average, a larger proportion of Caucasian/white gang members than any other racial/ethnic group (Howell, Egley, & Gleason, 2002). In short, the demographic composition of gangs is an extension of the characteristics of the larger community.

Is female gang involvement increasing?

Youth gang membership among girls is much more common (Moore & Hagedorn, 2001) and is documented more widely by law enforcement (Egley & Ritz, 2006) than in the past. During early adolescence, roughly one-third of all youth gang members are female (Esbensen & Winfree, 1998; Gottfredson & Gottfredson, 2001), but studies show that females leave the gang at an earlier age than males (Gottfredson & Gottfredson, 2001; Thornberry, Krohn, Lizotte, Smith, & Tobin, 2003). Gender-mixed gangs are also more commonly reported now than in the past. In 2000, 42 percent of all gang-problem jurisdictions in the NYGS reported that a majority of their gangs had female members (Egley & Ritz, 2006). Furthermore, emerging research has also documented that the gender composition of the gang is importantly associated with gang delinquency rates. In one study, females in all- or majority-female gangs exhibited the lowest delinquency rates, and males and females in majority-male gangs exhibited the highest delinquency rates (including higher rates than males in all-male gangs) (Peterson, Miller, & Esbensen, 2001).

What proportion of serious and violent crime is attributable to gang members?

Studies of large urban samples show that gang members are responsible for a large proportion of all violent offenses committed during the adolescent years. Rochester gang members (30 percent of the sample) self-reported committing 68 percent of all adolescent violent offenses; in Seattle, gang

(Continued)

(Continued)

members (15 percent of the sample) self-reported committing 85 percent of adolescent robberies; and in Denver, gang members (14 percent of the sample) self-reported committing 79 percent of all serious violent adolescent offenses (Thornberry, 1998; Thornberry et al., 2004).

What are the major risk factors for gang membership?

Risk factors that predispose many youths to gang membership are also linked to a variety of adolescent problem behaviors, including serious violence and delinquency. The major risk factor domains are individual characteristics, family conditions, school experiences and performance, peer group influences, and the community context. Risk factors predictive of gang membership include prior and/or early involvement in delinquency, especially violence and alcohol/drug use; poor family management and problematic parent-child relations; low school attachment and achievement and negative labeling by teachers; association with aggressive peers and peers who engage in delinquency; and neighborhoods in which large numbers of youth are in trouble and in which drugs and firearms are readily available (Howell & Egley, 2005; see also Esbensen, 2000; Hill, Lui, & Hawkins, 2001; Thornberry, 1998; Wyrick & Howell, 2004). The accumulation of risk factors greatly increases the likelihood of gang involvement, just as it does for other problem behaviors. The presence of risk factors in multiple risk-factor domains appears to increase the likelihood of gang involvement even more (Thornberry et al., 2003).

What are the consequences of gang membership?

Prolonged gang involvement is likely to take a heavy toll on youths' social development and life-course experiences. The gang acts as "a powerful social network" in constraining the behavior of members, limiting access to prosocial networks and cutting individuals off from conventional pursuits (Thornberry et al., 2003; Thornberry et al., 2004). These effects of the gang tend to produce precocious, off-time, and unsuccessful transitions that bring disorder to the life course in a cascading series of difficulties, including school dropout, early parenthood, and unstable employment. For some gang members, the end result of this foreclosure of future opportunities is continued involvement in criminal activity throughout adolescence and into adulthood.

Despite the apparent popular belief among youth that joining a gang will afford protection, in reality the opposite is true. Youth are far more likely to be violently victimized while in a gang than when they are not (Peterson, Taylor, & Esbensen, 2004). This relationship holds irrespective of the primary reason for joining a gang (i.e., whether for protection or not). Furthermore, in two studies, involvement in gang fights more than doubled or tripled the odds of serious injury (Loeber, Kalb, & Huizinga, 2001).

What can be done about youth gangs?

Over reliance on one strategy or another is unlikely to produce fundamental changes in the scope and severity of a community's gang problem (Curry & Decker, 2003; Wyrick & Howell, 2004). A balance of prevention, intervention, and suppression strategies and programs is likely to be far more effective (see Esbensen, 2000). For example, the Gang Resistance Education And Training (G.R.E.A.T.) prevention program (http://www.great-online.org/) could serve youth in gang-problem areas, while an Intervention Team (National Youth Gang Center, 2002) could work with active gang members, and a Gang Suppression Unit could target the most violent gangs and gang members.

SOURCE: National Youth Gang Center (2008).

Many Americans mistakenly believe that minority youth are involved in more violent crime than is actually the case due, at least in part, to how serious crimes are highlighted by the news media (Dorfman & Schiraldi, 2001) and because of their disproportionate involvement in violent crimes. The majority of juveniles arrested in 2006 were either White (67.1%) or Black (30.3%); American Indians or Alaskan Natives and Asian or Pacific Islanders each represented around 1% of juvenile arrestees. Blacks outnumbered Whites and others arrested for murder, robbery, prostitution (and commercial vice), and gambling. Black juveniles accounted for 51% of arrests for violent crime, whereas White juveniles accounted for 66.3% of juveniles arrested for property crime. As previously stated, most juveniles were arrested for Part I larceny-theft, followed by other assaults, disorderly conduct, and drug abuse violations, Part II crimes. Among American Indians or Alaskan Natives, more juveniles were arrested for liquor law violations (2.6%), whereas Asians or Pacific Islanders were more likely to be arrested for running away (4.7%) (Federal Bureau of Investigation, 2006; see Table 9.2).

Youth involvement in crimes, especially violent index crimes is troubling. According to the National Center for Injury Prevention and Control (located in the Centers for Disease Control and Prevention), homicide was the fourth leading cause of death for children ages 1 to 11 in 2002 (Snyder & Sickmund, 2008). The UCR Supplemental Homicide Reports provide information on murder victims, and the OJJDP provides the juvenile homicide victim data online. Highlight Box 9.2 was constructed using the online data to present juvenile homicide victimizations between 2000 and 2005. There were 9,312 juvenile murder victims during this time period, many of them very young. During this period, the number of victims fluctuated over time and steadily decreased between 2003 and 2005 for the youngest victims. The number of victims between the ages of 12 to 14 and 15 to 17 fluctuated between 2000 and 2005 and increased between 2004 and 2005 from 127 to 149 (12 to 14) and 697 to 736 (15 to 17) (Snyder, Finnegan, & Kang, 2007). Data on juvenile victims of nonfatal crimes are reported in victimization studies discussed next.

Highlight Box 9.2

Juvenile Victims of Homicide: 2000–2005

Count	0 to 5	6 to11	12 to14	15 to 17	Total
2000	612	119	147	661	1,539
2001	674	137	122	680	1,613
2002	604	147	125	664	1,540
2003	634	120	109	654	1,517
2004	591	108	127	697	1,523
2005	580	115	149	736	1,580
Total	3,695	746	779	4,092	9,312

Snyder, H., Finnegan, T., and Kang, W. (2007). "Easy Access to the FBI's Supplementary Homicide Reports: 1980–2005." Online. Available: http://ojjdp.ncjrs.gov/ojstatbb/ezashr/

Data source: Federal Bureau of Investigation. Supplementary Homicide Reports 1980–2005 (machine-readable data files).

How would you explain the high number of 0- to 5-year-old juvenile homicide victims? The 15- to 17-year-old juvenile homicide victims?

SOURCE: Snyder, Finnegan, & Kang (2007).

Table 9.2a Arrest Statistics by Race, 2006 Arrests Under 18 Years of Age

Offense charged	Total	White	Black	American Indian or Alaskan Native	Asian or Pacific Islander	Percent Distribution[1] Total	White	Black	American Indian or Alaskan Native	Asian or Pacific Islander
TOTAL	1,621,167	1,088,376	490,838	18,592	23,361	100.0	67.1	30.3	1.1	1.4
Murder and non-negligent manslaughter	950	374	566	6	10	100.0	39.1	59.2	0.6	1.0
Forcible rape	2,507	1,592	863	30	22	100.0	63.5	34.4	1.2	0.9
Robbery	26,060	8,074	17,569	116	301	100.0	31.0	67.4	0.4	1.2
Aggravated assault	44,314	24,697	18,652	439	526	100.0	55.7	42.1	1.0	1.2
Burglary	60,987	40,426	19,221	657	683	100.0	66.3	31.5	1.1	1.1
Larceny-theft	205,201	138,577	60,545	2,275	3,804	100.0	67.5	29.5	1.1	1.9
Motor vehicle theft	25,285	13,576	10,965	265	479	100.0	53.7	43.4	1.0	1.9
Arson	5,868	4,646	1,075	49	98	100.0	79.2	18.3	0.8	1.7
Violent crime[2]	73,837	34,737	37,650	591	859	100.0	47.0	51.0	0.8	1.2
Property crime[2]	297,341	197,225	91,806	3,246	5,064	100.0	66.3	30.9	1.1	1.7
Other assaults	181,288	106,785	70,639	1,899	1,965	100.0	58.9	39.0	1.0	1.1
Forgery and counterfeiting	2,568	1,892	617	21	38	100.0	73.7	24.0	0.8	1.5
Fraud	5,656	3,623	1,938	34	61	100.0	64.1	34.3	0.6	1.1
Embezzlement	1,036	624	391	6	15	100.0	60.2	37.7	0.6	1.4
Stolen property; buying, receiving, possessing	15,574	8,750	6,522	123	179	100.0	56.2	41.9	0.8	1.1
Vandalism	85,850	67,487	16,417	854	1,092	100.0	78.6	19.1	1.0	1.3

[1]Because of rounding, the percentages may not add to 100.0.

[2]Violent crimes are offenses of murder and nonnegligent manslaughter, forcible rape, robbery, and aggravated assault. Property crimes are offenses of burglary, larceny-theft, motor vehicle theft, and arson.

Table 9.2b Crime in the United States

Offense charged	Total	White	Black	American Indian or Alaskan Native	Asian or Pacific Islander	Percent Distribution				
						Total	White	Black	American Indian or Alaskan Native	Asian or Pacific Islander
Weapons; carrying, possessing, etc.	34,611	21,142	12,745	285	439	100.0	61.1	36.8	0.8	1.3
Prostitution and commercialized vice	1,207	533	651	0	13	100.0	44.2	53.9	0.8	1.1
Sex offenses (except forcible rape and prostitution)	11,465	8,185	3,086	80	114	100.0	71.4	26.9	0.7	1.0
Drug abuse violations	143,267	97,800	43,080	1,126	1,261	100.0	68.3	30.1	0.8	0.9
Gambling	1,620	140	1,471	1	8	100.0	8.6	90.8	0.1	0.5
Offenses against the family and children	3,621	2,737	823	26	35	100.0	75.6	22.7	0.7	1.0
Driving under the influence	14,225	13,328	506	232	159	100.0	93.7	3.6	1.6	1.1
Liquor laws	102,230	93,368	4,987	2,635	1,240	100.0	91.3	4.9	2.6	1.2
Drunkenness	12,035	10,764	977	200	94	100.0	89.4	8.1	1.7	0.8
Disorderly conduct	152,869	88,420	61,438	1,799	1,212	100.0	57.8	40.2	1.2	0.8
Vagrancy	3,734	2,872	823	17	22	100.0	76.9	22.0	0.5	0.6
All other offenses (except traffic)	278,902	200,742	70,771	3,104	4,285	100.0	72.0	25.4	1.1	1.5
Suspicion	316	205	108	0	3	100.0	64.9	34.2	0.0	0.9
Curfew and loitering law violations	114,166	69,624	42,496	814	1,232	100.0	61.0	37.2	0.7	1.1
Runaways	83,749	57,393	20,896	1,489	3,971	100.0	68.5	25.0	1.8	4.7

SOURCE: Federal Bureau of Investigation (2006, Table 43).

The NCVS includes reported violent victimizations (rape, sexual assault, robbery, aggravated assault, and simple assault) and property victimizations (attempted and completed theft, household burglary, and motor vehicle theft). According to the NCVS, in 1995, the Black juvenile (12 to 15 years old) victimization rate was higher (120.4) than the White juvenile victimization rate (106.8) for crimes of violence, although the victimization rate for 16- to 19-year-olds was higher for Whites (110.5) than Blacks (100.0) (Klaus & Matson, 2000). When gender is included, victimization rates are higher for 12- to 15-year-olds and 16- to 19-year-old White males, compared with Black males, although Black female rates for both age groups were much greater than White female victimization rates (Klaus & Matson, 2000). More recently, Blacks were victims of violent crimes (rape, robbery, aggravated assault) at higher rates than Whites, Hispanics, and others. Youth 16 to 19 years of age had the highest prevalence rate for sexual assault victimizations (Kilpatrick, Saunders, & Smith, 2003).

According to the NCVS (2005), youth between the ages of 12 and 15 had high rates of violent crime victimizations between 2002–2003 (48.1) and 2004–2005 (46.9), and there was a slight decrease (–2.5%) in victimization rates. In the 16- to 19-year-old category, violent victimizations fell from 55.6 to 45.09 between 2002–2003 and 2004–2005, a 19% decrease (Catalano, 2006). According to Snyder and Sickmund (2008) other important facts about juvenile victimizations include:

- Violent victimization is related to individual, family, and community characteristics (p. 30).
- A youth's risk of being a violent crime victim is most likely due to family and community characteristics, not race (p. 30).
- Juveniles are as likely to be victims of suicide as they are to be victims of homicide (p. 25).
- More than half of all victims of child maltreatment were White (p. 54).

Information about juvenile drug use is not included in the NCVS and is collected in other surveys such as the Monitoring the Future Series, the National Survey on Drug Use and Health, and the Youth Risk Behavior Surveillance System. Juvenile justice data on juveniles arrested and processed for drug offense are also available. Substance abuse is widespread among American youth; in 2004, 51.1% of high school seniors reported trying an illicit drug by the time they graduated, and 38.8% had used an illicit drug in the past year. In 2006, the percentages decreased to 46.8% and 35.9%, respectively (Johnston, O'Malley, & Bachman, 2004; National Institute on Drug Abuse, 2008). Alcohol and marijuana continue to be the drugs of choice of 8th graders, 10th graders, and 12th graders. In 1997, 38.5% of 12th graders reported using marijuana in the past 12 months, compared with 31.5% in 2006. One of the surprising findings from annual Monitoring the Future Surveys is the subgroup differences for Whites, African Americans, and Hispanics. African Americans have lower percentages of marijuana use than do Whites and Hispanics, although there was a slight increase for Blacks in reported use in 2006 (Johnston, O'Malley, Bachman, & Schulenberg, 2007; see Table 9.3).

Table 9.3 Trends in Marijuana Annual Prevalence of Use by Subgroups

Percentage who used in last twelve months

									Class of:							
	1991	1992	1993	1994	1995	1996	1997	1998	1999	2000	2001	2002	2003	2004	2005	2006
Approx. N =	15,000	15,800	16,300	15,400	15,400	14,300	15,400	15,200	13,600	92,800	12,800	12,900	14,600	14,600	14,700	14,200
Total	23.9	21.9	26.0	30.7	34.7	35.8	38.5	37.5	37.8	36.5	37.0	36.2	34.9	34.3	33.6	31.5
Race (2-year average):[b]																
White	28.2	24.9	25.9	30.2	34.2	36.4	38.7	39.9	39.1	38.2	38.5	38.7	37.9	37.3	36.6	34.8
Black	11.4	11.5	14.2	20.7	26.8	30.2	30.4	30.0	30.4	30.0	29.0	27.3	26.3	25.5	26.3	27.7
Hispanic	23.6	24.7	23.5	25.7	29.7	32.3	36.4	37.2	37.8	40.5	37.6	34.6	31.1	29.5	29.6	28.7

SOURCE: Johnston et al. (2007, Table D-9, p. 274).

[a]Parental education is an average score of mother's education and father's education. See Appendix B (in original source) for details.

[b]To derive percentages for each racial subgroup, data for the specified year and the previous year have been combined to increase subgroup sample sizes and the more stable estimates. For the data beginning in 2005, see the race/ethnicity note at the end of Appendix B.

Drug and alcohol use are more common among juvenile offenders than among students (Belenko, Sprott, & Petersen, 2004). Juvenile drug behavior is often correlated with delinquency, although the nature of the relationship is unclear. Despite decades of research, we still do not know whether drug use causes delinquency or vice versa. We do know that most youthful offenders are involved with drugs or alcohol at an early age and that this involvement "increases the likelihood of chronic contact with the juvenile justice system" (Belenko et al., 2004, p. 4).

UCR estimates of arrests of juveniles for drug abuse violations by race between 2001 and 2006 are presented in Table 9.4. In 2001, there were an estimated 113,695 juvenile arrests for drug abuse violations, compared with an estimated 143,267 arrests in 2006 (Federal Bureau of Investigation, 2001, 2006). Most arrestees were White, and total arrests increased from 2003 to 2006 and steadily increased for Blacks during that time period.

In the past, the National Institute of Justice's Arrestee Drug Abuse Monitoring Program provided information on drug use among juveniles detained for committing crimes. Although African Americans have lower levels of drug use, minority arrestees have a higher prevalence of positive drug tests. Fifty-six percent of African Americans, 55% of Hispanics, and 48% of Whites tested positive for any drug (Belenko et al., 2004). In 2002, the majority of male detainees in five jurisdictions—Birmingham, Alabama; Cleveland, Ohio; Phoenix, Arizona; San Antonio, Texas; and San Diego, California— tested positive for drugs in their urine, most often marijuana. Although fewer females tested positive for any drug, they too tested positive for marijuana most often (Arrestee Drug Abuse Monitoring Program, 2002). The implications for the prevention and control of drug-involved youth in juvenile justice are presented later in the chapter.

Data on Youth in the Juvenile Justice System

Data on youth in the juvenile justice system are neither as recent nor as readily available as arrest and victimization data. The primary sources of information are the OJJDP Web

Table 9.4   Juvenile Arrests by Race for Drug Abuse Violations, 2001–2006

Year	Total Arrests	White	Black	American Indian or Alaskan Native	Asian or Pacific Islanders
2001	113,695	77,744	33,947	892	1,112
2002	133,494	97,766	33,208	1,152	1,368
2003	137,052	98,849	35,638	1,225	1,340
2004	138,110	98,485	37,257	1,193	1,175
2005	139,776	96,207	41,076	1,301	1,192
2006	143,267	97,800	43,080	1,126	1,261

SOURCE: Federal Bureau of Investigation, UCR (2001–2006, Table 43).

site, Fact Sheets on Juvenile Court Statistics, and the Census of Juveniles in Residential Placement series. Similar to the pattern in arrest data, most delinquency cases involved males, and most youth are referred to juvenile court for property offenses (Puzzanchera, Stahl, Finnegan, Tierney, & Snyder, 2003b). According to OJJDP, there were approximately 1.66 million delinquency cases in U.S. juvenile courts in 2004; the majority of cases were for property offenses, followed by public order offenses such as obstruction of justice, disorderly conduct, and weapons offenses. The number of delinquency cases processed in juvenile courts decreased between 1995 and 2004, as did the cases involving males. Delinquency cases involving females increased during the decade from 398,600 to 452,500. White youth comprised 66% and Black youth comprised 31% of delinquency cases in 2004 (Stahl, 2008a). Status offense cases in juvenile courts numbered an estimated 159,400 in 2004, and 60% of the cases involved younger juveniles (under age 16). For Blacks, Whites, and Asian/Pacific Islander, most referrals were for truancy; for AI/AN, most referrals were for liquor law violations (Stahl, 2008b)

Estimates of person offense delinquency cases by race processed in 2002 are presented in Figure 9.2. Black youth are more likely to be formerly petitioned to juvenile court, although less likely to be adjudicated delinquent, than White youth or youth of other races. Once adjudicated, Black youth and youth of other races are also more likely to be placed outside the home than are White youth. Minority youth also are overrepresented in residential placement. According to the most recent data, 61% of juveniles in residential placement are minorities and 39% are White. Of the minorities, 38% are Black and 19% are Hispanic (Snyder & Sickmund, 2008). These 2003 data are slightly lower than earlier figures (63% of juveniles in residential placement were minority, and 37% were White reported by OJJDP in 2004). The overrepresentation of Black youth might seem to make sense in that Black youth are disproportionately arrested for violent crimes, especially murder and robbery. However, as stated earlier, fewer juveniles are arrested for violent crimes, and most youth arrested overall are White. In 2002, juveniles waived to criminal court comprised about 1% of petitioned delinquency cases. Black youth were more likely to be waived for drug offenses, and White youth were more likely to be waived of property offenses; youth of other races were waived most often for person offenses (Snyder & Sickmund, 2008). When we take into consideration that many of the most serious violent, property, drug, and public order offenders are waived to adult court, the overrepresentation of Black and other minority youth in juvenile justice case processing is less clear. Several scholars have called attention to this issue, focusing primarily on the disproportionate number of minorities in confinement mentioned at the beginning of the chapter.

Contemporary Issues in Race and Juvenile Justice

DISPROPORTIONATE MINORITY CONFINEMENT

Broadly speaking, DMC refers to the problem of overrepresentation of minority youth at different stages in the juvenile justice system (Snyder & Sickmund, 2008). The

Figure 9.2   Person offense case processing by race, 2002.

For person offense cases in 2002, juvenile courts ordered sanctions after adjudication at similar rates for white youth (369 of 1,000) and black youth (362 of 1,000 cases)

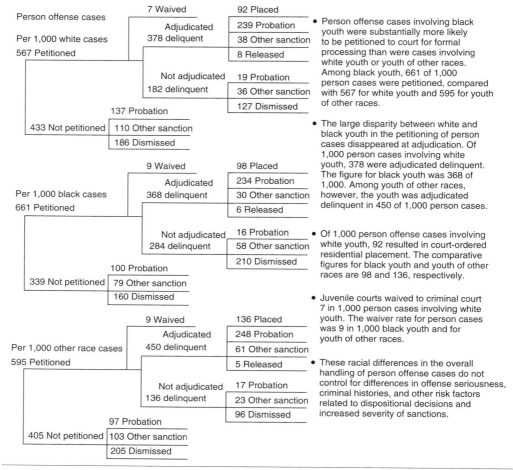

- Person offense cases involving black youth were substantially more likely to be petitioned to court for formal processing than were cases involving white youth or youth of other races. Among black youth, 661 of 1,000 person cases were petitioned, compared with 567 for white youth and 595 for youth of other races.

- The large disparity between white and black youth in the petitioning of person cases disappeared at adjudication. Of 1,000 person cases involving white youth, 378 were adjudicated delinquent. The figure for black youth was 368 of 1,000. Among youth of other races, however, the youth was adjudicated delinquent in 450 of 1,000 person cases.

- Of 1,000 person offense cases involving white youth, 92 resulted in court-ordered residential placement. The comparative figures for black youth and youth of other races are 98 and 136, respectively.

- Juvenile courts waived to criminal court 7 in 1,000 person cases involving white youth. The waiver rate for person cases was 9 in 1,000 black youth and for youth of other races.

- These racial differences in the overall handling of person offense cases do not control for differences in offense seriousness, criminal histories, and other risk factors related to dispositional decisions and increased severity of sanctions.

SOURCE: Snyder and Sickmund, *Juvenile Offenders and Victims: 2006 National Report* (2008) p. 179

concept of DMC has been expanded to include disproportionate contact as well as confinement. DMC exists in most states, at all decision points; is greater for African Americans than Hispanics; and is often greater in states with smaller minority populations (Leiber, 2002). DMC research has yielded varying results; in the earlier studies, some found no effects, whereas others found either direct and/or indirect race effects during intake (Leiber & Mack, 2003), detention (Wordes, Bynum, & Corley, 1994), probation (Bridges & Steen, 1998), confinement (Bridges, Conley, Engen, & Price-Spratlen, 1995), and in the child welfare system (Lau et al., 2003). More recent research has sought to identify effective strategies for reducing DMC (Cabaniss, Frabutt, Kendrick, & Arbuckle, 2007; Hsia, Bridges, & McHale, 2004; Short & Sharp, 2005) and examine DMC in the context of race and gender (Carr, Hudson, Hanks, & Hunt, 2008; Guevara, Herz, & Spohn, 2006; Leiber & Mack, 2003), at multiple stages instead of just one

(Guevara, Herz, & Spohn, 2006), and inclusive of Blacks, Latino/as, and American Indians (Rodriguez, 2007).

Why is the proportion of juvenile minorities (especially African Americans, Native Americans, and Hispanics) at several stages in juvenile justice greater than their proportion of the general population? There are several plausible explanations. First, juvenile justice has always been a racialized system (Ward, 2001). Feld (1999) cited race and the macrostructural transformation of cities as two societal-level factors important for understanding juvenile justice policies and practices. More specifically, Feld was referring to the racial segregation in urban areas and the deindustrialization of cities that has occurred during the last few decades. Feld agreed with Sampson and Wilson's (1995) structural/cultural approach, which takes into consideration a community's structural disorganization, cultural isolation, and the concentration effects of poverty that foster criminal involvement (see Chap. 3). Feld (1999) stated, "As African Americans became urban Americans and the public attributed increases in crime primarily to urban Black youth, race and crime intersected to produce more punitive juvenile justice policies" (p. 6). Another explanation is that within a racialized system, police and other justice practitioners hold stereotypical views of minority youth that often result in differential treatment. Finally, minority youth's disproportionate involvement in violent crimes warrants stiffer penalties.

Even if discrimination and more punitive policies directed at Black and other minority youth account for some of the overrepresentation, their involvement in delinquency in some communities is also a factor. For some youth, delinquency and crime are accepted behaviors. Some explanations for acceptance of and involvement in delinquency include alienation, reactions to discrimination, and "the street" factor. According to the colonial model (see Chap. 3), some youth turn to crime in response to their alienation (Tatum, 1994). Others are reacting to their perceptions of inequality, oppression, and discrimination (S. Johnson, 1996; Vega & Gil, 1998). There is often a belief that crime and delinquency are acceptable because of the lack of access to legitimate means of economic gain that many attribute to racism. Relatedly, the "street factor" often requires involvement in crime to prove one's toughness; violence is viewed favorably, often as a way of gaining status (E. Anderson, 1999), and is therefore more likely to be tolerated and perpetuated. S. Johnson (1996) coined the term *subcultural backlash* to refer to empathy for involvement in criminal behavior by members of an oppressed group.

The U.S. Congress formerly addressed DMC in 1988 by amending the Juvenile Justice and Delinquency Prevention (JJDP) Act of 1974. Any state receiving federal funding through formula grants was required to determine whether DMC existed and, if necessary, address the problem. In 1991, OJJDP began to assist states in addressing DMC issues (Devine et al., 1998). Congress made DMC a core requirement of the JJDP Act in 1992, which mandated compliance as a condition of funding. OJJDP developed an equation for determining the extent of minority overrepresentation, known as the "DMC index." The DMC index is a ratio of the percentage of the confined minority juvenile population divided by the percentage of minority juveniles in the general population. According to Leiber (2002), OJJDP has adopted a tentative approach to DMC in light of the politics of race, crime, racial bias, state resistance, and other practical considerations. Leiber stated, "OJJDP has adopted a judicious approach to implementation of DMC, which appears to follow the 'spirit' of the mandate and attempts to make inroads—'to get something done' rather than accomplishing 'nothing at all' " (p. 16). OJJDP has developed a "problem-solving process" and provides technical assistance to the states.

The DMC problem-solving process usually involves several activities, which include assigning organizational responsibility, identifying the extent to which DMC exists, assessing the reasons for DMC if it exists, developing an intervention plan, evaluating the effectiveness of strategies to address DMC, and monitoring DMC trends over time (Devine et al., 1998). Unfortunately, there are several challenges to reducing DMC that include lack of resources, inadequate information systems, development of intervention strategies, and transition from planning (state) to implementation (local).

In a follow-up study to the Pope and Feyerherm (1990) review of the research literature discussed at the beginning of this chapter, Pope, Lovell, and Hsia (2002) reviewed DMC studies published between 1989 and 2001 and concluded, "The majority of studies continue to provide evidence of race effects, direct or indirect, at certain stages of juvenile justice processing and in certain jurisdictions" (p. 6). Relatedly, Ward (2001) noted,

> Despite these efforts among researchers and policy makers, nearly a decade later the DMC problem is only more entrenched, and its significance amplified in the wake of a retributive turn in the philosophy and organization of juvenile justice administration. . . . The limited impact of research and policy in this area extends in part from conceptual and empirical problems which have characterized efforts to understand and address the race-effect in juvenile justice. (p. 2)

Ward (2001) identified the following problems:

- Conceptualization of race as a category of social meaning and organization is ahistorical and theoretically narrow. (p. 2)
- Race [is] measured as an individual case-level characteristic. (p. 2)
- There has been virtually no effort to determine how race may operate at broader cultural and structural levels to affect outcomes in the administration of juvenile justice. (p. 2)
- DMC mandates assume that the existence of disparity can be summarily determined by whether or not minority youth are overrepresented among juveniles confined in institutions. This approach overlooks disparities in the types of institutions to which juveniles are committed as well as differential practices in the handling of juveniles within a given institution. (p. 3)
- Policies derived from a proportional measure of racial inequality in juvenile justice may therefore oversimplify the processes and forms through which race operates to produce categories of difference and structures of opportunity in juvenile justice administration. (p. 3)
- There is little empirical evidence to support the assumption that a decision maker's racial background is related to their orientation. (p. 3)

- Inability of the DMC index score to capture important qualitative manifestations of inequality in juvenile confinement, which may be masked by low levels of disproportionality. (p. 28)
- The DMC Index measure severely homogenizes the practice and experience of confinement, ignoring important social-contextual factors which differentiate the meaning of institutionalization in juvenile justice, and its relevance to racial inequality. (p. 29)

The JJDP Act of 2002 modified previous DMC provisions that focused solely on confinement to include contact, emphasizing the overrepresentation of minority youth at all points in the justice process. Additionally, states that fail to address DMC may forfeit 20% of their federal funding (Office of Juvenile Justice and Delinquency Prevention, 2004). After nearly three decades of DMC research and reform efforts, "Although some progress is evident, there are no agreed-on strategies nor even complete accord on how best to explain the inequities" (Kempf-Leonard, 2007, p. 82). Some progress has been made in identifying effective strategies for reducing DMC that include data review and decision point mapping, cultural competency training, increasing community-based detention alternatives, reducing decision-making subjectivity, reducing barriers to family involvement, and developing state leadership to legislate system-level changes (Cabaniss et al., 2007).

THE FUTURE OF DMC

Future efforts to reduce DMC will require federal, state, and local officials to continue to recognize the importance of this problem for youth, families, communities, and juvenile justice. It requires a cadre of persons with a background in addressing DMC. One example is the W. Haywood Burns Institute, which assists communities in developing strategies to reduce the number of minority youth in detention (J. Bell, 2005). During this decade, several methodological issues important in DMC research have been identified. First, the limitations of the DMC Index are now more readily known. OJJDP has developed the DMC Relative Rate Index (RRI) to measure disparity at varying decision points (Snyder & Sickmund, 2008). The RRI is designed to "quantify the nature of the decisions at each decision point for each racial group and then compare these decisions" (Snyder & Sickmund, 2008, p. 190). The information can be presented in a RRI matrix that might be useful for understanding disparity within jurisdictions, although it does not explain disparities. Second, DMC research is more comprehensive, focusing on multistages instead of one stage (Guevara, Herz, & Spohn, 2006). Third, DMC researchers also recognize that defining one's racial/ethnic status is more complex than knowing one's skin color and that culture, class, and country of origin also are important (Kempf-Leonard, 2007). Fourth, DMC research has expanded to include both race and gender (Carr, Hudson, Hanks, & Hunt, 2008; Guevara, Herz, & Spohn, 2006; Leiber & Mack, 2003). Future research will more directly address the issues presented earlier, as well as examine the treatment of mixed race youth in juvenile justice.

As Pope and Feyerherm (1990) noted, the DMC problem and race effects will not end until structural and economic factors that contribute to youth involvement in delinquency are recognized and addressed. It also is important to prevent delinquency and address the problem of youth with favorable attitudes about delinquency.

MINORITY FEMALE DELINQUENCY

Girls' delinquency is not a new phenomenon, although it receives much more attention today than in the past. This is due, at least in part, to an increase in female arrests and court referrals for serious crimes. Arrests for two offenses, aggravated assaults and other assaults, increased more for both males and females than other violent crimes during the 1980s and early 1990s (Chesney-Lind & Shelden, 2004). It is important to keep in mind that violent crimes are a small portion of youth arrests, that they have declined in recent years, and that more males than females are arrested for these offenses. For example, in 2006, 40 females and 724 males under age 18 were arrested for murder (Federal Bureau of Investigation, 2006). According to Chesney-Lind and Shelden (2004), patterns of female juvenile arrests have not changed much during the past 25 years: "Females have typically been arrested for the following offenses: running away, larceny-theft, liquor law violations, curfew violations, disorderly conduct, other assaults, and the catch all category 'all other offenses' " (p. 18). Although patterns of female arrests have not changed much over time, the proportion of females referred to juvenile court for delinquency cases increased from 19% in 1989 to 24% in 1998. Those referred for offenses against the person increased from 20% in 1989 to 28% by 1998 (Puzzanchera et al., 2003a). Female person offense cases increased 22% between 1994 and 2004. In 2004, females accounted for 17% of delinquency, 20% of drug violation, 17% of property, 18% of public order, and 30% of person offense cases in juvenile courts (Stahl, 2008a). Females comprised 44% of petitioned status offense cases, primarily for running away (62%) (Stahl, 2008b).

Interestingly, although the extent of female delinquency is less than male delinquency, patterns of behavior and risk factors for both groups are quite similar. For example, poverty, family problems, academic failure, dropping out of school, and substance abuse are risk factors for both sexes (Chesney-Lind & Shelden, 2004; Deschenes & Esbensen, 1999; Howell, 2003). Unlike males, female delinquents experience more physical and sexual abuse; experience pregnancy and adolescent motherhood; have lower self-esteem; and have family and school relationship issues that require programs that focus on their unique risk factors as well as those they share with males (American Bar Association and National Bar Association, 2001; Hubbard & Pratt, 2002). For years, researchers have considered whether females are treated differently in the juvenile justice system (see Chesney-Lind & Shelden, 2004). Carr et al. (2008) used a gendered organization framework to examine official responses to delinquency in two single-sex minimum-security residential programs and a coed aftercare program in an Alabama county. They concluded that "the girls treatment facility confined more girls for less serious offenses than the boys' program" (p. 39).

Are minority female youth more delinquent than White females? This question is extremely difficult to answer because there is no specific data set that addresses this issue. Taylor, Biafora, Warheit, and Gil (1997) studied family factors and deviance in adolescent girls in Miami public schools. They found that 37.5% of the respondents engaged in serious delinquency, and Black respondents participated in more delinquent behaviors. They also reported that family factors influenced girls' delinquency differently in Hispanic, African American, and White non-Hispanics. In a study of minority girls, Walker-Barnes, Arrue, and Mason (1998) found that friends were perceived to be a greater risk factor for gang involvement than family, neighborhood, and self. In a study of girls' sexual development in the inner city, Dunlap, Golub, and Johnson (2003) found that many girls were compelled to have sex by the age of 13, which often resulted in various forms of independent sexual behavior, such as prostitution and teen pregnancy.

According to Chesney-Lind and Shelden (2004), both Black and White girls are more likely to be arrested for traditional female offenses, like running away and prostitution. They cite a study by Snyder and Sickmund (1999) that reported higher alcohol consumption, cigarette smoking, and marijuana use among White and Hispanic females and higher percentages of assaults and sex for Black females. The number of girls in gangs is believed to be increasing (see Highlight Box 9.1). Deschenes and Esbensen (1998) found that most participants in their study were not gang members. Among those who were, Hispanic and the "Other" category of females were more likely to be gang members than were African American and White females. Girls in gangs were more likely to engage in violent behavior and to be victims of violent crime. Chesney-Lind and Shelden (2004) concluded that, although there are differences between Black and White girls offending, they are not as pronounced as some might expect.

We do know that, according to the most recent data available, minority females comprise a disproportionate share of females in residential placement (American Bar Association and the National Bar Association, 2001; Office of Juvenile Justice and Delinquency Prevention, 2004; Snyder & Sickmund, 2008). A few recent studies have examined race and gender effects on juvenile justice decision making (Guevara, Herz, & Spohn, 2006; Leiber & Mack, 2002). Utilizing data from case files in two midwestern juvenile courts collected between 1990 and 1994, Guevara et al. (2006) reported that White females were more likely to receive out-of-home placements than non-White females. They concluded that non-White females were either treated no differently or more leniently than White females. In 2001, the American Bar Association and the National Bar Association issued a historic report on girls and juvenile justice. The report notes that, over the past two decades, the number of girls in the system has increased, and the system is not prepared to meet their special needs. Delinquent girls' experiences are believed to require different interventions than those usually found in community and institutional settings. There is now a focus on providing gender-specific programs that take into consideration the experiences and risks that girls face. Bloom, Owen, Deschenes, and Rosenbaum (2002) made several policy and program recommendations that focus on the needs of girls and young women for the state of California and policymakers in other states. Absent from their recommendations, and what is not known, is whether implementing gender-specific programs will take race and ethnic differences into account.

JUVENILES, RACE, AND THE DEATH PENALTY

A detailed discussion of the death penalty was presented in Chapter 7. Two international treaties expressly forbid the execution of juveniles: the International Covenant on Civil and Political Rights and the United Nations Convention on the Rights of Child. Today there are no juveniles under sentence of death as a result of the 2005 U.S. Supreme Court decision in *Roper v. Simmons*. In 2004, the Supreme Court agreed to hear *Roper v. Simmons* (2005) after the Supreme Court of Missouri held that the death penalty for juveniles violates the U.S. Constitution's prohibition against cruel and unusual punishment (*Simmons v. Roper*, 2003). *Roper v. Simmons* (2005) required that the Supreme Court decide whether a lower court can reject a standard set by the Supreme Court and whether the death penalty for a 17-year-old offender violated the Eighth Amendment (see Highlight Box 9.3). In a 5–4 decision, the Supreme Court ruled that executing juveniles is unconstitutional. As a result, the death sentence of Christopher Simmons and other juveniles with death sentences are now invalid. The close decision in this case is indicative of the lack of consensus on this issue.

Should juveniles be executed? In the past all juveniles who receive death sentences had not been tried as adults and usually have killed. Prior to the *Roper v. Simmons* (2005) decision, the U.S. Supreme Court had decided several juvenile death penalty cases, including *Eddings v. Oklahoma* (1982), *Thompson v. Oklahoma* (1988), *Stanford v. Kentucky* (1989), *Wilkins v. Missouri* (1989), and *Atkins v. Virginia* (2002). In the companion cases of *Stanford* and *Wilkins*, the Court ruled that the death penalty for juveniles who were at least 16 years old at the time of their offense did not violate the Eighth Amendment. At that time, the Court left it up to the states to decide whether they would execute juveniles at age 16 or 17 because there was no national consensus on the issue (see Highlight Box 9.3).

Throughout U.S. history, there have been 12 known cases where juveniles were executed for crimes committed before the age of 14. The first juvenile executed in the United States was Thomas Graunger, executed in Plymouth, Massachusetts, for committing the crime of bestiality when he was 16 years old, in 1642. The youngest known person to be executed in the United States was James Arcene, a Cherokee child who was 10 years old when he was hanged by the federal government in 1885 (Hakins, 2004). Since World War II, the youngest known person to be executed in the United States was George Stinney, a 14-year-old African American boy who was executed in South Carolina in 1944 for murdering two White girls, ages 8 and 11. Since 1976, 22 men have been executed for crimes committed as juveniles. Three were executed in Texas in 2002, and one was executed in Oklahoma in 2003.

Highlight Box 9.3

The U.S. Supreme Court, Juveniles, and the Death Penalty

Eddings v. Oklahoma, 455 U.S. 104 (1982) Judges are required to consider mitigating circumstances, including family life and emotional abuse, during the sentencing phase of capital cases.

Thompson v. Oklahoma, 487 U.S. 815 (1988) Sentencing juveniles to death who are age 15 and younger at the time of their crime is prohibited by the Eighth Amendment to the U.S. Constitution.

Stanford v. Kentucky and Wilkins v. Missouri, 492 U.S. 361 (1989) The Eighth Amendment to the United States Constitution does not prohibit the death penalty for crimes committed at ages 16 or 17.

Atkins v. Virginia, 536 U.S. 304 (2002) The U.S. Constitution prohibits executions of mentally retarded criminals.

In re Stanford, 537 U.S. 968 (2002) The U.S. Supreme Court decided not to take the case, over a strong dissent by Justice Stevens (joined by Justices Breyer, Ginsburg, and Souter). On December 8, 2003, the Kentucky governor granted clemency to Kevin Stanford, changing his death sentence to life in prison without parole.

Roper v. Simmons, 543 U.S. 551 (2005) The U.S. Supreme Court declared that the execution of juveniles violates the Eighth Amendment's prohibition against cruel and unusual punishment.

SOURCE: Death Penalty Information Center (2008).

Prior to the *Simmons* decision, all juvenile offenders under sentence of death were males, and most committed their offenses at the age of 17. Forty-eight offenders (65%) were minorities—1 American Indian, 2 Asians, 30 Blacks, and 15 Latinos—and 25 were Whites (Streib, 2004). According to the National Coalition to Abolish the Death Penalty (2004), two of three children sent to death row were people of color, and two of three people executed for crimes they committed as children have been African American. In the past, the federal government had imposed the death penalty against American Indian children for crimes they committed as young as 10 years old.

Even before the recent landmark decision, opposition to the execution of juveniles seemed to be increasing. Numerous organizations including the American Bar Association, the American Psychiatric Association, the Child Welfare League of America, and the Children's Defense Fund are opposed to the death penalty for juveniles (American Bar Association, 2004). Sixty-nine percent of Americans responding to a Gallup Poll survey opposed the execution of juveniles (Death Penalty Information Center, 2004; National Coalition to Abolish the Death Penalty, 2004), and several Supreme Court justices are opposed to the death penalty as well (Greenberger, 2002). Although the era of executing juveniles in the United States is over, reducing juvenile involvement in serious and heinous crimes requires prevention as well as confinement approaches. Delinquency prevention strategies are discussed next.

DELINQUENCY PREVENTION

The federal government has provided billions of dollars to state and local governments to assist them in their efforts to prevent crime (Sherman, 1997). Yet the amount spent on prevention programs pales in comparison to what is spent on punishment and placement of youth in secure confinement. During this decade, numerous delinquency programs have emerged, many of them targeting minority youth and funded by OJJDP. According to Taylor Greene and Penn (2005), the pendulum might be shifting from punishment to prevention in juvenile justice due, at least in part, to the high cost of "get tough" policies that do not necessarily work. Howell (2003) identified several juvenile justice programs and strategies that do not work, including drug abuse resistance education, zero-tolerance policies, shock incarceration, and incarceration of juveniles in adult prisons (see Highlight Box 9.4).

Although we know more about what does not work today, we still do not know enough about what does work. More than a decade ago, Sherman et al. (1997) conducted a study of factors that relate to juvenile crime and the effect of prevention programs on youth violence. The study was in response to a mandate from Congress to the attorney general to evaluate the effectiveness of crime prevention programs. Sherman (1997) concluded that, although some programs work, some do not, and others are promising, there is a need to identify what works in areas of concentrated poverty where homicides are rampant. It is also important to know what works in delinquency prevention in any community where there is a heightened problem of fear, violence, and victimization.

Highlight Box 9.4

Juvenile Justice Programs and Strategies That Clearly Do Not Work

- D.A.R.E.
- Zero-tolerance policies
- Curfew laws
- Punishment without treatment and rehabilitation services
- Removal of antisocial youth from their families, schools, neighborhoods, and communities for treatment
- Bed-driven treatment for mental health, delinquency, and drug abuse problems
- Large congregate, custodial corrections facilities
- Shock incarceration, "scared straight" programs, and boot camps
- Programs involving large groups of antisocial adolescents
- Piecemeal solutions
- Transfer of juveniles to the criminal justice system
- Incarceration of juveniles in adult prisons

SOURCE: Howell (2003, p. 147).

The Center for the Study and Prevention of Violence has been identifying and evaluating violence prevention efforts since 1996. It selects "Blueprint Model Programs" based on several criteria for effectiveness, including evidence of a deterrent effect with a strong research design, sustained effect, and replication elsewhere (Center for the Study and Prevention of Violence, 2008). "Promising Programs" are required to meet only the first criteria. Currently, there are 11 Blueprint Model Programs and 17 Promising Programs. The Blueprint Model Programs include Big Brothers Big Sisters of America (mentoring), Multisystemic Therapy, Olweus Bullying Prevention Program, Project Towards No Drug Abuse, and Life Skills Training (Center for the Study and Prevention of Violence, 2008).

Howell (2003) offered the "comprehensive strategy framework" for integrating the delinquency prevention and juvenile justice fields. The key components of this strategy are prevention, effective early intervention with at-risk children and families, and graduated sanctions for youth in the juvenile justice system (Howell, 2003). The comprehensive strategy is research based and flexible, and it has been implemented in several jurisdictions. It utilizes a developmental prevention approach that focuses on risk and protective factors in the family, school, peer group, community, and the individual.

Are any of the strategies mentioned earlier effective for minority youth who are locked into urban and rural environments plagued by poverty and disorder? Will they counter the pressures and temptations of drugs and delinquency? Do they take into consideration the pressures placed on some youth to engage in delinquent behaviors in order to survive in their neighborhoods? Do they acknowledge the relationship between violence and victimization? These and other questions about delinquency prevention programs for minority youth remain unanswered.

Taylor Greene and Penn (2005) noted that identifying programs that work for minority youth is difficult. First, just because programs like the Blueprint Models, for example, have proved effectiveness based on their deterrent effects, research design, effects, and replication does not necessarily mean that they work for minority youth. If we could determine that more than 50% of the study samples of effective programs are minority youth, then we could believe that they are effective with these youth. One Blueprint Model Program that probably works with minority youth is the Boys & Girls Clubs of America mentoring programs because this organization has traditionally serviced minorities in communities with concentrated effects of social disorganization and poverty.

Culturally specific programs such as the "rites of passage" programs that emerged decades ago have not been adequately evaluated. These programs provide training and rites-of-passage ceremonies for males and females and are used in both community and institutional settings (Taylor Greene & Penn, 2005). "Afrocentric prevention programs" emphasize traditional African values, including collectiveness and spirituality. Most culturally specific programs utilize a developmental approach and focus on encouraging youth to lead positive lifestyles. Many culturally relevant programs are locally funded and grassroots efforts that do not follow youth who complete their programs into adulthood. Although some specifically target delinquency prevention, others do not. Relatedly, there are few evaluations of culturally specific programs. Two studies of a community treatment program for juvenile felons in Cincinnati, Ohio, found little support for the Afrocentric program (see King, Holmes, Henderson, & Latessa, 2001; Wooldredge, Hartman, Latessa, & Holmes, 1994). Some support has been found for drug prevention programs that are culturally specific and focus on Afrocentric values (Belgrave, Townsend, Cherry, & Cunningham, 1997), Native American culture (Van Stelle, Allen, & Moberg, 1998), and Asian American culture (Zane, Aoki, Ho, Huang, & Jang, 1998).

Understanding the relationship between violence and victimization is also important to developing effective prevention programs for minority youth. Youth who witness violence in their homes and in their communities are more vulnerable to involvement in delinquency. For some, the behavior is viewed as acceptable, and for others, it is required to have what E. Anderson (1999) described as "juice" or status. McGee (2003) and McGee and Baker (2002) found that direct victimization as a measure of exposure to violence was a predictor of problem behaviors. Victimization was linked to both internalizing and externalizing behaviors. McGee (2003) suggested that violence prevention programs must take into consideration the specific needs of students exposed to danger and the importance of developing problem- and emotion-focused coping strategies. Relatedly, programs that emphasize resilience also are important.

Conclusion

This chapter traced the historical legacy of race in the juvenile justice system. It also presented information on the extent of juvenile crime and victimization and juveniles in the juvenile system. The issue of DMC was examined to shed light on the ongoing problem of minority youth in the system. When considered historically, it is not surprising that this problem exists, and it is likely to continue, if not worsen, unless society addresses the social conditions that foster delinquency and the racial attitudes that still taint the treatment of minority youth in the system. It was noted that Feld (1999) described social and demographic changes in the 1970s that produced macrostructural conditions that resulted in escalating youth violence in the 1980s. By the 1990s, the panic over violent juvenile crime adversely impacted urban Black males, who unfortunately were the perpetrators of the most violent of crimes: murder (Feld, 1999). The hybrid mix of rehabilitation and punishment that evolved in juvenile justice has had both positive and negative results. On the positive side, prevention has reemerged as a cost-effective approach to delinquency, and juvenile homicides and other violent crimes have decreased. However, the punitive era in juvenile justice has proved to be quite costly and not necessarily effective. More troubling is that DMC has only improved slightly during the past decade.

Three other contemporary issues were also presented in this chapter: female delinquency, juveniles and the death penalty, and delinquency prevention. It is easy to lose sight of the fact that most of the more than 72 million juveniles in our country are not delinquent regardless of their race, class, and gender. Parents, teachers, other individuals, and numerous community organizations are dedicated to the development of American youth. It makes much better sense to invest in education, health, and delinquency prevention than it does to invest in correctional institutions. This will require the identification of prevention programs that work, can be replicated, and are adequately funded. Research on the effectiveness of culturally relevant programs targeting minority youth also must receive more attention.

Discussion Questions

1. How relevant is the historical treatment of minority youth in juvenile justice to the DMC issue today?

2. Do you think programs for delinquent youth should vary by race/ethnicity?

3. Do you think females are delinquent for different reasons than males?

4. Is executing a juvenile more "cruel and unusual" than executing an adult?

5. Does juvenile victimization in the home or the community lead to delinquency or future criminality?

Internet Exercise

Use the National Youth Gang Center Web site, www.irr.com/nygc/, to summarize the most recent data on the extent of youth gangs in the United States. Include two demographic characteristics (race, ethnicity, gender, age) in your summary.

Internet Sites

For more information on juveniles in the juvenile justice system, visit:

Office of Juvenile Justice and Delinquency Prevention Web page and *Statistical Briefing Book*: http://ojjdp.ncjrs.org, http://ojjdp.ncjrs.org/ojstatbb/index.html.

Juvenile Justice Evaluation Center Disproportionate Minority Confinement: http://www.jrsa.org/jjec/programs/dmc/intervention.html

Center for the Study of Violence Prevention: http://www.colorado.edu/cspv/blueprints/index.html

National Center for Juvenile Justice State Profiles: http://www.ncjj.org/stateprofiles/

National Youth Gang Center: http://www.iir.com/nygc/

Conclusion

We set out to examine race and crime by incorporating a more historical approach, providing overviews of components of the criminal justice system (police, courts, and corrections), and examining numerous contemporary issues. Throughout the book, we reviewed historical and contemporary research, and we presented legal and sociohistorical factors that are important to understanding race and crime. In addition to the ongoing debate about the real meaning of the concepts of race and crime, our historical approach found that, unfortunately, little has changed over the past two centuries. Even so, a few things do stand out. First, as we noted in Chapter 1, concerns regarding race/ethnicity are not new. They have existed since the founding of America. Beginning with the overstated "criminal aggression" of Native Americans, and more recently with the "superpredator" concerns expressed concerning African American and Latino youth, there has always been some group or groups that have garnered special attention, although it was not always warranted. Over time, however, most White ethnic immigrants and some Latinos have been able to assimilate into "Whiteness" and escape such continued scrutiny. Although some Blacks have "passed" as Whites, they have never received as much attention. Changes in the classification of race and ethnicity in the 2000 Census of the Population shed light on citizens who classify themselves as "mixed" instead of the other traditional racial categories. How we define race has implications for the study of race and crime because it is not clear how mixed races fit into cultural notions about race. In the past, individuals known to be of mixed heritage or foreign-born and those with any characteristics that differed from the (White) nativists were not viewed the same as Whites. Today, the distinctions are less clear especially for those that appear to be White. Race and ethnicity continue to maintain either a cultural or subcultural context, often based on physical characteristics.

Second, in most instances, legislation has been the response to the perceived criminality of racial groups. Early on, legislation such as the slave codes, Black codes, the Indian Removal Act, and Chinese Exclusion Act was direct and clearly targeted specific groups. Later, legislation (i.e., death penalty and drug statutes) that purported to be race neutral continued to differentially impact racial minorities. Thus, the role of the law cannot be overstated. In fact, it is clear that, since the founding of America, there have been laws that were either explicitly or implicitly designed for certain racial/ethnic groups. More often than not, the "beneficiaries" of such laws have been racial minorities. Therefore, on the one hand, although racial and ethnic minorities have been overrepresented in many criminal justice indices because they have been recorded, legal and socioeconomic factors have undoubtedly contributed to this finding. On the other hand, we are left with no definitive evidence as to why elevated levels of crimes have persisted in minority communities. Our review of the seemingly endless theoretical perspectives on the subject pointed to a variety of potential explanations, but more research still needs to be done.

Our final thoughts relate to prospects for the future of both the study of race and crime and the plight of those in the criminal and juvenile justice systems. We are encouraged about the increased attention that race and crime has received in the discipline during the past decade. At this point in our history, it is difficult to know whether this trend will continue without thoroughly examining indicators such as integration in scholarly research and textbooks and the number of minority and majority scholars in the field that are involved in such research. Despite the increased attention, there is still so much we do not know. The study of race and crime requires more attention to historical, methodological, and theoretical issues.

Historical research is necessary to better understand the present. There continue to be gaps in information about race and crime for certain groups and during certain periods. Future historical research must include a more comprehensive analysis of the role of politics, and the ideologies of politicians and political parties, in order to better understand the political context of race and crime that goes beyond facts about legislation, case law, and key players in the administration of justice during different time periods. Researchers have examined race, crime, and politics, although more information is needed to understand how racist sentiments are connected to anticrime sentiments in political rhetoric and policy, often making crime a symbol for racism (see Ehrlichman, 1982; Marion & Oliver, 2006; Tonry, 1995).

In the future, theoretical research must recognize diversity not only within racial categories, but also within individuals. Explanations of race and crime should continue to contextualize race and ethnic disparities, as well as recognize that, although there is no general theory of race and crime, some theoretical concepts are more generalizable than others (such as class and labeling) and that a more global perspective can be instructive (Gabbidon, 2007a). Theories of race and crime that are based on data collection and statistics must acknowledge the limitations of data and strive to improve them. A discussion about whether crime statistics should continue to be reported and recorded by race is essential. In light of the historical legacy of segregation and discrimination in justice agencies, it seems that recording racial categories is important, although some consistency from one source to another is required. Recent changes in the reporting of juvenile justice data by race/ethnicity make it easier to understand DMC at all stages.

Overall, we are not optimistic about the overrepresentation of racial minorities (primarily African Americans, Latinos, and, more recently, Native Americans) in the arrest, delinquency, victimization, and correctional figures. Unfortunately, more of the same also equates to more minority youth taken into custody and more minorities being placed on death row, being wrongly accused and incarcerated (and in the worst case scenarios executed), and, as a consequence of such trends, in the coming years, more minorities may be disenfranchised. Yet, there has been some progress, including: the move toward the reduction/elimination of the crack/cocaine sentencing disparity, more attention being paid to wrongful convictions, abolition of the death penalty for juveniles, recognition of the challenge of prisoner reentry, efforts to reduce DMC, efforts to reduce racial profiling in traffic stops, reductions in juvenile drug use, and recognition of the importance of prevention. A shift to evidence-based research is also promising. At the same time, there are still challenges, including the possibility of profiling Latino/as in efforts to curb illegal immigration, continued injustice in the treatment of juveniles where racial issues might be more hidden, and failure to seek alternatives to incarceration that will be less harmful to individuals, families, and communities.

If our dire forecast for the future is to change, a few things need to happen as we move into the next decade. First, racial minorities must continue to look beyond governmental institutions to provide the impetus for change. By now it should be clear that the government offers no panacea and that such expectations are unrealistic. Further, drawing on our historical observations of racial and ethnic groups, we have noted that educational achievement holds a central place in changing the course of most minority involvement in crime. Second, for those legislators interested in tackling race and crime, they might want to heed the advice of Herbert Packer when seeking to determine what should be against the law. In his classic work, *The Limits of the Criminal Sanction* (1968), which is known for its delineation of the crime control and the due process models of justice, Packer also lays out the following criteria that serve as guidelines for determining what should be against the law:

(1) the conduct is prominent in most people's view of socially threatening behavior, and is not condoned by any significant segment of society;

(2) subjecting it to the criminal sanction is not inconsistent with the [societal] goals of punishment;

(3) suppressing it will not inhibit socially desirable conduct;

(4) it may be dealt with through even-handed and nondiscriminatory enforcement;

(5) controlling [the conduct] through the criminal process will not expose that process to severe qualitative or quantitative strain; and

(6) no reasonable alternatives to the criminal sanction exist for dealing with the behavior. (p. 298)

If legislators followed most of Packer's guidelines, it would go a long way toward ending the centuries-old trend of criminal justice legislation unfairly impacting racial minorities.

Third, legislators need to be open to more thoughtful approaches to investigating and handling certain offenses. For example, since September 11, 2001, police agencies across the country are devoting more attention to preventing terrorism. Many American law enforcement agencies have adopted the British model for preventing terrorism that advocates increased public surveillance (Klinger & Grossman, 2002).

Although most Americans are willing to give up some rights in the interest of protecting the country from terrorism, the dangers of the abuse of powers are of concern, particularly to American minorities. The focus on homeland security during the current era must not diminish the importance of police and the citizenry working together to solve problems of crime and disorder. Relatedly, manpower and financial resources that have been diverted to fighting terrorism and homeland security could have a negative impact on both street crimes and progress made in the last decade in the extent of reported crime (see Lehrer, 2007).

Finally, there must be acknowledgment of the failure of both research and policy to ameliorate the historical and socioeconomic conditions that have contributed to the nexus of race, crime, and the administration of justice. Legislators must understand that the development of a nation's youth is an important component of national security. It is more fiscally sound to invest in the welfare of children, youth, and young adults than to ignore the adverse socioeconomic, biological, and psychological conditions that often lead to criminality. The cost of incarceration that does not work (for most offenders) is much more than the cost of investing in developmental prevention. If there is any hope for reducing the disproportionate involvement in crime by minorities in this century, legislation that adequately funds prevention research and programs is critical.

Appendix

Table A.1 Arrests by Race, 1933–2006

Year	Total	White	Negro	Mexican	Indian	Chinese	Japanese	Other	Unknown
1933	312,360	236,125	76,224	10,418	1,233	1,040	176	7,813	11
1934	343,582	247,753	80,618	11,820	1,699	971	164	2,190	
1935	392,251	284,236	91,171	16,456	2,592	1,057	243	2,312	
1936	461,589	333,922	104,998	16,897	2,787	1,120	228	2,291	
1937	520,153	383,306	113,524	16,028	2,651	837	238	2,080	
1938	554,376	411,679	120,863	17,638	3,029	942	330	1,822	
1939	576,920	427,158	126,001	23,184	3,647	1,032	440	2,269	
1940	609,013	439,695	138,746	23,127	3,624	775	570	2,078	
1941	630,568	452,275	148,119	21,559	4,688	731	431	1,493	
1942	585,988	431,908	146,737	18,384	5,438	499	102	1,132	
1943	490,764	358,254	125,339	17,817	6,084	554	135	1,275	
1944	488,979	351,609	129,322	20,062	5,820	544	81	1,521	
1945	543,852	390,315	145,571	19,793	5,700	432	140	1,776	
1946	645,431	78,211	59,172	20,330					
1947	734,041	536,695	187,781		6,040	423	154	2,948	
1948	759,698	557,125	191,921		6,846	653	309	2,844	
1949	792,029	582,447	198,596		6,881	743	302	3,060	
1950	793,671	576,422	205,576		7,334	842	285	3,212	
1951	831,288	598,722	218,823		8,953	862	233	3,695	
1952	1,110,675	808,357	281,442		17,908	223	119	2,626	
1953	1,791,160	1,270,466	481,095		32,084	407	144	6,964	
1954	1,688,555	1,206,110	439,762		33,212	363	194	8,914	
1955	1,861,764	1,310,481	510,228		38,032	256	253	4,514	

Year	Total	White	Negro	Indian	Chinese	Japanese	Other
1956	2,070,794	1,390,944	634,022	40,554	236	114	4,924
1957	2,068,677	1,405,967	616,028	37,715	267	273	8,427
1958	2,340,004	1,583,070	696,209	43,126	1,252	296	16,051
1959	2,612,704	1,742,399	788,799	56,555	1,486	3,000	20,465
1960	3,498,926	2,320,635	1,064,814	71,662	2,066	5,570	34,179
1961	3,608,317	2,424,631	1,073,491	79,716	1,725	3,428	25,326
1962	3,923,465	2,602,011	1,195,796	86,027	2,003	3,116	34,512
1963	4,259,463	2,943,148	1,186,870	101,253	1,817	2,640	23,735
1964	4,381,419	3,053,818	1,194,377	99,195	1,097	2,488	30,444
1965	4,743,123	3,235,386	1,347,994	113,398	1,293	2,970	42,082
1966	4,797,741	3,329,475	1,315,796	108,489	1,482	3,626	38,873
1967	5,265,302	3,630,787	1,462,556	121,398	1,726	3,490	45,345
1968	5,349,450	3,700,012	1,471,730	119,265	1,666	4,186	52,591
1969	5,576,705	3,842,895	1,558,740	115,645	1,426	3,613	54,386
1970	6,257,104	4,373,157	1,688,389	130,981	2,582	3,882	58,113
1971	6,626,085	4,623,891	1,791,474	138,677	3,274	3,832	64,937
1972	6,706,950	4,664,220	1,847,566	130,375	4,234	1,821	58,734
1973	6,248,286	4,458,567	1,636,237	110,433	3,049	2,134	37,866
1974	5,853,060	4,112,443	1,561,781	86,725	4,033	4,827	83,251
1975	7,671,230	5,538,890	1,935,422	115,554	4,629	5,817	70,918
1976	7,383,960	5,336,889	1,870,206	104,797	2,998	3,925	65,145
1977	8,972,109	6,428,993	2,308,429	108,520	5,999	5,240	114,928
1978	9,687,995	6,792,934	2,562,454	105,881	8,212	5,802	212,712
1979	9,467,502	6,849,179	2,342,664	102,392	6,089	7,668	159,510

(Continued)

(Continued)

Year	Total	White	Black	American Indian/ Alaskan Native	Asian/ Pacific Islander
1980	9,683,672	7,145,763	2,375,204	109,480	53,225
1981	10,264,187	7,482,012	2,619,463	104,261	58,451
1982	10,000,078	7,070,374	2,777,145	91,845	60,714
1983	10,247,859	7,291,129	2,796,038	93,736	66,956
1984	8,890,662	6,528,686	2,216,299	89,873	55,804
1985	10,239,478	7,337,681	2,721,144	111,459	69,194
1986	10,335,942	7,370,777	2,788,934	106,926	69,305
1987	10,750,309	7,386,639	3,168,129	116,916	78,625
1988	10,067,447	6,903,070	2,977,266	105,050	82,061
1989	11,224,528	7,559,138	3,459,177	113,777	92,436
1990	11,151,368	7,712,339	3,224,060	122,586	92,383
1991	10,516,399	7,251,862	3,049,299	115,345	99,893
1992	11,876,204	8,030,171	3,598,259	130,770	117,004
1993	11,741,751	7,855,287	3,647,174	126,017	113,273
1994	11,846,833	7,894,414	3,705,713	126,503	120,203

SOURCE: FBI, *Crime in the United States, 1933–2006.*

Table A.2 Total Arrests and Arrests for Murder by Race 1950–2006

Year	Total	White	Black	American Indian/ Alaskan Native	Asian/ Pacific Islander
1995	11,386,627	7,607,522	3,523,409	129,843	125,853
1996	11,072,832	7,404,170	3,400,338	139,290	129,034
1997	10,516,707	7,061,803	3,201,014	132,734	121,156
1998	10,225,920	6,957,337	3,033,710	122,879	111,994
1999	9,100,050	6,283,294	2,600,510	113,705	102,541
2000	9,068,977	6,324,006	2,528,368	112,192	104,411
2001	9,306,587	6,465,887	2,617,669	119,281	103,750
2002	9,797,385	6,923,390	2,633,632	130,636	109,727
2003	9,529,469	6,723,093	2,570,770	125,438	110,168
2004	9,940,671	7,042,510	2,660,770	130,545	106,846
2005	10,189,691	7,117,040	2,830,778	135,877	105,996
2006	10,437,620	7,270,214	2,924,724	130,589	112,093

SOURCE: FBI, *Crime in the United States, 1950–2006.*

Table A.3 Reported Murders and Arrests by Race, 1950–2006

Year	Total Arrests	Murders	White	Negro	Indian	Chinese	Japanese	Other
1950	793,671	6,336	3,372	2,889	36	2	2	35
1951	831,288	6,522	3,407	3,029	46	7	4	29
1952	1,110,675	1,288	444	829	—	—	—	15
1953	1,791,160	2,199	883	1,306	5	—	—	5
1954	1,688,555	1,706	630	1,064	9	1	—	2
1955	1,861,764	2,073	807	1,241	17	2	1	5
1956	2,070,794	2,028	683	1,336	2	4	1	2
1957	2,068,677	2,007	761	1,225	4	1	1	15
1958	2,340,004	2,303	840	1,427	8	2	2	24
1959	2,612,704	2,610	975	1,596	13	2	1	23
1960	3,498,926	4,120	1,536	2,511	29	4	4	36
1961	3,608,317	3,694	1,493	2,154	18	4	2	23
1962	3,923,465	4,404	1,672	2,665	24	1	1	41
1963	4,259,463	5,338	2,288	2,948	53	1	3	45
1964	4,381,419	5,442	2,310	3,041	44	1	3	43
1965	4,743,123	6,509	2,675	3,704	46	3	5	76
1966	4,797,741	7,114	2,911	4,068	66	1	4	64
1967	5,265,302	8,218	3,200	4,883	49	2	9	75
1968	5,349,450	9,458	3,536	5,699	93	2	2	126
1969	5,576,705	10,412	3,743	6,444	63	3	5	154
1970	6,257,104	11,847	4,503	7,097	76	6	—	165
1971	6,626,085	13,302	4,716	8,276	88	7	4	211
1972	6,706,950	13,806	5,145	8,347	110	14	4	186

Year	Total Arrests	Murders	White	Negro	Indian	Chinese	Japanese	Other
1973	6,248,286	12,913	5,236	7,478	118	18	3	60
1974	5,853,060	12,464	4,897	7,122	104	25	6	310
1975	7,671,230	15,173	6,581	8,257	143	18	11	163
1976	7,383,960	12,875	5,792	6,886	100	6	4	87
1977	8,972,109	17,122	7,866	8,731	155	16	7	347
1978	9,687,995	18,698	8,703	9,243	153	16	10	573
1979	9,467,502	18,238	9,010	8,693	155	16	8	356

Year	Total Arrests	Murders	White	Black	American Indian/ Alaskan Native	Asian or Pacific Islander
1980	7,663,682	16,987	8,533	8,199	136	119
1981	10,264,187	20,404	10,129	9,998	133	144
1982	10,000,078	18,475	9,008	9,174	141	152
1983	10,247,859	18,027	8,836	8,935	114	142
1984	8,890,662	13,656	7,339	6,133	91	93
1985	10,239,478	15,612	7,817	7,562	109	124
1986	10,355,942	15,953	8,028	7,659	146	120
1987	10,750,309	16,678	7,642	8,746	131	159
1988	10,067,447	16,090	7,243	8,603	99	145
1989	11,224,528	17,944	7,567	10,118	120	139
1990	11,151,368	18,190	7,942	9,952	132	164
1991	10,516,399	18,096	7,861	9,924	143	168
1992	11,876,204	19,463	8,466	10,728	107	162

(Continued)

(Continued)

Year	Total Arrests	Murders	White	Black	American Indian/ Alaskan Native	Asian or Pacific Islander
1993	11,741,751	20,243	8,243	11,656	131	213
1994	11,846,833	18,475	7,705	10,420	126	224
1995	11,386,627	16,691	7,245	9,074	134	238
1996	11,072,832	14,439	6,176	7,928	119	216
1997	10,516,707	12,759	5,345	7,194	94	126
1998	10,225,920	12,318	5,478	6,580	139	121
1999	9,100,050	9,716	4,460	5,029	105	122
2000	9,068,977	8,683	4,231	4,238	87	127
2001	9,306,587	9,416	4,561	4,585	122	148
2002	9,797,385	10,099	4,814	5,047	115	123
2003	9,529,469	9,063	4,454	4,395	101	113
2004	9,940,671	9,983	4,935	4,760	122	166
2005	10,189,691	10,083	4,955	4,898	109	121
2006	10,437,620	9,801	4,595	4,990	110	106

References

Abbott, E. (Ed.). (1931). *Crime and the foreign born* (National Commission on Law Observance and Enforcement, Report No. 10). Washington, DC: Government Printing Office.

ABC News Poll. (March 7, 2002). Retrieved March 17, 2004, from the Polling the Nations database at http://www.orspub.com/

ABC/*Washington Post* Poll. (2003). Retrieved March 2, 2003, from Polling the Nations database at http://www.orspub.com/

Abril, J. C. (2007a). Cultural conflict and crime: Violation of Native American Indian cultural values. *International Journal of Criminal Justice Sciences, 2,* 44–62.

Abril, J. C. (2007b). Perceptions of crime seriousness, cultural values, and collective efficacy among Native American Indians and non-Indians within the same reservation community. *Applied Psychology in Criminal Justice, 3,* 172–196.

Abu-Jamal, M. (1995). *Live from death row.* New York: Addison-Wesley.

Adams, K. (1999). What we know about police use of force. In *Use of force by police: Overview of national and local data* (NCJ 176330, pp. 1–14). Washington, DC: U.S. Department of Justice, Office of Justice Programs.

Adkins, S. (2002, December 27). Judges' characteristics are a factor. *York Daily Record.* Retrieved July 29, 2003, from http://ydr.com/story/justice/4962/printer/

Agnew, R. (1992). Foundation for a general strain theory of crime and delinquency. *Criminology, 30,* 47–87.

Agozino, B. (2003). *Counter-colonial criminology: A critique of imperialist reason.* London: Pluto Press.

Akers, R. L. (2000). *Criminological theories: Introduction, evaluation, and application.* Los Angeles: Roxbury Publishing.

Allen, H. E., Simonsen, C. E., & Latessa, E. J. (2004). *Corrections in America: An introduction.* Upper Saddle River, NJ: Prentice Hall.

Allen, T. W. (1994). *The invention of the White race: Vol. 1. Racial oppression and social control.* New York: Verso.

American Bar Association. (2004). *Christopher Simmons.* Retrieved June 15, 2004, from http://www .abanet.org/crimjust/juvjus/simmons.html

American Bar Association and the National Bar Association. (2001). *Justice by gender.* Washington, DC: Author.

American Civil Liberties Union. (1999). *Driving while Black: Racial profiling on our nation's highways.* New York: Author.

Anderson, C. (1994). *Black labor, White wealth.* Edgewood, MD: Duncan & Duncan.

Anderson, E. A. (1994, May). The code of the streets. *Atlantic Monthly,* pp. 81–94.

Anderson, E. A. (1999). *The code of the streets: Decency, violence, and the moral life of the inner city.* New York: Norton.

Anderson, J. (2001). What to do about "much ado" about drug courts? *International Journal of Drug Policy, 12,* 469–475.

Anderson, P. R., & Newman, D. J. (1998). *Introduction to criminal justice* (6th ed.). New York: McGraw-Hill.

Aptheker, B. (1971). The social functions of the prisons in the United States. In A. Y. Davis (Ed.), *If they come in the morning* (pp. 39–48). New Rochelle, NY: Third Press.

Aptheker, H. (1993). *American Negro slave revolts.* New York: International Publishers. (Original work published 1943)

Argersinger v. Hamlin, 407 U.S. 25 (1972).

Arrestee Drug Abuse Monitoring Program. (2002). *Preliminary data on drug use and related matters among adult arrestees and juvenile detainees.* Washington, DC: National Institute of Justice. Retrieved June 15, 2004, from http://www.ojp.usdoj.gov/nij/adam/ADAMPrelim2002.pdf

Arthur, J. A. (1998). Proximate correlates of Blacks, support for capital punishment. *Journal of Crime and Justice, 21,* 159–172.

Atkins v. Virginia, 536 U.S. 304 (2002).

Austin, J., & Irwin, J. (2001). *It's about time: America's imprisonment binge* (3rd ed.). Belmont, CA: Wadsworth.

Austin, R. (1983). The colonial model, subcultural theory, and intragroup violence. *Journal of Criminal Justice, 11,* 93–104.

Baadsager, P., Sims, B., Baer, J., & Chambliss, W. J. (2000). The overrepresentation of minorities in America's imprisonment binge. *Corrections Management Quarterly, 4,* 1–7.

Bachman, R. (1991). An analysis of American Indian homicide: A test of social disorganization and economic deprivation at the reservation county level. *Journal of Research in Crime and Delinquency, 28,* 456–471.

Baker, D. V. (2003). The racist application of capital punishment to African Americans. In M. Free (Ed.), *Racial issues in criminal justice: The case of African Americans* (pp. 177–201). Westport, CT: Greenwood.

Baker, D. V. (2007). American Indian executions in historical context. *Criminal Justice Studies, 20,* 315–373.

Baker, D. V. (2008). Black female executions in historical context. *Criminal Justice Review, 33,* 64–88.

Baldus, D. C., & Woodworth, G. (2003). Race discrimination and the death penalty: An empirical and legal overview. In J. Acker, R. Bohm, & C. S. Lanier (Eds.), *America's experiment with capital punishment* (2nd ed., pp. 501–551). Durham, NC: Carolina Academic Press.

Baldus, D. C., Woodworth, G., & Pulaski, C. A. (1990). *Equal justice and the death penalty: A legal and empirical analysis.* Boston: Northeastern University Press.

Banner, S. (2002). *The death penalty: An American history.* Cambridge, MA: Harvard University Press.

Barak, G., Flavin, J., & Leighton, P. (2006). *Class, race, gender, and crime* (2nd ed.). New York: Rowman & Littlefield.

Barkan, S. F., & Cohn, S. F. (1994). Racial prejudice and support for the death penalty by Whites. *Journal of Research in Crime and Delinquency, 31,* 202–209.

Barker, T., & Carter, D. (1991). *Police deviance* (2nd ed.). Cincinnati, OH: Anderson.

Barlow, D. E., & Barlow, M. H. (2000). *Police in a multicultural society.* Prospect Heights, IL: Waveland Press.

Barlow, D. E., & Barlow, M. H. (2002). Racial profiling: A survey of African American police officers. *Police Quarterly, 5,* 334–358.

Barro, R. J. (1999, September 27). Does abortion lower the crime rate? *Business Week,* p. 30.

Bastian, L. (1990). *Hispanic victims.* Washington, DC: U.S. Department of Justice, Bureau of Justice Statistics.

Bastian, L. (1992). *Criminal victimization in the United States: 1973–1990 trends.* Washington, DC: U.S. Department of Justice, Office of Justice Programs.

Batson v. Kentucky, 476 U.S. 79, 108 (1986).

Battle v. Anderson, 376 F. Supp. 402 (1974).

Baumer, E., Horney, J., Felson, R., & Lauritsen, J. (2003). Neighborhood disadvantage and the nature of violence. *Criminology, 41,* 39–71.

Baumer, E., Messner, S., & Rosenfeld, R. (2003). Explaining spatial variation in support for capital punishment: A multilevel analysis. *American Journal of Sociology, 108,* 844–875.

Bavon, A. (2001). The effect of the Tarrant County drug court project on recidivism. *Evaluation and Program Planning, 24,* 13–22.

Beard, A. (2007). *Judge finds Duke prosecutor in contempt.* Retrieved March 8, 2008, from www .abcnews.go.com/sports.

Beaumont, G., & Tocqueville, A. (1964). *On the penitentiary system in the United States and its application in France.* Carbondale: Southern Illinois University Press. (Original work published 1833)

Beck, E. M., & Tolnay, S. E. (1995). Violence toward African Americans in the era of the White lynch mob. In D. F. Hawkins (Ed.), *Ethnicity, race, and crime: Perspectives across time and place* (pp. 121–144). Albany, NY: State University of New York Press.

Becker, S. (2004). Assessing the use of profiling in searches by law enforcement personnel. *Journal of Criminal Justice, 32,* 183–193.

Beckett, K., & Sasson, T. (2000). *The politics of injustice: Crime and punishment in America.* Thousand Oaks, CA: Pine Forge Press.

Bedau, H. A., & Radelet, M. L. (1987). Miscarriage of justice in potentially capital cases. *Stanford Law Review, 40,* 21–179.

Behrens, A., Uggen, C., & Manza, J. (2003). Ballot manipulation and the "menace of Negro domination": Racial threat and felon disenfranchisement in the United States, 1850–2002. *American Journal of Sociology, 109,* 559–605.

Belenko, S., Sprott, J. B., & Petersen, C. (2004). Drug and alcohol involvement among minority and female juvenile offenders: Treatment and policy issues. *Criminal Justice Policy Review, 15,* 3–36.

Belgrave, F. Z., Townsend, T. G., Cherry, V. R., & Cunningham, D. M. (1997). The influence of an Africentric worldview and demographic variables on drug knowledge, attitudes, and use among African American youth. *Journal of Community Psychology, 25,* 421–433.

Bell, D. (1960). *The end of ideology.* Glencoe, IL: Free Press.

Bell, J. (2005). Solvable problem: Reducing the disproportionality of youths of color in juvenile detention facilities. *Corrections Today, 67,* 80–83.

Benekos, P. J., & Merlo, A. V. (2008). Juvenile justice: The legacy of punitive policy. *Youth Violence and Juvenile Justice, 6,* 28–46.

Bernard, T. J. (1992). *The cycle of juvenile justice.* New York: Oxford University Press.

Blakely v. Washington, 542 U.S. 296 (2004).

Blalock, H. M. (1967). *Toward a theory of minority group relations.* New York: Wiley.

Blauner, R. (1969). Internal colonialism and ghetto revolt. *Social Problems, 16,* 393–408.

Blauner, R. (1972). *Racial oppression in America.* New York: Harper & Row.

Bloom, B., Owen, B., Deschenes, E. P., & Rosenbaum, J. (2002). Moving toward justice for female juvenile offenders in the new millennium. *Journal of Contemporary Criminal Justice, 18,* 37–56.

Blumer, M. (1984). *The Chicago school of sociology: Institutionalization, diversity, and the rise of sociological research.* Chicago: University of Chicago Press.

Blumstein, A. (1982). On the racial disproportionality of the United States' prison populations. *Journal of Criminal Law & Criminology, 73,* 1259–1281.

Blumstein, A. (1993). Racial disproportionality of U.S. prison populations revisited. *University of Colorado Law Review, 63*. Retrieved August 25, 2002, from Lexis-Nexis database.

Blumstein, A. (1996). *Youth violence, guns, and the illicit drug markets* (Research preview). Washington, DC: National Institute of Justice.

Blumstein, A. (2002). Prisons: A policy challenge. In J. Q. Wilson & J. Petersilia (Eds.), *Crime: Public policies for crime control* (pp. 451–483). Oakland, CA: Institute for Contemporary Studies.

Blumstein, A., & Wallman, J. (Eds.). (2005). *The crime drop in America* (2nd ed.). New York: Cambridge University Press.

Bobo, L. D., & Johnson, D. (2004). A taste for punishment: Black and White Americans' views on the death penalty and the war on drugs. *Du Bois Review, 1,* 151–180.

Bohm, R. M. (1991). American death penalty opinion, 1936–1986: A critical examination of Gallup polls. In R. M. Bohm (Ed.), *The death penalty in America: Current research* (pp. 113–145). Cincinnati, OH: Anderson.

Bohm, R. M. (1999). *Deathquest: An introduction to the theory and practice of capital punishment.* Cincinnati, OH: Anderson.

Bohm, R. M. (2001). *A primer on crime and delinquency theory* (2nd ed.). Belmont, CA: Wadsworth/ Thomson Learning.

Bohm, R. M. (2007). *Death Quest III: An introduction to the theory and practice of capital punishment in the United States, 3rd edition.* Cincinnati, OH: Anderson Publishing Company.

Bohm, R. M., Clark, L., & Aveni, A. (1991). Knowledge and death penalty opinion: A test of the Marshall hypotheses. *Journal of Research in Crime and Delinquency, 28,* 360–387.

Bond-Maupin, L. (1998). Self-determination? Juvenile justice in one American Indian community. *Journal of Contemporary Criminal Justice, 14,* 26–41.

Bonger, W. A. (1943). *Race and crime.* New York: Columbia University Press.

Borchard, E. M. (1932). *Convicting the innocent: Sixty-five actual errors of criminal justice.* Garden City, NY: Doubleday.

Bosworth, M., & Flavin, L. (Eds.). (2007). *Race, gender, & punishment: From colonialism to the war on terror.* New Brunswick, NJ: Rutgers University Press.

Bowers, W. J., Sandys, M. R., & Steiner, B. D. (1998). Juror predispositions: Guilt-trial experience, and premature decision making. *Cornell Law Review, 83,* 1476–1556.

Bowers, W. J., Steiner, B. D., & Sandys, M. R. (2001). Death sentencing in Black and White: An empirical analysis of the role of jurors' race and jury composition. *The University of Pennsylvania Journal of Constitutional Law, 3,* 171–274.

Bradley, M. (1978). *The iceman inheritance: Prehistoric sources of Western man's racism, sexism, and aggression.* New York: Kayode.

Braga, A. A., Weisburd, D. L., Waring, E. J., Mazerolle, L. G., Spelman, W., & Gajewski, F. (1999). Problem-oriented policing in violent crime places: A randomized controlled experiment. *Criminology, 37,* 541–580.

Braman, D. (2002). Families and incarceration. In M. Mauer & M. Chesney-Lind (Eds.), *Invisible punishment: The collateral consequences of mass imprisonment* (pp. 117–135). New York: New Press.

Brandon, R., & Davies, C. (1973). *Wrongful imprisonment: Mistaken convictions and their consequences.* London: Archon Books.

Brewer, T. W. (2004). Race and jurors' receptivity to mitigation in capital cases: The effect of jurors,' defendants,' and victims' race in combination. *Law and Human Behavior, 28,* 529–545.

Brezina, T., Agnew, R., Cullen, F. T., & Wright, J. P. (2004). The code of the street: A quantitative assessment of Elijah Anderson's subculture of violence thesis and its contribution to youth violence research. *Youth Violence and Juvenile Justice, 2,* 303–328.

Bridges, G. S., Conley, D., Engen, R. L., & Price-Spratlen, T. (1995). Social contexts of punishment: Effects of crime and community social structure on racial disparities in the administration of juvenile justice. In K. Kempf Leonard, C. Pope, & W. Feyerherm (Eds.), *Minorities in Juvenile Justice,* (pp. 128–152). Thousand Oaks, CA: Sage Publications.

Bridges, G., & Steen, S. (1998). Racial disparities in official assessments of juvenile offenders: Attributional stereotypes as mediating mechanisms. *American Sociological Review, 63,* 554–570.

Brodkin, K. (1999). *How Jews became White folks and what that says about race in America.* New Brunswick, NJ: Rutgers University Press.

Brodkin Sacks, K. (1997). How did Jews become White folks? In R. Delgado & J. Stefancic (Eds.), *Critical White studies: Looking behind the mirror* (pp. 395–401). Philadelphia: Temple University Press.

Brown v. Board of Education of Topeka et al., 347 US 483 (1954).

Brown, M. C., & Warner, B. D. (1992). Immigrants, urban politics and policing in 1900. *American Sociological Review, 57,* 293–305.

Brownstein, H. H. (1996). *The rise and fall of a violent crime wave: Crack cocaine and the social construction of a crime problem.* Guilderland, NY: Harrow & Heston.

Brunson, R. K. (2007). "Police don't like Black people": African-American young men's accumulated police experiences. *Criminology & Public Policy, 6,* 71–102.

Brunson, R. K., & Stewart, E. A. (2006). Young African American women, the street code, and violence: An exploratory analysis. *Journal of Crime & Justice, 29,* 1–19.

Bureau of Indian Affairs. (1974). *The American Indians: Answers to 101 questions.* Washington, DC: U.S. Department of the Interior.

Bureau of Indian Affairs. (1976). *Law enforcement services annual report 1976.* Washington, DC: U.S. Department of the Interior.

Bureau of Justice Assistance. (2007). *2005 national gang threat assessment.* Retrieved February 22, 2008, from http://www.ojp.usdoj.gov/BJA/what/2005_threat_assesment.pdf

Bureau of Justice Statistics. (1992). *Criminal victimization in the United States: 1973–1990 trends.* Washington, DC: U.S. Department of Justice, Office of Justice Programs.

Bureau of Justice Statistics. (2001). *Hate crimes reported in NIBRS, 1997–99.* Washington, DC: U.S. Department of Justice, Office of Justice Programs.

Bureau of Justice Statistics. (2003). *Over half of the increase in state prison populations since 1995 is due to an increase in the prisoners convicted of violent offenses.* Washington, DC: Department of Justice. Retrieved March 22, 2004, from http://www.ojp.usdoj.gov/bjs/glance/corrtyp.htm

Bureau of Justice Statistics. (2006). *Compendium of federal justice statistics, 2004.* Washington, DC: Department of Justice.

Bureau of Justice Statistics. (2008). *Demographic trends in jail populations.* Retrieved July 21, 2008, from http://www.ojp.usdoj.gov/bjs/glancetables/jailracetab.htm.

Burgess, E. W. (1925). The growth of the city: An introduction to a research project. In R. Park & E. W. Burgess (Eds.), *The city* (pp. 47–62). Chicago: University of Chicago Press.

Bushway, S., Stoll, M., & Weiman, D. F. (Eds.). (2007). *Barriers to reentry? The labor market for released prisoners in post-industrial America.* New York: Russell Sage.

Butler, P. (1995). Racially based jury nullification: Black power in the criminal justice system. *The Yale Law Journal, 105,* 677–725.

Cabaniss, E. R., Frabutt, J. M., Kendrick, M. H., & Arbuckle, M. B. (2007). Reducing disproportionate minority contact in the juvenile justice system: Promising practices. *Aggression and Violent Behavior: A Review Journal, 12,* 393–401.

Cahalan, M., & Parsons, L. (1986). *Historical corrections in the United States, 1850–1984.* Washington, DC: Bureau of Justice Statistics.

Camp, G. M., & Camp, C. G. (1985). *Prison gangs: Their extent, nature, and impact on prisons.* Washington, DC: U.S. Department of Justice.

Cao, L., Adams, A. T., & Jensen, V. J. (2000). The empirical status of the Black subculture-of-violence thesis. In M. W. Markowitz & D. Jones Brown (Eds.), *The system in Black and White: Exploring the connections between race, crime, and justice* (pp. 47–61). Westport, CT: Praeger.

Carey, S. M., Finigan, M., Crumpton, D., & Waller, M. (2006). California drug courts: Outcomes, costs, and promising practices: An overview of Phase II in a statewide study. *Journal of Psychoactive Drugs, 3,* 345–356.

Carlson, P. M. (2001). Prison interventions: Evolving strategies to control security threat groups. *Corrections Management Quarterly, 5,* 10–22.

Carr, N. T., Hudson, K., Hanks, R. S., & Hunt, A. N. (2008). Gender effects along the juvenile justice system: Evidence of a gendered organization. *Feminist Criminology, 3,* 25–43.

Carroll, C. (1900). *The Negro a beast.* St. Louis, MO: American Book and Bible House.

Carroll, L. (1974). *Hacks, Blacks, and cons: Race relations in a maximum security prison.* Prospect Heights, IL: Waveland Press.

Carter, D. (1969). *Scottsboro: A tragedy of the American South.* Baton Rouge: Louisiana State University Press.

Carter, D. L. (1983). Hispanic interaction with the criminal justice system in Texas: Experiences, attitudes, and perceptions. *Journal of Criminal Justice, 11,* 213–227.

Castelle, G., & Loftus, E. F. (2002). Misinformation and wrongful convictions. In S. Westervelt & J. A. Humphrey (Eds.), *Wrongly convicted: Perspectives on failed justice* (pp. 17–35). New Brunswick, NJ: Rutgers University Press.

Catalano, S. M. (2006). *Criminal victimization, 2005.* Washington, DC: U.S. Department of Justice, Office of Justice Programs, Bureau of Justice Statistics.

Catterall, H. (Ed.). (1968). *Judicial cases concerning American slavery and the Negro* (Vol. 1). New York: Negro University Press. (Original work published 1926)

Cernkovich, S. A., Giordano, P. C., & Rudolph, J. L. (2000). Race, crime, and the American dream. *Journal of Research in Crime and Delinquency, 37,* 131–170.

Chabrán, R., & Chabrán, R. (1996). *The Latino encyclopedia.* Tarrytown, NY: Marshall Cavendish Corporation.

Chambliss, W. (1964). A sociological analysis of the law of vagrancy. *Social Problems, 12,* 67–77.

Chambliss, W. (Ed.). (1969). *Crime and the legal process.* New York: McGraw-Hill.

Chambliss, W. J. (1991). *Trading textbooks for prison cells.* Baltimore: National Center for Institutions and Alternatives.

Chamlin, M. B., Myer, A. J., Sanders, B. A., & Cochran, J. K. (2008). Abortion as crime control: A cautionary tale. *Criminal Justice Policy Review, 19,* 135–152.

Chapin, B. (1983). *Criminal justice in colonial America, 1606–1660.* Athens: University of Georgia Press.

Chesney-Lind, M., & Shelden, R. G. (2004). *Girls, delinquency, and juvenile justice.* Belmont, CA: Wadsworth/Thomson Learning.

Chicago Commission on Race Relations. (1922). *The Negro in Chicago: A study of race relations and a race riot in 1919.* Chicago: Author.

Chilton, B. (2004). Regional variations in lethal and nonlethal assaults. *Homicide Studies, 8,* 40–56.

Chircos, T. G., & Crawford, C. (1995). Race and imprisonment: A contextual assessment of the evidence. In D. Hawkins (Ed.), *Ethnicity, race, and crime* (pp. 281–309). Albany: State University of New York Press.

Christianson, S. (1981). Our Black prisons. *Crime and Delinquency, 27,* 364–375.

Christianson, S. (1998). *With liberty for some: 500 years of imprisonment in America.* Boston: Northeastern University Press.

Christianson, S. (2004). *Innocent: Inside wrongful conviction cases.* New York: New York University Press.

Cintron, M. (2005). Latino delinquency: Defining and counting the problem. In E. Penn, H. Taylor Greene, & S. Gabbidon (Eds.), *Race and juvenile justice* (pp. 27–45). Durham, NC: Carolina Academic Press.

Clarke, H. J. (1992). *Christopher Columbus and the African holocaust.* Brooklyn, NY: A & B.

Clarke, J. W. (1998). "Without fear or shame": Lynching, capital punishment and the subculture of violence in the American South. *British Journal of Political Science, 28,* 269–289.

Clear, T. R. (2007). *Imprisoning communities: How mass incarceration makes disadvantaged neighborhoods worse.* New York: Oxford University Press.

Clear, T. R., & Cole, G. F. (2000). *American corrections* (5th ed.). Belmont, CA: Wadsworth.

Clear, T. R., & Cole, G. F. (2003). *American corrections* (6th ed.). Belmont, CA: Wadsworth.

Clear, T. R., & Dammer, H. R. (2003). *The offender in the community* (2nd ed.). Belmont, CA: Wadsworth.

Clear, T. R., Rose, D. R., & Ryder, J. A. (2001). Incarceration and the community: The problem of removing and returning offenders. *Crime & Delinquency, 47,* 335–351.

Clear, T. R., Rose, D. R., Waring, E., & Scully, K. (2003). Coercive mobility and crime: A preliminary examination of concentrated incarceration and social disorganization. *Justice Quarterly, 20,* 33–64.

Clement, P. T. (1993). The incorrigible child: Juvenile delinquency in the United States from the 17th through the 19th centuries. In A. G. Hess & P. F. Clement (Eds.), *History of juvenile delinquency: A collection of essays on crime committed by young offenders, in history and in selected countries* (pp. 453–490). Aalen, Germany: Scientia Verlag.

Cloward, R. A., & Ohlin, L. E. (1960). *Delinquency and opportunity: A theory of delinquent gangs.* New York: The Free Press.

Cochran, J. K., & Chamlin, M. B. (2005). Can information change public opinion? Another test of the Marshall hypotheses. *Journal of Criminal Justice, 33,* 573–584.

Cochran, J. K., & Chamlin, M. B. (2006). The enduring racial divide in death penalty support. *Journal of Criminal Justice, 34,* 84–99.

Cochran, J. K., Sanders, B., & Chamlin, M. B. (2006). Profiles in change: An alternative look at the Marshall hypotheses. *Journal of Criminal Justice Education, 17,* 205–226.

Cohen, A. K. (1955). *Delinquent boys: The culture of the gang.* New York: The Free Press.

Cohen, F. (1971). *Handbook of federal Indian law.* Albuquerque: University of New Mexico Press.

Cohen, S. (2003). *The wrong man: America's epidemic of wrongful death row convictions.* New York: Carroll & Graf.

Coker v. Georgia, 429 U.S. 815 (1977).

Cole, D. (1999). *No equal justice.* New York: New Press.

Collins, C. F. (1997). *The imprisonment of African American women.* Jefferson, NC: McFarland.

Colvin, M. (1997). *Penitentiaries, reformatories, and chain gangs.* New York: St. Martin's Press.

Conley, D. J., & Debro, J. (2002). Black Muslims in California prisons: The beginning of a social movement for Black prisoners in the United States. In C. E. Reasons, D. C. Conley, & J. Debro (Eds.), *Race, class, gender, and justice in the United States* (pp. 278–291). Boston: Allyn & Bacon.

Cooper, C. (2001/2002). Subjective states of mind & custodial arrest: Race-based policing. *The Journal of Intergroup Relations, 38,* 3–18.

Cooper, C. S. (2003). Drug courts: Current issues and future perspectives. *Substance Use & Misuse, 38,* 1671–1711.

Coppa, F. J., & Curran, T. J. (1976). From the Rhine to the Mississippi: The German emigration to the United States. In F. J. Coppa & T. J. Curran (Eds.), *The immigrant experience in America* (pp. 44–62). Boston: Twayne Publishers.

Cose, E. (2005, March 14). Long after the alarm went off. *Newsweek,* p. 37.

Courtwright, D. T. (2001). *Dark paradise.* Cambridge, MA: Harvard University Press.

Covington, J. (1995). Racial classification in criminology: The reproduction of racialized crime. *Sociological Forum, 10,* 547–568.

Covington, J. (2003). The violent Black male: Conceptions of race in criminological theories. In D. F. Hawkins (Ed.), *Violent crime: Assessing race & ethnic differences* (pp. 254–279). Cambridge, UK: Cambridge University Press.

Cox, O. C. (1945). Lynching and the status quo. *Journal of Negro Education, 14,* 576–588.

Crank, J. P. (1998). *Understanding police culture.* Cincinnati, OH: Anderson.

Crawford, C. (2000). Gender, race, and habitual offender sentencing in Florida. *Criminology, 38,* 263–280.

Crawford, C., Chircos, T., & Kleck, G. (1998). Race, racial threat, and sentencing of habitual offenders. *Criminology, 36,* 481–511.

Cresswell, L., & Deschenes, E. (2001). Minority and non-minority perceptions of drug court program severity and effectiveness. *Journal of Drug Issues, 31,* 259–292.

Crow, M. C., & Johnson, K. A. (2008). Race, ethnicity, and habitual-offender sentencing: A multilevel analysis of individual and contextual threat. *Criminal Justice Policy Review, 19,* 63–83.

Crutchfield, R. D. (2004). Warranted disparity? Questioning the justification of racial disparity in criminal justice processing. *Columbia Human Rights Law Review, 36,* 15–40.

Curran, D. J., & Renzetti, C. M. (2001). *Theories of crime.* Boston: Allyn & Bacon.

Curry, G. D., & Decker, S. H. (2003). *Confronting gangs: Crime and community* (2nd ed.). Los Angeles, CA: Roxbury.

D'Alessio, S. J., Eitle, D., & Stolzenberg, L. (2005). The impact of serious crime, racial threat, and economic inequality on private police size. *Social Science Research, 34,* 267–282.

Daniels, R. (1988). *Asian America: Chinese and Japanese in the United States since 1850.* Seattle: University of Washington Press.

Darwin, C. (1859). On the origin of species by means of natural selection, or the *preservation of favoured races in the struggle for life.* London: John Murray.

Darwin, C. (1871). The descent of man and selection in relation to sex. London: John Murray.

Davidson, R. T. (1974). *Chicano prisoners: The keys to San Quentin.* Prospect Heights, IL: Waveland Press.

Davis, A. Y. (1981). *Women, race & class.* New York: Random House.

Davis, A. Y. (1997). Race and criminalization: Black Americans and the punishment industry. In W. Lubiano (Ed.), *The house that race built: Black Americans, U.S. terrain* (pp. 264–279). New York: Pantheon.

Davis, A. Y. (2000). From the convict lease system to the super-max prison. In J. James (Ed.), *States of confinement: Policing, detention, and prisons* (pp. 60–74). New York: St. Martin's Press.

Davis, A. Y. (2003). *Are prisons obsolete?* New York: Seven Stories Press.

De Las Casas, B. (1993). *The devastation of the Indies: A brief account.* Baltimore: Johns Hopkins University Press. (Original work published 1552)

Death Penalty Information Center. (2004). *U.S. Supreme Court: Roper v. Simmons.* Retrieved June 6, 2004, from http://www.deathpenaltyinfo.org

Death Penalty Information Center. (2008). *Juveniles and the death penalty.* Retrieved March 26, 2008, from http://www.deathpenaltyinfo.org/article.php?did=205&scid=27 3/26/2008.

Dedman, B. (2004, January 21). Profiling study cites dozens of locals. *Boston Globe.* Retrieved December 15, 2004, from http://www.boston.com/news/local/massachusetts/articles/2004/01/21/profiling_study_cites_dozens_of_locales/

DeFrances, C. J. (2001). *State-funded indigent defense services, 1999.* Washington, DC: U.S. Bureau of Justice Statistics.

DeFrances, C. J., & Litras, M. F. X. (2000). *Indigent defense services in large counties, 1999.* Washington, DC: Bureau of Justice Statistics.

del Carmen, A. (2008). *Racial profiling in America.* Upper Saddle River, NJ: Pearson/Prentice-Hall.

Delgado, R., & Stefancic, J. (2001). *Critical race theory: An introduction.* New York: New York University Press.

Delisi, M., & Regoli, R. (1999). Race, conventional crime, and criminal justice: The declining importance of skin color. *Journal of Criminal Justice, 27,* 549–557.

Dempsey, J. S. (1999). *An introduction to policing.* Belmont, CA: Wadsworth.

Demuth, S., & Steffensmeier, D. (2004). The impact of gender and race-ethnicity in the pretrial release process. *Social Problems, 51,* 222–242.

Denov, M. S., & Campbell, K. M. (2005). Criminal injustice: Understanding the causes, effects, and responses to wrongful convictions in Canada. *Journal of Contemporary Criminal Justice, 21,* 224–249.

Deschenes, E. P., & Esbensen, F. A. (1999). Violence among girls: Does gang membership make a difference? In M. Chesney-Lind & J. M. Hagedorn (Eds.), *Female gangs in America* (pp. 277–294). Chicago: Lakeview Press.

Desmond v. Blackwell, 235 F. Supp. 246 (1964).

Devine, P., Coolbaugh, K., & Jenkins, S. (1998). *Disproportionate minority confinement: Lessons from five states.* Washington, DC: Office of Juvenile Justice and Delinquency Prevention.

Dickey, G. (2004). *Downtown opium dens attracted many.* El Paso Community College Local History Project. Retrieved December 16, 2004, from http://www.epcc.edu/ftp/Homes/monicaw/ borderlands/21_opium.htm

DiIulio, J. J., Jr. (1995, November 27). The coming of the super-predators. *Weekly Standard,* p. 23.

DiIulio, J. J., Jr. (1996, Spring). They're coming: Florida's youth crime bomb. *Impact,* pp. 25–27.

Dillon, R. R., Robbins, M. S., Szapocznik, J., & Pantin, H. (2008). Exploring the role of parental monitoring of peers on the relationship between family functioning and delinquency in the lives of African American and Hispanic adolescents. *Crime & Delinquency, 54,* 65–94.

DiLulio, J. (1996). My Black crime problem, and ours. *City Journal, 6,* 14–28.

Dinnerstein, L., & Reimers, D. M. (1982). *Ethnic Americans* (2nd ed.). New York: Harper & Row.

Donohue, J. J., & Levitt, S. D. (2001). The impact of legalized abortion on crime. *The Quarterly Journal of Economics, CXVI,* 379–420.

Donohue, J. J., & Levitt, S. D. (2004). Further evidence that legalized abortion lowered crime: A reply to Joyce. *Journal of Human Resources, 39,* 29–49.

Donohue, J. J., & Levitt, S. D. (2006). Measurement error, legalized abortion and the decline in crime: A response to Foote and Goetz (2005). *National Bureau of Economic Research Working Paper,* no. 11987.

Donzinger, S. R. (Ed.). (1996). *The real war on crime: The report of the national criminal justice commission.* New York: Harper Perennial.

Dorfman, L., & Schiraldi, V. (2001). *Off balance: Youth, race, and crime in the news.* Washington, DC: Building Blocks for Youth.

Douglas, J. E., & Olshaker, M. (1996). *Mind hunter.* New York: Pocket Star Books.

Drug Policy Alliance. (2008). *Tulia, Texas.* Retrieved March 18, 2008, from http://www.drugpolicy .org/law/police/tulia/index.cfm

Du Bois, W. E. B. (1891). *Enforcement of the slave trade laws* (American Historical Association, Annual Report). Washington, DC: Government Printing Office.

Du Bois, W. E. B. (1899, May 18). The Negro and crime. *The Independent, 51.*

Du Bois, W. E. B. (1920). Crime. *The Crisis, 19,* 172–173.

Du Bois, W. E. B. (1996). *The Philadelphia Negro: A social study.* Philadelphia: The University of Pennsylvania Press. (Original work published 1899)

Du Bois, W. E. B. (1901). The spawn of slavery: The convict lease system in the South. *Missionary Review of the World, 14,* 737–745. (Reprinted in S. L. Gabbidon, H. Taylor Greene, & V. Young, 2002, *African American classics in criminology and criminal justice,* pp. 83–88, Thousand Oaks, CA: Sage.)

Dugdale, R. L. (1877). *"The Jukes": A study in crime, pauperism, disease, and heredity.* New York: G. P. Putnam's Sons.

Dulaney, M. R. (1879). *The origin of races and color.* Philadelphia: Harper & Brothers.

Dulaney, W. M. (1996). *Black police in America.* Bloomington: Indiana University Press.

Dunlap, E., Golub, A., & Johnson, B. D. (2003). Girls' sexual development in the inner city: From compelled childhood sexual contacts to sex-for-things exchanges. *Journal of Child Sexual Abuse, 12,* 73–96.

Durose, M. R. (2007). *State court sentencing of convicted felons 2004–Statistical tables.* Washington, DC: Bureau of Justice Statistics.

Durose, M. R., & Langan, P. (2007). *Felony sentences in state courts, 2004.* Washington, DC: Bureau of Justice Statistics.

Durose, M. R., Smith, E. L., & Langan, P. A. (2007). *Contacts between police and the public, 2005.* Washington, DC: U.S. Department of Justice.

Dyer, J. (2000). *The perpetual prisoner machine: How America profits from crime.* Boulder, CO: Westview Press.

Dyson, M. E. (2006). *Come hell or high water: Hurricane Katrina and the color of disaster.* New York: Basic Civitas Books.

Eberheart v. Georgia, 433 U.S. 917 (1977).

Eddings v. Oklahoma, 455 U.S. 104 (1982).

Egley, A., Jr. (2002). *National Youth Gang Survey trends from 1996 to 2000* (Fact Sheet no. 2002–2003). Washington, DC: U.S. Department of Justice, Office of Juvenile Justice and Delinquency Prevention.

Egley, A., Jr., Howell, J. C., & Major, A. K. (2004). Recent patterns of gang problems in the United States: Results from the 1996–2002 National Youth Gang Surveys. In F. A. Esbensen, S. G. Tibbetts, & L. Gaines (Eds.), *American youth gangs at the millennium* (pp. 90–108). Long Grove, IL: Waveland Press.

Egley, A., Jr., Howell, J. C., & Major, A. K. (2006). *National Youth Gang Survey: 1999–2001.* Washington, DC: U.S. Department of Justice, Office of Juvenile Justice and Delinquency Prevention.

Egley, A., Jr., & Ritz, C. E. (2006). *Highlights of the 2004 National Youth Gang Survey* (Fact Sheet no. 2006–01). Washington, DC: U.S. Department of Justice, Office of Juvenile Justice and Delinquency Prevention.

Eitle, D., & Turner, R. J. (2003). Stress exposure, race, and young male adult crime. *Sociological Quarterly, 44,* 243–269.

Ellis, L. (1997). Criminal behavior and *r/K* selection: An extension of gene-based evolutionary theory. *Deviant Behavior, 8,* 148–176.

Ellis, L., & Walsh, A. (1997). Gene-based evolutionary theories in criminology. *Criminology, 35,* 229–275.

Ellis, L., & Walsh, A. (1999). Criminologists' opinions about causes and theories of crime and delinquency. *The Criminologist, 24,* 1, 4–6.

Ellis, L., & Walsh, A. (2000). *Criminology: A global perspective.* Needham Heights, MA: Allyn & Bacon.

Emsley, C. (1983). *Policing and its context, 1750–1870.* New York: Schocken Books.

Engel, R. S., & Calnon, J. M. (2004). Examining the influence of driver's characteristics during traffic stops with police: Results from a national survey. *Justice Quarterly, 21,* 49–90.

Engen, R. L., Gainey, R. R., Crutchfield, R. D., & Weis, J. G. (2003). Discretion and disparity under sentencing guidelines: The role of departures and structured sentencing alternatives. *Criminology, 41,* 99–130.

Epps, E. G. (1967). Socioeconomic status, race, level of aspiration and juvenile delinquency: A limited empirical test of Merton's conception of deviance. *Phylon, 28,* 16–27.

Esbensen, F., & Deschenes, E. P. (1998). A multisite examination of gang membership: Does gender matter? *Criminology, 36,* 799–827.

Esbensen, F. A. (2000). *Preventing adolescent gang involvement* (Youth Gang Series). Washington, DC: U.S. Department of Justice, Office of Juvenile Justice and Delinquency Prevention.

Esbensen, F. A., & Winfree, L. T., Jr. (1998). Race and gender differences between gang and non-gang youth: Results from a multi-site survey. *Justice Quarterly, 15*, 505–525.

Escobar, E. J. (1999). *Race, police, and the making of a political identity: Mexican Americans and the Los Angeles Police Department, 1900–1945.* Berkeley: University of California Press.

Escobedo v. Illinois, 378 U.S. 478, 84 S. Ct. 1758 (1964).

Faust, A. B. (1927). *The German element in the United States* (Vol. I). New York: The Steuben Society of America.

Feagin, R. F., & Booher Feagin, C. (1996). *Racial and ethnic relations* (5th ed.). Upper Saddle River, NJ: Prentice Hall.

Feagin, R. F., & Booher Feagin, C. (2008). *Racial and ethnic relations* (8th ed.). Upper Saddle River, NJ: Prentice Hall.

Federal Bureau of Investigation. (1930–2002). *Crime in the United States.* Washington, DC: Government Printing Office. Retrieved from http://www.fbi.gov/ucr/ucr.htm

Federal Bureau of Investigation. (1998). *National Incident-Based Reporting System.* Washington, DC: Government Printing Office.

Federal Bureau of Investigation. (2001). *Crime in the United States, 2000.* Washington, DC: Government Printing Office.

Federal Bureau of Investigation. (2002). *Hate crime statistics, 1995–2001.* Washington, DC: Government Printing Office. Retrieved July 15, 2004, from http://www.fbi.gov/ucr/ucr.htm#hate

Federal Bureau of Investigation. (2003). *Crime in the United States, 2003 Section IV arrests.* Retrieved from http://www.fbi.gov/ucr/cius_03/pdf/03sec4.pdf

Federal Bureau of Investigation. (2004). *Crime in the United States, 2002.* Washington, DC: Government Printing Office. Retrieved from http://www.fbi.gov/ucr/cius_02/pdf/4section four.pdf

Federal Bureau of Investigation. (2005). *Crime in the United States, 2005 Section IV arrests.* Retrieved from http://www.fbi.gov/ucr/05cius/arrests/index.html

Federal Bureau of Investigation. (2006a). *Crime in the United States, 2006.* Retrieved from http://www.fbi.gov/ucr/cius2006/arrests/index.html.

Federal Bureau of Investigation. (2006b). *Hate crime statistics, 2006.* Washington, DC: Government Printing Office. Retrieved February 2, 2008, from http://www.fbi.gov/ucr/hc2006/index.html.

Federal Bureau of Investigation. (2006c). *Law enforcement officers feloniously killed and assaulted, 2006. Officers feloniously killed.* Retrieved March 18, 2008, from http://www.fbi.gov/ucr/killed/2006/feloniouslykilled.html.

Federal Bureau of Investigation. (2007). *Crime in the United States, 2006.* Washington, DC: Government Printing Office. Retrieved from http://www.fbi.gov/ucr/cius2006/index.html

Feld, B. (1999). *Bad kids: Race and the transformation of the juvenile court.* New York: Oxford University Press.

Fellner, J., & Mauer, M. (1998). *Losing the vote: The impact of felony disenfranchisement laws in the United States.* Washington, DC: Human Rights Watch and the Sentencing Project.

Fernandez, K. E., & Bowman, T. (2004). Race, political institutions, and criminal justice: An examination of the sentencing of Latino offenders. *Columbia Human Rights Law Review, 36*, 41–70.

Fielding, J., Tye, G., Ogawa, P., Imam, I., & Long, A. (2002). Los Angeles County drug court programs: Initial results. *Journal of Substance Abuse Treatment, 23*, 217–224.

Finger, B. (1959). *Concise world history.* New York: Philosophical Library.

Finkelhor, D., & Ormrod, R. (2004). *Prostitution of juveniles: Patterns from NIBRS.* Washington, DC: Office of Justice Programs, Office of Juvenile Justice and Delinquency Prevention.

Finkelstein, N. H. (2007). *American Jewish history.* Philadelphia: The Jewish Publication Society.

Fischer, B. (2003). Doing good with a vengeance: A critical assessment of the practices, effects and implications of drug treatment courts in North America. *Criminal Justice, 3,* 227–248.

Fixico, D. L. (2000). *The urban Indian experience in America.* Albuquerque: University of New Mexico Press.

Flanagan, T. J. (1996). Reform or punish: Americans' views of the correctional system. In T. J. Flanagan & D. R. Longmire (Eds.). *Americans view crime and justice: A national public opinion survey* (pp. 75–92). Thousand Oaks, CA: Sage Publications.

Fleisher, M. S., & Decker, S. H. (2001). An overview of the challenge of prison gangs. *Corrections Management Quarterly, 5,* 1–9.

Flowers, R. B. (1988). *Minorities and criminality.* Westport, CT: Greenwood Press.

Fluery-Steiner, B. (2009). Capital jury project. In H. T. Greene & S. L. Gabbidon (Eds.), *Encyclopedia of race and crime.* Thousand Oaks, CA: Sage Publications.

Fluery-Steiner, B., & Argothy, V. (2004). Lethal "borders": Elucidating jurors' racialized discipline to punish in Latino defendant death cases. *Punishment & Society, 6,* 67–84.

Fogelson, R. (1977). *Big-city police.* Cambridge, MA: Harvard University Press.

Foote, C. L., & Goetz, C. F. (2006). Testing economic hypotheses with state-level data: A comment on Donohue and Levitt (2001). *Federal Reserve Bank of Boston Working Paper,* no. 05–15.

Forst, B. (2003, July 28–30). *Managing errors in the new era of the policing.* Paper presented at the annual conference on Criminal Justice Research and Evaluation, Washington, DC.

Fox, C., & Huddleston, W. (2003). Drug courts in the U.S. *Issues of Democracy: The Changing Face of U.S. Courts, 8,* 13–19.

Fox, J. A., & Zawitz, M. W. (2003). *Homicide trends in the United States: 2000 update* (Crime data brief). Washington, DC: U.S. Department of Justice, Bureau of Justice Statistics.

Franklin, J. H., & Moss, A. A. (2000). *From slavery to freedom: A history of African Americans* (8th ed.). New York: McGraw-Hill.

Frazier, E. F. (1939). Rebellious youth. In *The Negro family in the United States* (pp. 268–280). Chicago: University of Chicago Press.

Frazier, E. F. (1949). Crime and delinquency. In *The Negro in the United States* (pp. 638–653). New York: Macmillan.

Fredrickson, D. D., & Siljander, R. P. (2002). *Racial profiling.* Springfield, IL: Charles C Thomas.

Free, M. D. (2002a). Race and presentencing decisions in the United States: A summary and critique of the research. *Criminal Justice Review, 27,* 203–232.

Free, M. D. (2002b). Racial bias and the American criminal justice system: Race and presentencing revisited. *Critical Criminology, 10,* 195–223.

Frey, C. P. (1981). The house of refuge for colored children. *Journal of Negro History, 66,* 10–25.

Friedman, L. M. (1993). *Crime and punishment in American history.* New York: Basic Books.

Fulwood v. Clemmer, 206 F. Supp. 370 (1962).

Furman v. Georgia, 408 U.S. 238 (1972).

Fyfe, J. J. (1981). Race and extreme police-citizen violence. In R. L. McNeely & C. E. Pope (Eds.), *Race, crime and criminal justice.* (Reprinted in J. F. Fyfe, Ed., 1982, *Readings on police use of deadly force,* pp. 173–194. Washington, DC: Police Foundation.)

Fyfe, J. J. (Ed.). (1982). *Readings on police use of deadly force.* Washington, DC: Police Foundation.

Gabbidon, S. L. (1999). W. E. B. DuBois on crime: American conflict theorist. *The Criminologist, 24,* 1, 3, 20.

Gabbidon, S. L. (2001). W. E. B. Du Bois: Pioneering American criminologist. *Journal of Black Studies, 31,* 581–599.

Gabbidon, S. L. (2003). Racial profiling by store clerks and security personnel in retail establishments: An exploration of "shopping while Black." *Journal of Contemporary Criminal Justice, 19,* 345–364.

Gabbidon, S. L. (2004). Crime prevention in the African American community: Lessons learned from the Nation of Islam. *Souls: A Critical Journal of Black Politics, Culture, and Society, 6,* 42–54.

Gabbidon, S. L. (2007a). *Criminological perspectives on race and crime.* New York: Routledge.

Gabbidon, S. L. (2007b). *W.E.B. Du Bois on crime and justice: Laying the foundations of sociological criminology.* Aldershot, UK: Ashgate Publications.

Gabbidon, S. L., Craig, R., Okafo, N., Marzette, L. N., & Peterson, S. A. (2008). The consumer racial profiling experiences of Black students at historically Black colleges and universities: An exploratory study. *Journal of Criminal Justice, 36,* 354–61.

Gabbidon, S. L., & Higgins, G. E. (2007). Consumer racial profiling and perceived victimization: A phone survey of Philadelphia area residents. *American Journal of Criminal Justice, 32,* 1–11.

Gabbidon, S. L., Kowal, L., Jordan, K. L., Roberts, J. L., & Vincenzi, N. (2008). Race-based peremptory challenges: An empirical analysis of litigation from the U.S. Court of Appeals, 2002–2006. *American Journal of Criminal Justice, 33,* 59–68.

Gabbidon, S. L., Marzette, L. N., & Peterson, S. A. (2007). Racial profiling and the courts: An empirical analysis of federal litigation, 1991 to 2006. *Journal of Contemporary Criminal Justice, 23,* 226–238.

Gabbidon, S. L., & Taylor Greene, H. (2001). The presence of African-American scholarship in early American criminology texts (1918–1960). *Journal of Criminal Justice Education, 12,* 301–310.

Gabbidon, S. L., & Taylor Greene, H. (Eds.). (2005). *Race, crime, and justice: A reader.* New York: Routledge.

Gabbidon, S. L., Taylor-Greene, H., & Young, V. D. (Eds.). (2002). *African American classics in criminology and criminal justice.* Thousand Oaks, CA: Sage Publications.

Gaes, G. G., Wallace, S., Gilman, E., Klein-Saffran, J., & Suppa, S. (2002). The influence of prison gang affiliation on violence and other prison misconduct. *The Prison Journal, 82,* 359–385.

Gallagher, C. A. (Ed.). (1997). *Rethinking the color line: Readings in race and ethnicity.* Mountain View, CA: Mayfield.

Gallegos-Castillo, A., & Patino, V. (2006). *Bridging community, research, and action: An emerging center on Latino youth development.* San Francisco, CA: National Center on Crime and Delinquency.

Galloway, A. L., & Drapella, L. A. (2006). Are effective drug courts an urban phenomenon? Considering their impact on recidivism among a nonmetropolitan adult sample in Washington State. *International Journal of Offender Therapy and Comparative Criminology, 50,* 280–293.

Gallup Organization. (2007). *Sixty-nine percent of Americans support death penalty: Majority say death penalty is applied fairly.* Retrieved February 14, 2008, from Gallup Brain database.

Gans, H. J. (2005). Race as class. *Contexts, 4,* 17–21.

Geis, G. (1972). Statistics concerning race and crime. In C. E. Reasons & J. L. Kuykendall (Eds.), *Race, crime, and justice* (pp. 61–78). Palisades, CA: Goodyear.

Georges-Abeyie, D. (1989). Race, ethnicity, and the spatial dynamic. *Social Justice, 16,* 35–54.

Gerber, J., & Engelhardt-Greer, S. (1996). Just and painful: Attitudes toward sentencing criminals. In T. J. Flanagan & D. R. Longmire (Eds.), *Americans view crime and justice: A national public opinion survey* (pp. 62–74). Thousand Oaks, CA: Sage Publications.

Gideon v. Wainwright, 372 U.S. 335 (1963).

Glaze, L. E., & Bonczar, T. P. (2007). *Probation and parole in the United States, 2006.* Washington, DC: Bureau of Justice Statistics.

Goldkamp, J. S. (1982). Minorities as victims of police shootings: Interpretations of racial disproportionality and police use of deadly force. In J. Fyfe (Ed.), *Readings on police use of deadly force* (pp. 128–151). Washington, DC: Police Foundation.

Goldkamp, J. S., & Weiland, D. (1993). *Assessing the impact of Dade County's felony drug court.* Washington, DC: U.S. Department of Justice.

Goldman, S. (1989, April–May). Reagan's judicial legacy: Completing the puzzle and summing up. *Judicature, 72,* 320–321, Table 1.

Goldman, S. (2003, May/June). W. Bush remaking the judiciary: Like father like son? *Judicature, 86,* 304.

Goldstein, H. (1979). Improving policing: A problem oriented approach. *Crime and Delinquency, 25,* 236–258.

Gossett, T. (1963). *Race: The history of an idea in America.* Dallas: Southern Methodist University Press.

Gottfredson, D., Najaka, S., & Kearley, B. (2003). Effectiveness of drug treatment courts: Evidence from a randomized trial. *Criminology & Public Policy, 2,* 171–198.

Gottfredson, G. D., & Gottfredson, D. C. (2001, October). *Gang problems and gang programs in a national sample of schools.* Ellicott City, MD: Gottfredson Associates.

Gould, L. A. (2000). White male privilege and the construction of crime. In The Criminal Justice Collective of Northern Arizona University (Eds.), *Investigating difference: Human and cultural relations in criminal justice* (pp. 27–43). Boston: Allyn & Bacon.

Gould, S. J. (1996). *The mismeasure of man.* New York: Norton.

Graham, O. L., Jr. (2004). *Unguarded gates: A history of America's immigration crisis.* Lanham, MD: Rowman & Littlefield.

Green, B. L., Furrer, C., Worcel, S., Burrus, S., & Finighan, M. W. (2007). How effective are family treatment drug courts? Outcomes from a four-site national study. *Child Maltreatment, 12,* 43–59.

Greenberg, D. (1976). *Crime and law enforcement in the colony of New York, 1691–1776.* Ithaca, NY: Cornell University Press.

Greenberger, R. S. (2002, October 11). The economy: Supreme Court narrowly refuses to consider death penalty plea. *The Wall Street Journal,* p. A2.

Greene, J. R., Piquero, A. R., Collins, P., & Kane, R. (1999). Doing research in public housing: Implementation issues from Philadelphia's 11th Street Corridor Community Policing Program. *Justice Research and Policy, 1,* 67–95.

Greenfield, L. A., & Smith, S. K. (1999). *American Indians and crime.* Washington, DC: U.S. Department of Justice, Office of Justice Programs.

Greenleaf, R. G., Skogan, W. G., & Lurigio, A. J. (2008). Traffic stops in the Pacific Northeast: Competing hypotheses about racial disparity. *Journal of Ethnicity in Criminal Justice, 6,* 3–22.

Greenwald, F. (1993). Treatment of behavioral problems of children and youth by early indigenous Americans. In A. G. Hess & P. F. Clement (Eds.), *History of juvenile delinquency: A collection of essays on crime committed by young offenders, in history and in selected countries* (pp. 735–756). Aalen, Germany: Scientia Verlag.

Gregg v. Georgia, 428 U.S. 153 (1976).

Griffin, M. L., & Hepburn, J. R. (2006). The effect of gang affiliation on violent misconduct among inmates during the early years of confinement. *Criminal Justice and Behavior, 33,* 419–448.

Grimke, A. H. (1915). *The ultimate criminal.* Washington, DC: American Negro Academy.

Grimshaw, A. D. (Ed.). (1969). *Racial violence in the United States.* Chicago: Aldine.

Growette Bostaph, L. M. (2008). Repeat citizens in motor vehicle stops: A Black experience. *Journal of Ethnicity in Criminal Justice, 6,* 41–64.

Guevara, L., Herz, D., & Spohn, C. (2006). Gender and decision making: What role does race play? *Feminist Criminology, 1,* 258–282.

Hagan, F. E. (2002). *Introduction to criminology: Theories, methods, and criminal behavior.* Belmont, CA: Wadsworth.

Hakins, S. (2004). *Too young to die.* Retrieved December 15, 2004, from http://www.fortunesociet .org/fa110205.htm

Hallett, M. A. (2006). *Private prisons in America: A critical race perspective.* Urbana, IL: University of Illinois Press.

Hanrahan, N. (2001). *All things censored: Mumia Abu-Jamal*. New York: Seven Stories Press.

Harer, M. D., & Steffensmeier, D. J. (1996). Race and prison violence. *Criminology, 34,* 323–355.

Harjo, S. S. (2002). Redskins, savages, and other Indian enemies: A historical overview of American media coverage of Native peoples. In C. R. Mann & M. S. Zatz (Eds.), *Images of color, images of crime* (2nd ed., pp. 56–70). Los Angeles, CA: Roxbury.

Harlow, C. W. (2005). *Hate crime reported by victims and police*. Washington, DC: U.S. Department of Justice, Office of Justice Programs.

Harmon, T. R. (2001). Guilty until proven innocent: An analysis of post-*Furman* capital errors. *Criminal Justice Policy Review, 12,* 113–139.

Harmon, T. R. (2004). Race for your life: An analysis of the role of race in erroneous capital convictions. *Criminal Justice Review, 29,* 76–96.

Harrell, A. (2003). Judging drug courts: Balancing the evidence. *Criminology & Public Policy, 2,* 207–212.

Harrell, E. (2007). *Black victims of violent crime*. Washington, DC: U.S. Department of Justice, Office of Justice Programs.

Harris Poll. (2004). Retrieved March 2, 2003, from Polling the Nations database at http://www.orspub.com/

Harris, D. A. (2002). *Profiles in injustice*. New York: New Press.

Hart, T. C., & Rennison, C. (2003). *Reporting crime to the police, 1992–2000*. Washington, DC: U.S. Department of Justice, Office of Justice Programs.

Hawkins, D. F. (1983). Black and White homicide differentials: Alternatives to an inadequate theory. *Criminal Justice and Behavior, 10,* 407–440.

Hawkins, D. F. (1986). Race, crime type, and imprisonment. *Justice Quarterly, 3,* 251–269.

Hawkins, D. F. (1987). Beyond anomalies: Rethinking the conflict perspective on race and capital punishment. *Social Forces, 65,* 719–745.

Hawkins, D. F., & Hardy, K. A. (1989). Black-White imprisonment rates: A state-by-state analysis. *Social Justice, 16,* 75–94.

Hawkins, D. F., & Kempf-Leonard, K. (2005). *Our children, their children: Confronting racial and ethnic differences in American juvenile justice*. Chicago, IL: The University of Chicago Press.

Hay, C., & Evans, M. M. (2006). Has *Roe v. Wade* reduced U.S. crime rates? Examining the link between mothers' pregnancy intentions and children's later involvement in law-violating behavior. *Journal of Research in Crime and Delinquency, 43,* 36–66.

Hayner, N. (1933). Delinquency areas in the Puget Sound region. *American Journal of Sociology, 39,* 314–328.

Hayner, N. (1938). Social factors in oriental crime. *American Journal of Sociology, 43,* 908–919.

Hayner, N. (1942). Variability in the criminal behavior of American Indians. *American Journal of Sociology, 47,* 602–613.

Healey, J. F. (2003). *Race, ethnicity, gender, and class: The sociology of group conflict and change* (3rd ed.). Thousand Oaks, CA: Pine Forge Press.

Healey, J. F. (2004). *Diversity and society: Race, ethnicity, and gender*. Thousand Oaks, CA: Pine Forge Press.

Healey, J. F. (2006). *Race, ethnicity, gender, and class: The sociology of group conflict and change* (4th ed.). Thousand Oaks, CA: Pine Forge Press.

Healey, J. F. (2007). *Diversity and society: Race, ethnicity, and gender* (2nd ed.). Thousand Oaks, CA: Pine Forge Press.

Henderson, C. R. (1901). *Introduction to the study of dependents, defective, and delinquent classes*. Boston: D.C. Heath.

Hernandez v. New York, U.S. 352 (1991).

Herrnstein, R. J., & Murray, C. (1994). *The bell curve: Intelligence and class structure in American life*. New York: The Free Press.

Herz, D. C., & Walsh, J. E. (2004). Faith-based programs for reentry courts: A summary of issues and recommendations. *Juvenile and Family Court Journal, 55,* 15–25.

Hester, T., & Feeney, T. (2007). Assembly votes to abolish the death penalty. *Star-Ledger.* Retrieved July 29, 2008, from ww.nj.com/news/index.ssf/2007/12/assembly_begins_debate_on_deat.html

Hickey, T. (1998). *Criminal procedure.* New York: McGraw-Hill.

Hickman, L. J., & Suttorp, M. J. (2008). Are deportable aliens a unique threat to public safety? Comparing the recidivism of deportable and nondeportable aliens. *Criminology & Public Policy, 7,* 59–82.

Hickman, M. J. (2003). *Tribal law enforcement, 2000.* Washington, DC: U.S. Department of Justice.

Hickman, M. J. (2006). *Citizen complaints about police use of force.* Washington, DC: U.S. Department of Justice.

Hickman, M. J., & Reaves, B. A. (2003). *Sheriffs' offices 2000.* Washington, DC: U.S. Department of Justice.

Higginbotham, A. L. (1978). *In the matter of color:* Race *and the American legal process: The colonial period.* Oxford, UK: Oxford University Press.

Higginbotham, A. L. (1996). *Shades of freedom: Racial politics and the presumptions of the American legal process.* Oxford, UK: Oxford University Press.

Higginbotham, A.L., & Jacobs, A. F. (1992). The law as an enemy: The legitimization of racial powerlessness through the Colonial and Antebellum criminal laws of Virginia. *North Carolina Law Review, 70,* 969–1070.

Higgins, G. E., & Gabbidon, S. L. (2008). Perceptions of consumer racial profiling and negative emotions: An exploratory study. *Criminal Justice and Behavior,* available at OnlineFirst at doi:10.1177/0093854808325686.

Higgins, G. E., Vito, G. F., & Walsh, W. F. (2008). Searches: An understudied area of racial profiling. *Journal of Ethnicity in Criminal Justice, 6,* 23–40.

Hill, K. G., Lui, C., & Hawkins, J. D. (2001). *Early precursors of gang membership: A study of Seattle youth* (Bulletin, Youth Gang Series). Washington, DC: U.S. Department of Justice, Office of Justice Programs, Office of Juvenile Justice and Delinquency Prevention.

Hinton, W. J., Sims, P. L., Adams, M. A., & West, C. (2007). Juvenile justice: A system divided. *Criminal Justice Policy Review, 18,* 466–483.

Hirsch, J. S. (2002). *Riot and remembrance: The Tulsa race riot and its legacy.* Boston: Houghton Mifflin.

Hirschi, T., & Hindelang, M. (1977). Intelligence and delinquency: A revisionist review. *American Sociological Review, 42,* 571–587.

Hoffman, F. L. (1896). *Race traits and tendencies of the American Negro.* New York: Macmillan.

Hogan, L. J. (1998). *The Osage Indian murders.* Frederick, MD: Amlex.

Holden-Smith, B. (1996). Inherently unequal justice: Interracial rape and the death penalty. *Journal of Criminal Law & Criminology, 86,* 1571–1583.

Hooton, E. A. (1939a). *Crime and the man.* Cambridge, MA: Harvard University Press.

Hooton, E. A. (1939b). *The American criminal: An anthropological study. Volume 1. The native white criminal of native parentage.* Cambridge, MA: Harvard University Press.

Howell, J. C. (2003). *Preventing and reducing juvenile delinquency: A comprehensive framework.* Thousand Oaks, CA: Sage Publications.

Howell, J. C., & Egley, A., Jr. (2005). Moving risk factors into developmental theories of gang membership. *Youth Violence and Juvenile Justice, 3*(4), 334–354.

Howell, J. C., Egley, A., Jr., & Gleason, D. K. (2002). *Modern day youth gangs* (Bulletin, Youth Gang Series). Washington, DC: U.S. Department of Justice, Office of Justice Programs, Office of Juvenile Justice and Delinquency Prevention.

Howell, J. C., Moore, J. P., & Egley, A., Jr. (2002). The changing boundaries of youth gangs. In C. R. Huff (Ed.), *Gangs in America III* (pp. 3–18). Thousand Oaks, CA: Sage Publications.

Hsia, H. M., Bridges, G. S., & McHale, R. (2004). *Disproportionate minority confinement: Year 2002 update.* Washington, DC: US Department of Justice, Office of Juvenile Justice and Delinquency Prevention.

Hubbard, D. J., & Pratt, T. C. (2002). Meta-analysis of the predictors of delinquency among girls. *Journal of Offender Rehabilitation, 34,* 1–13.

Hudson v. McMillian, 503 U.S. 1 (1992).

Huebner, B. M. (2003). Administrative determinates of inmate violence. *Journal of Criminal Justice, 31,* 107–117.

Huey, J., & Lynch, M. J. (1996). The image of Black women in criminology: Historical stereotypes as theoretical foundation. In M. J. Lynch & E. B. Patterson (Eds.), *Justice with prejudice: Race and criminal justice in America* (pp. 72–88). Guilderland, NY: Harrow & Heston.

Huff, C. R. (2002). Wrongful convictions and public policy: The American society of criminology 2001 presidential address. *Criminology, 40,* 1–18.

Huff, C. R. (2004). Wrongful convictions: The American experience. *Canadian Journal of Criminology and Criminal Justice, 46,* 107–120.

Huff, C. R., Rattner, A., & Sagarin, E. (1996). *Convicted but innocent: Wrongful conviction and public policy.* Thousand Oaks, CA: Sage Publications.

Hughes, T. A., James Wilson, D., & Beck, A. J. (2001). *Trends in state parole, 1990–2000.* Washington, DC: Bureau of Justice Statistics.

Huling, T. (2002). Building a prison economy in rural America. In M. Mauer & M. Chesney-Lind (Eds.), *Invisible punishment: The collateral consequences of mass imprisonment* (pp. 197–213). New York: New Press.

Ignatiev, N. (1996). *How the Irish became White.* New York: Routledge.

Illinois suspends death penalty: Governor calls for review of "flawed" system. (2000). Retrieved March 5, 2003, from http://www.CNN.com

Inciardi, J. A. (1999). *Criminal justice* (6th ed.). New York: Harcourt Brace.

Institute of Texan Cultures. (1998). *The El Paso Chinese colony.* Retrieved December 16, 2004, from http://www.texancultures.utsa.edu/txtext/chinese/chineseelpaso.htm

Institute on Race and Poverty. (2003). *Minnesota statewide racial profiling report: All participating jurisdictions.* Minneapolis, MN: Institute on Race and Poverty.

International Association of Chiefs of Police. (2003). *Police use of force in America 2001.* Retrieved December 16, 2004, from http://www.theiacp.org

Ioimo, R., Tears, R. S., Meadows, L. A., Becton, J. B., & Charles, M. T. (2007). The police view of biased-based policing. *Police Quarterly, 10,* 270–287.

Iorizzo, L. J., & Mondello, S. (2006). *The Italian Americans* (3rd ed.). Youngstown, NY: Cambria Press.

Jackson, G. (1970). *Soledad brother: The prison letters of George Jackson.* Chicago: Lawrence Hill Books.

Jackson, J., Sr., & Jackson, J., Jr. (1996). *Legal lynching: Racism, injustice, and the death penalty.* New York: Marlowe.

Jackson, P. I. (1989). *Minority group threat, crime, and policing.* New York: Praeger.

James, J. (Ed.). (2003). *Imprisoned intellectuals: America's political prisoners write on life, liberation, and rebellion.* Lanham, MD: Rowman & Littlefield.

Jang, S. J., & Johnson, B. R. (2003). Strain, negative emotions, and deviant coping among African Americans: A test of general strain theory. *Journal of Quantitative Criminology, 19,* 79–105.

Jang, S. J., & Johnson, B. R. (2005). Gender, religiosity, and reactions to strain among African Americans. *The Sociological Quarterly, 46,* 323–357.

Jang, S. J., &, Lyons, J. A. (2006). Strain, social support, and retreatism among African Americans. *Journal of Black Studies, 37,* 251–274.

Jenkins, P. (1994). The ice age: The social construction of a drug panic. *Justice Quarterly, 11,* 7–31.

Johnson v. California, 543 U.S. 499 (2005)

Johnson, B. D. (2003). Racial and ethnic disparities in sentencing departures across modes of conviction. *Criminology, 41,* 449–489.

Johnson, D. R. (1981). *American law enforcement: A history.* St. Louis, MO: Forum Press.

Johnson, G. B. (1941). The Negro and crime. *Annals of the American Academy of Political and Social Sciences, 217,* 93–104.

Johnson, S. L. (1996). *Subcultural backlash: A new variable in the explanation of the over-representation of African-Americans in the criminal justice system.* Unpublished doctoral dissertation, The Pennsylvania State University.

Johnson Listwan, S., Sundt, J., Holsinger, A., & Latessa, E. (2003). The effect of drug court programming on recidivism: The Cincinnati experience. *Crime and Delinquency, 49,* 389–411.

Johnston, L. D., O'Malley, P. M., & Bachman, J. G. (2004). *Monitoring the future: National results on adolescent drug use overview of key findings, 2002.* Washington, DC: National Institute on Drug Abuse.

Johnston, L. D., O'Malley, P. M., Bachman, J. G., & Schulenberg, J. E. (2007). *Monitoring the Future national survey results on drug use, 1975–2006. Volume I: Secondary school students* (NIH Publication No. 07-6205). Bethesda, MD: National Institute on Drug Abuse.

Jones v. Willingham, 248 F. Supp. 791 (1965).

Jones, J. (2003). *Support for the death penalty remains high at 74%: Slight majority prefers death penalty to life imprisonment as punishment for murder.* Retrieved March 2, 2003, from www.gallup.com/content/print.asp

Jones, T. (1977). The police in America: A Black viewpoint. *The Black Scholar,* pp. 22–39.

Jones-Brown, D. (2007). Forever the symbolic assailant: The more things change, the more they stay the same. *Criminology & Public Policy, 6,* 103–122.

Jordan, L. J. (2008, July 2). *Justice department seeks to legalize racial profiling.* New York: Associated Press.

Joyce, T. (2004a). Did legalized abortion lower crime? *Journal of Human Resources, 39,* 1–38.

Joyce, T. (2004b). Further tests of abortion and crime. *National Bureau of Economic Research Working paper,* no. 10564.

Kahane, L. H., Paton, D., & Simmons, R. (2008). The abortion-crime link: Evidence from England and Wales. *Economica, 75,* 1–21.

Kanellos, N. (1977). *The Hispanic-American almanac.* Detroit, MI: Gale Research.

Katz, W. L. (1986). *The invisible empire: The Ku Klux Klan impact on history.* Seattle, WA: Open Hand.

Keil, T. J., & Vito, G. F. (2006). Capriciousness or fairness? Race and prosecutorial decisions to seek the death penalty in Kentucky. *Journal of Ethnicity in Criminal Justice, 4,* 27–50.

Keith, M. (1996). Criminalization and racialization. In J. Muncie, E. McLaughlin, & M. Langan (Eds.), *Criminological perspectives: A reader* (pp. 271–283). London: Sage Publications.

Kelling, L., & Moore, M. H. (1988). *The evolving strategy of policing.* Washington, DC: National Institute of Justice.

Kempf-Leonard (2007). Minority youths and juvenile justice: Disproportionate minority contact after nearly 20 years of reform efforts. *Youth Violence and Juvenile Justice, 5,* 71–87.

Kennedy, R. (1997). *Race, crime, and the law.* New York: Pantheon Books.

Kilpatrick, D. G., Saunders, B. E., & Smith, D. W. (2003). *Youth victimization: Prevalence and implications.* Washington, DC: National Institute of Justice.

Kim, H. (2001). The Filipino Americans. *Journal of American Ethnic History, 20,* 135–137.

Kim, H. C. (1999). *Koreans in the hood: Conflict with African Americans.* Baltimore, MD: Johns Hopkins University Press.

Kimbrough v. United States, 128 S. Ct. 558 (2007).

Kindermann, C., Lynch, J., & Cantor, D. (1997). *Effects of the redesign on victimization estimates.* Washington, DC: U.S. Department of Justice, Office of Justice Programs.

King, R. D. (2007). The context of minority group threat: Race, institutions, and complying with hate crime law. *Law & Society Review, 41,* 189–224.

King, W. R., Holmes, S. T., Henderson, M. L., & Latessa, E. J. (2001). Community corrections partnership: Examining the long-term effects of youth participation in an Afrocentric diversion program. *Crime & Delinquency, 47,* 558–572.

Kirk, D. S. (2008). The neighborhood context of racial and ethnic disparities in arrest. *Demography, 45,* 55–77.

Klaus, P., & Maston, C. (2000). *Criminal victimization in the United States, 1995.* Washington, DC: U.S. Department of Justice.

Klein, M. W. (1995). Street gang cycles. In J. Q. Wilson & J. Petersilia (Eds.), *Crime* (pp. 217–236). San Francisco, CA: Institute for Contemporary Studies.

Klinger, D. A., & Grossman, D. (2002). Who should deal with foreign terrorists on U.S. soil? Socio-legal consequences of September 11 and ongoing threat of terrorist attacks in America. *Harvard Journal of Law and Public Policy, 25,* 815–835.

Knepper, P. (2001). *Explaining criminal conduct: Theories and systems in criminology.* Durham, NC: Carolina Academic Press.

Knepper, P. E. (1996). Race, racism, and crime statistics. *Southern University Law Review, 24,* 71–112.

Knepper, P. E., & Potter, D. M. (1998). Crime, politics, and minority populations: Use of official statistics in the United States and Japan. *International Journal of Comparative and Applied Criminal Justice, 22,* 145–155.

Krauss, E., & Schulman, M. (1997). The myth of Black jury nullification: Racism dressed up in jurisprudential clothing. *Cornell Journal of Law and Public Policy, 7,* 57–76.

Krisberg, B. (1975). *Crime and privilege: Toward a new criminology.* Englewood Cliffs, NJ: Prentice Hall.

Krivo, L. J., & Peterson, R. D. (1996). Extremely disadvantaged neighborhoods and urban crime. *Social Forces, 75,* 619–650.

Krivo, L. J., & Peterson, R. D. (2000). The structural context of homicide: Accounting for racial differences in process. *American Sociological Review, 65,* 547–559.

Kroeger, B. (2004, March 20). *When a dissertation makes a difference.* Retrieved March 22, 2004, from http://www.nytimes.com/

Kubrin, C. E. (2005). Gangstas, thugs, and hustlas: Identity and the code of the streets in rap music. *Social Problems, 52,* 360–378.

Kusow, A., Wilson, L. C., & Martin, D. E. (1997). Determinants of citizen satisfaction with the police: The effects of residential location. *Policing, 20,* 655–664.

Kuykendall, M. A., & Burns, D. E. (1980). The Black police officer: An historical perspective. *Journal of Contemporary Criminal Justice, 1,* 4–12.

LaFree, G. (1995). Race and crime trends in the United States, 1946–1990. In D. F. Hawkins (Ed.), *Ethnicity, race, and crime* (pp. 169–193). Albany: State University of New York Press.

LaFree, G., & Russell, K. (1993). The argument for studying race and crime. *Journal of Criminal Justice Education, 4,* 273–289.

Lambert, E., & Clarke, A. (2001). The impact of information on an individual's support of the death penalty: A partial test of the Marshall hypothesis among college students. *Criminal Justice Policy Review, 12,* 215–234.

Lane, R. (1967). *Policing the city: Boston, 1822–1885.* Cambridge, MA: Harvard University Press.

Langbein, J. H. (1976). The historical origins of the sanction of imprisonment for serious crimes. *Journal of Legal Studies, 5,* 35–60.

Lanier, C., & Huff-Corzine, L. (2006). American Indian homicide: A county level analysis utilizing social disorganization theory. *Homicide Studies, 10,* 181–194.

Lanier, M. M., & Henry, S. (1998). *Essential criminology.* Boulder, CO: Westview.

Lattimore, P. K. (2007). The challenge of reentry. *Corrections Today, 69,* 88–91.

Lau, A. S., McCabe, K. M., Yeh, M., Garland, A. F., Hough, R. L., & Landsverk, J. (2003). Race/ethnicity and rates of self-reported maltreatment among high-risk youth in public sectors of care. *Child Maltreatment, 8,* 183–194.

Lee, C. (2007). Hispanics and the death penalty: Discriminatory charging practices in San Joaquin County, California. *Journal of Criminal Justice, 35,* 17–27.

Lee, M. T., & Martinez, R. (2002). Social disorganization revisited: Mapping the recent immigration and black homicide relationship in northern Miami. *Sociological Focus, 35,* 363–380.

Lee, N. (1995). Culture conflict and crime in Alaskan native villages. *Journal of Criminal Justice, 23,* 177–189.

Lehrer, E. (2007). Crime's up: An old issue is about to resurface. *The Weekly Standard, 12,* 23–26.

Leiber, M. J. (2002). Disproportionate minority confinement (DMC) of youth: An analysis of state and federal efforts to address the issue. *Crime & Delinquency, 48,* 3–45.

Leiber, M. J., & Blowers, A. N. (2003). Race and misdemeanor sentencing. *Criminal Justice Policy Review, 14,* 464–485.

Leiber, M. J., & Mack, K. Y. (2003). The individual and joint effects of race, gender, and family status on juvenile justice decision-making. *Journal of Research in Crime and Delinquency, 40,* 34–70.

Leo, R. A. (2002). False confessions: Causes, consequences, and solutions. In S. Westervelt & J. A. Humphrey (Eds.), *Wrongly convicted: Perspectives on failed justice* (pp. 36–54). New Brunswick, NJ: Rutgers University Press.

Leo, R. A. (2005). Rethinking the study of miscarriages of justice: Developing a criminology of wrongful conviction. *Journal of Contemporary Criminal Justice, 21,* 201–223.

Leonard, K. K., Pope, C., & Feyerherm, W. H. (1995). *Minorities in juvenile justice.* Thousand Oaks, CA: Sage Publications.

Levitt, S. D., & Dubner, S. J. (2005). *Freakonomics: A rogue economist explores the hidden side of everything.* New York: William Morrow.

Light, I. (1977). The ethnic vice industry, 1880–1944. *American Sociological Review, 42,* 464–479.

Lilly, R. J., Cullen, F. T., & Ball, R. A. (2001). *Criminological theory: Context and consequences.* Thousand Oaks, CA: Sage Publications.

Linderman, T. F., & Innerst, S. (2007, October 10). *Study finds racial profiling of shoppers is real, but it goes unreported.* Pittsburgh, PA: Pittsburgh Post-Gazette.

Loeber, R., Kalb, L., & Huizinga, D. (2001). *Juvenile delinquency and serious injury victimization* (Bulletin). Washington, DC: U.S. Department of Justice, Office of Justice Programs, Office of Juvenile Justice and Delinquency Prevention.

Lombroso, C. (1911). *Criminal man.* New York: Putnam. (Original work published 1876)

Longmire, D. R. (1996). Americans' attitudes about the ultimate weapon: Capital punishment. In T. J. Flanagan & D. R. Longmire (Eds.), *Americans view crime and justice: A national public opinion survey* (pp. 93–108). Thousand Oaks, CA: Sage Publications.

Lopez, J. (2000). Political incarceration. In J. James (Ed.), *States of confinement: Policing, detention, and prisons* (pp. 303–321). New York: St. Martin's Press.

Lopez, R., Roosa, M. W., Tein, J. T., & Dinh, K. T. (2004). Accounting for Anglo-Hispanic differences in school misbehavior. *Journal of Ethnicity in Criminal Justice, 2,* 27–46.

Lotke, E. (1996). *The prison-industrial complex.* Baltimore: National Center on Institutions and Alternatives.

Lotke, E. (1998). Hobbling a generation: Young African American men in Washington, D.C.'s, criminal justice system—five years later. *Crime and Delinquency, 44,* 355–366.

Lott, J. R., & Whitley, J. (2007). Abortion and crime: Unwanted children and out-of-wedlock births. *Economic Inquiry, 45,* 304–324.

Luna-Firebaugh, E. M. (2003, July 28–30). *Tribal law enforcement in P.L. 280 states.* Paper presented at the annual conference on Criminal Justice Research and Evaluation, Washington, DC.

Lundman, R. J., & Kaufman, R. L. (2003). Driving while Black: Effects of race, ethnicity, and gender on citizen self-reports of traffic stops and police actions. *Criminology, 41,* 195–220.

Lynn, R. (2002). Skin color and intelligence in African Americans. *Population and Environment, 23,* 365–375.

MacDonald, H. (2003). *Are cops racist?: How the war against the police harms Black Americans.* Chicago: Ivan R. Dee.

MacDonald, H. (2008). Is the criminal-justice system racist? *City Journal.* Retrieved May 6, 2008, from www.city-journal.org/.

Madriz, E. (1997). Latina teenagers: Victimization, identity, and fear of crime. *Social Justice, 24,* 39–55.

Maguire, K. (Ed.). (1995). Reported confidence in the police: Table 2.11. In *Sourcebook of criminal justice statistics* (p. 147). Bernan Associates.

Maguire, K. (Ed.). (2002). Reported confidence in the police: Table 2.16. In *Sourcebook of criminal justice statistics* (p. 108). Washington, DC: Author.

Maguire, K., & Pastore, A. L. (Eds.). (1996). Reported confidence in the police: Table 2.12. In *Sourcebook of criminal justice statistics* (p. 133). Diane Publishing.

Maguire, K., & Pastore, A. L. (Eds.). (1997). Reported confidence in the police: Table 2.11. In *Sourcebook of criminal justice statistics* (p. 119). Claitor.

Maguire, K., & Pastore, A. L. (Eds.). (1998). Reported confidence in the police: Table 2.17. In *Sourcebook of criminal justice statistics* (p. 107). Diane Publishing.

Maguire, K., & Pastore, A. L. (Eds.). (2000). Reported confidence in the police: Table 2.18. In *Sourcebook of criminal justice statistics* (p. 102). Washington, DC: Author.

Maguire, K., & Pastore, A. L. (Eds.). (2002). *Sourcebook of criminal justice statistics.* Retrieved from http://www.albany.edu/sourcebook/

Maguire, K., & Pastore, A. L. (Eds.). (2003). *Sourcebook of criminal justice statistics.* Retrieved from http://www.albany.edu/sourcebook/pdf/t213.pdf

Maguire, K., & Pastore, A. L. (Eds.). (2004). Reported confidence in the police: Table 2.13. In *Sourcebook of criminal justice statistics.* Retrieved from www.albany.edu/sourcebook/

Maguire, K., & Pastore, A. L. (Eds.). (2007). *Sourcebook of criminal justice statistics* [Online].

Males, M., & Macallair, D. (2000). *The color of justice: An analysis of juvenile adult court transfers in California.* Washington, DC: Building Blocks for Youth.

Maltz, M. (1977). Crime statistics: A historical perspective. *Crime & Delinquency, 23,* 32–40.

Manatu-Rupert, N. (2001). Media images and the victimization of Black women: Exploring the impact of sexual stereotyping on prosecutorial decision making. In D. Jones-Brown & M. Markowitz (Eds.), *The system in black and white* (pp. 181–196). Westport, CT: Praeger.

Mandak, J. (2003, June 12). *The Patriot-News* (AP), p. B7.

Mann, C. R. (1990a). Black female homicide in the United States. *Journal of Interpersonal Violence, 5,* 176–201.

Mann, C. R. (1990b). Random thoughts on the ongoing Wilbanks-Mann discourse. In B. D. Maclean & D. Milovanovic (Eds.), *Racism, empiricism, and criminal justice* (pp. 15–19). Vancouver, Canada: Collective Press.

Mann, C. R. (1993). *Unequal justice: A question of color.* Bloomington: Indiana University Press.

Manza, J., & Uggen, C. (2006). *Locked out: Felon disenfranchisement and democracy in America.* New York: Oxford University Press.

Manza, J., Brooks, C., & Uggen, C. (2004). Civil death or civil rights? Public attitudes towards felon disenfranchisement in the United States. *Public Opinion Quarterly, 68,* 276–287.

Mapp v. Ohio, 367 U.S. 1081, 81 S.CT (1961).

Marger, M. (Ed.). (1997). *Race and ethnic relations: American and global perspectives* (4th ed.). Belmont, CA: Wadsworth.

Marion, N. E., & Oliver, W. M. (2006). *The public policy of crime and criminal justice.* Upper Saddle River, NJ: Pearson/Prentice Hall.

Martin, D. L. (2002). The police role in wrongful convictions: An international comparative study. In S. Westervelt & J. A. Humphrey (Eds.), *Wrongly convicted: Perspectives on failed justice* (pp. 77–95). New Brunswick, NJ: Rutgers University Press.

Martinez, R. (2002). *Latino homicide.* New York: Routledge.

Martinez, R. (2003). Moving beyond black and white violence: African American, Haitian, and Latino homicides in Miami. In D. F. Hawkins (Ed.), *Violent crime: Assessing race and ethnic differences* (pp. 22–43). New York: Cambridge University Press.

Martinez, R. (2006). Coming to America: The impact of the new immigration on crime. In R. Martinez & A. Valenzuela (Eds.), *Immigration and crime: Race, ethnicity, and violence* (pp. 1–19). New York: New York University Press.

Martinez, R., Lee, M. T., & Nielsen, A. L. (2001). Revisiting the Scarface legacy: The victim/offender relationship and Mariel homicides in Miami. *Hispanic Journal of Behavioral Sciences, 23,* 37–56.

Martinez, R., & Valenzuela, A. (Eds.). (2006). *Immigration and crime: Race, ethnicity, and violence.* New York: New York University Press.

Martinson, R. (1974). What works? Questions and answers about prison reform. *Public Interest, 24,* 22–54.

Massey, D. S., & Denton, N. A. (1993). *American apartheid.* Cambridge, MA: Harvard University Press.

Mauer, M. (1990). *Young African American men and the criminal justice system: A growing national problem.* Washington, DC: The Sentencing Project.

Mauer, M. (1999). *Race to incarcerate.* New York: New Press.

Mauer, M. (2004). Extended view: Racial disparity and the criminal justice system: An assessment of causes and responses. *SAGE Race Relations Abstracts, 29,* 34–56.

Mauer, M. (2006). *Race to incarcerate* (rev. ed.). New York: New Press.

Mauer, M., & Huling, T. (1995). *Young African Americans and the criminal justice system: Five years later.* Washington, DC: The Sentencing Project.

Maupin, J. R., & Maupin, J. R. (1998). Juvenile justice decision making in a rural Hispanic community. *Journal of Criminal Justice, 26,* 373–384.

Mazerolle, L. G., & Terrill, W. (1997). Problem-oriented policing in public housing: Identifying the distribution of problem places. *Policing, 20,* 235–255.

McCamey, W. P., Scaramella, G. L., & Cox, S. M. (2003). *Contemporary municipal policing.* Boston: Allyn & Bacon.

McCleskey v. Kemp, 481 U.S. 279 (1987).

McCluskey, C. P. (2002). *Understanding Latino delinquency.* New York: LFB Scholarly Publishing.

McCluskey, C. P., Krohn, M. D., Lizotte, A. J., & Rodriguez, M. L. (2002). Early substance use and school achievement: An examination of Latino, White, and African-American youth. *Journal of Drug Issues, 32,* 921–944.

McCluskey, C. P., & Tovar, S. (2003). Family processes and delinquency: The consistency of relationships by ethnicity and gender. *Journal of Ethnicity in Criminal Justice, 1,* 37–61.

McDonald, H. (2003). *Are cops racist? How the war against the police harms Black Americans.* Chicago: Ivan R. Dee.

McGee, Z. T. (1999). Patterns of violent behavior and victimization among African American youth. *Journal of Offender Rehabilitation, 30,* 47–64.

McGee, Z. T. (2003). Community violence and adolescent development: An examination of risk and protective factors among African American youth. *Journal of Contemporary Criminal Justice, 19,* 293–314.

McGee, Z. T., & Baker, S. R. (2002). Impact of violence on problem behavior among adolescents. *Journal of Contemporary Criminal Justice, 18,* 74–93.

McGee, Z. T., Barber, A., Joseph, E., Dudley, J., & Howell, R. (2005). Delinquent behavior, violent victimization, and coping strategies among latino adolescents. *Journal of offender rehabilitation, 42*(3), 41–56.

McGuffee, K., Garland, T. S., & Eigenberg, H. (2007). Is jury selection fair? Perceptions of race and the jury selection process. *Criminal Justice Studies, 20,* 445–468.

McIntosh, P. (2002). White privilege, color, and crime: A personal account. In C. R. Mann & M. S. Zatz (Eds.), *Images of color, images of crime* (2nd ed., pp. 45–53). Los Angeles, CA: Roxbury.

McIntyre, C. C. L. (1992). *Criminalizing a race: Free Blacks during slavery.* New York: Kayode.

McKinney, K. (2003). *OJJDP's tribal youth initiatives*. Washington, DC: US Department of Justice, Office of Justice Programs, Office of Juvenile Justice and Delinquency Prevention.

McMahon, J., Garner, J., Davis, R., & Kraus, A. (2002). *How to correctly collect and analyze racial profiling data: Your reputation depends on it*. Washington, DC: US Department of Justice, Office of Community-Oriented Policing Services.

Meagher, T. J. (2005). *The Columbia guide to Irish American history*. New York: Columbia University Press.

Mears, D. (2002). Sentencing guidelines and the transformation of juvenile justice in the 21st century. *Journal of Contemporary Criminal Justice, 18*, 6–19.

Mears, D. P. (2007). Faith-based reentry programs: Cause for concern or showing promise? *Corrections Today, 69*, 30–33.

Mears, D. P., & Watson, J. (2006). Towards a fair and balanced assessment of supermax prisons. *Justice Quarterly, 23*, 232–270.

Meeks, K. (2000). *Driving while Black*. New York: Broadway Books.

Meier, A., & Rudwick, E. (1970). *From plantation to ghetto* (rev. ed.). New York: Hill & Wang.

Mennel, R. M. (1973). *Thorns and thistles*. Hanover: University of New Hampshire Press.

Menninger, K. (1966). *The crime of punishment*. New York: Viking Press.

Merton, R. K. (1938). Social structure and anomie. *American Sociological Review, 3,* 672–682.

Merton, R. K., & Montagu-Ashley, M. F. (1940). Crime and the anthropologist. *American Anthropologist, 42,* 384–408.

Messerschmidt, J. (1983). *The trial of Leonard Peltier*. Boston: South End Press.

Messner, S. F., Krohn, M. D., & Liska, A. E. (Eds.). (1989). *Theoretical integration in the study of deviance and crime: Problems and prospects*. Albany: State University of New York Press.

Miller, E. J. (2007). The therapeutic effects of managerial reentry courts. *Federal Sentencing Reporter, 20*, 127–135.

Miller, J. (2001). Breaking the individual back in: A commentary on Wacquant and Anderson. *Punishment & Society, 3,* 153–160.

Miller, J. G. (1992a). *Hobbling a generation: Young African American males in Washington, D.C.'s, criminal justice system*. Baltimore: National Center for Institutions and Alternatives.

Miller, J. G. (1992b). *Hobbling a generation: Young African American males in the criminal justice system of America's cities*. Baltimore: National Center for Institutions and Alternatives.

Miller, J. G. (1996). *Search and destroy: African American males in the criminal justice system*. New York: Cambridge University Press.

Miller, K. (1969). *Out of the house of bondage*. New York: Arno Press and *The New York Times*. (Original work published 1908)

Miller, W. (1958). Lower class culture as a generating milieu of gang delinquency. *Journal of Social Issues, 14,* 5–19.

Minnesota Lawyer Staff. (2008, January 14). *Minnesota advocates for human rights* express concern about immigration crackdown. Gale document no. A173409283.

Minton, T. D. (2006). *Jails in Indian country, 2004*. Washington, DC: Bureau of Justice Statistics.

Miranda v. Arizona, 384 U.S. 436,86 S.Ct. 1602 (1966).

Mirande, A. (1987). *Gringo justice*. Notre Dame, IN: Notre Dame University Press.

Mitchell, A. D. (2006). The effect of the Marshall hypothesis on attitudes towards the death penalty. *Race, Gender & Class, 13*, 221–239.

Mitchell, O. (2005). A meta-analysis of race and sentencing research: Examining the inconsistencies. *Journal of Quantitative Criminology, 21*, 439–466.

Monkkonen, E. H. (1981). *Police in urban America, 1860–1920*. Cambridge, NY: Cambridge University Press.

Moore, J. W., & Hagedorn, J. M. (2001). *Female gangs: A focus on research* (Bulletin. Youth Gang Series). Washington, DC: U.S. Department of Justice, Office of Juvenile Justice and Delinquency Prevention.

Morgan, K. D., & Smith, B. (2008). The role of race on parole decision-making. *Justice Quarterly, 25,* 411–435.

Morin, J. L. (2005). *Latino/a rights and justice in the United States: Perspectives and approaches.* Durham, NC: Carolina Academic Press.

Mosher, C. J., Miethe, T. D., & Phillips, D. M. (2002). *The mismeasure of crime.* Thousand Oaks, CA: Sage Publications.

Mumola, C. J. (2007). *Arrest-related deaths in the United States, 2003–2005.* Washington, DC: U.S. Department of Justice, Office of Justice Programs.

Munoz, E. A., & McMorris, B. J. (2002). Misdemeanor sentencing decisions: The cost of being Native American. *The Justice Professional, 15,* 239–259.

Munoz, E. A., McMorris, B. J., & DeLisi, M. J. (2005). Misdemeanor criminal justice: Contextualizing effects of Latino ethnicity, gender, and immigration status. *Race, Gender & Class, 11,* 112–134.

Myers, M. A. (1998). *Race, labor, & punishment in the New South.* Columbus: Ohio State University Press.

Myrdal, G. (1944). *An American dilemma: The Negro problem and modern democracy.* New York: Harper & Brothers.

National Advisory Commission on Civil Disorders. (1968). *Report of the National Advisory Commission on Civil Disorders.* Washington, DC: Government Printing Office.

National Center for Juvenile Justice. (2004). *State juvenile justice profiles.* Retrieved July 15, 2004, from http://www.ncjj.org/stateprofiles/

National Coalition to Abolish the Death Penalty. (2004). *America's shame: Killing kids.* Retrieved December 15, 2004, from http://www.ncadp.org/fact_sheet1.html

National Commission on Law Observance and Enforcement. (1931a). *Crime and the foreign born* (Report No. 10). Washington, DC: Government Printing Office.

National Commission on Law Observance and Enforcement. (1931b). *Report on lawlessness in law enforcement.* Washington, DC: U.S. Government Printing Office.

National Drug Control Strategy. (2008). Washington, DC: Office of the President.

National Institute on Drug Abuse. (2008*). InfoFacts—High School and Youth Trends.* Retrieved March 24, 2008, from http://www.nida.nih.gov/Infofacts/HSYouthtrends.html.

National Minority Advisory Council on Criminal Justice. (1979). *Racism in the criminal courts* (draft). Washington, DC: U.S. Department of Justice.

National Organization of Black Law Enforcement Executives. (2001). *A NOBLE perspective: Racial profiling: A symptom of biased-based policing.* Retrieved July 31, 2004, from http://www.noblenational.org/pdf/RacialProfiling901.pdf

National Urban League Poll. (2001). Retrieved March 17, 2004, from Polling the Nations database at http://www.orspub.com/

National Youth Gang Center. (2002). *Planning for implementation.* Tallahassee, FL: Author.

National Youth Gang Center. (2008). *Frequently asked questions regarding gangs.* Retrieved March 20, 2008, from http://www.iir.com/nygc/faq.htm#q3.

Neubauer, D. (2002). *America's courts & the criminal justice system* (7th ed.). Belmont, CA: Wadsworth.

Neverdon-Morton, C. (1989). *Afro-American women of the South and the advancement of the race, 1895–1925.* Knoxville: University of Tennessee Press.

Newport, F., & Ludwig, J. (2000). *Protests by Blacks over Amadou Diallo verdict not surprising given long-standing perceptions among Blacks that they are discriminated against in most areas of their daily lives.* Retrieved from http://www.gallup.com/poll/fromtheed/ed003.asp

New York State Special Commission on Attica. (1972). *Attica: The official report of the New York State Special Commission on Attica.* New York: Praeger.

Nolan, J. J. III, Akiyama, Y., & Berhanu, S. (2002). The Hate Crime Statistics Act of 1990: Developing a method for measuring the occurrence of hate violence. *American Behavioral Scientist, 46,* 136–153.

Northern v. Nelson, 315 F. Supp. 687 (1970).

Norton, H. K. (2004). Virtual museum of the City of San Francisco. *The Chinese.* Retrieved from http://www.sfmusuem.org

O'Connor, J. P. (2008). *The framing of Mumia-Abu Jamal.* Chicago: Lawrence Hill Books.

Office of Juvenile Justice and Delinquency Prevention. (2004). *Disproportionate minority contact. About DMC: Core requirement of JJDP Act.* Retrieved July 15, 2004, from http://www.ojjdp.ncjrs .org/dmc/about/core.html

Office of Juvenile Justice and Delinquency Prevention. (2008). *Statistical briefing book.* Retrieved March 20, 2008, from http://ojjdp.ncjrs.gov/ojstatbb/crime/qa05101.asp?qaDate=2006.

Ogletree, C. J., & Sarat, A. (Eds.). (2006). *From lynch mobs to the killing state: Race and the death penalty in America.* New York: New York University.

Oliver, W. (1984). Black males and the tough guy image: A dysfunctional compensatory adaptation. *The Western Journal of Black Studies, 8,* 199–203.

Oliver, W. (1989a). Black males and social problems: Prevention through Afrocentric socialization. *Journal of Black Studies, 20,* 15–39.

Oliver, W. (1989b). Sexual conquest and patterns of Black-on-Black violence: A structural-cultural perspective. *Violence and Victims, 4,* 257–273.

Oliver, W. (1994). *The violent social world of Black men.* New York: Lexington Books.

Oliver, W. (2003). The structural-cultural perspective: A theory of Black male violence. In D. F. Hawkins (Ed.), *Violent crime: Assessing race and ethnic differences* (pp. 280–318). Cambridge, UK: Cambridge University Press.

Oliver, W. (2006). "The streets": An alternative Black male socialization institution. *Journal of Black Studies, 36,* 918–937.

Onwudiwe, I. D., & Lynch, M. J. (2000). Reopening the debate: A reexamination of the need for a Black criminology. *Social Pathology: A Journal of Reviews, 6,* 182–198.

Oshinsky, D. M. (1996). *"Worse than slavery": Parchman farm and the ordeal of Jim Crow justice.* New York: The Free Press.

Packer, H. (1968). *The limits of the criminal sanction.* Palo Alto, CA: Stanford University Press.

Pager, D. I. (2007a). *Marked: Race, crime, and finding work in an era of mass incarceration.* Chicago: University of Chicago Press.

Pager, D. I. (2007b). Two strikes and you're out: The intensification of racial and criminal stigma. In D. Weiman, S. Bushway, & M. Stoll (Eds.), *Barriers to reentry? The labor market for released prisoners in post-industrial America* (pp. 151–173). New York: Russell Sage.

Palmer, J. W., & Palmer, S. E. (1999). *Constitutional rights of prisoners* (6th ed.). Cincinnati, OH: Anderson Publishing.

Parade Magazine Poll. (2002, February 10). Retrieved March 17, 2004, from Polling the Nations database at http://www.orspub.com/

Parenti, C. (1999). *Lockdown America: Police and prisons in the age of crisis.* New York: Verso.

Parker, K. F., Dewees, M. A., & Radelet, M. L. (2002). Racial bias and the conviction of the innocent. In S. Westervelt & J. A. Humphrey (Eds.), *Wrongly convicted: Perspectives on failed justice* (pp. 114–131). New Brunswick, NJ: Rutgers University Press.

Pastore, A. L., & Maguire, K. (Eds.). (1999). Reported confidence in the police: Table 2.18. In *Sourcebook of criminal justice statistics* (p. 105). Diane Publishing.

Pastore, A. L., & Maguire, K. (Eds.). (2001). Reported confidence in the police: Table 2.16. In *Sourcebook of criminal justice statistics* (p. 109). Washington, DC: Author.

Pastore, A. L., & Maguire, K. (Eds.). (2003). Reported confidence in the police: Table 2.13. In *Sourcebook of criminal justice statistics* (p. 116). Diane Publishing.

Pastore, A. L., & Maguire, K. (Eds.). (2007). *Sourcebook of criminal justice statistics* [Online]. Available from http://www.albany.edu/sourcebook.

Paternoster, R. (1983). Race of the victim and location of crime: The decision to seek the death penalty in South Carolina. *Journal of Criminal Law and Criminology, 74,* 701–731.

Patterson, W. L. (Ed.). (1970). *We charge genocide: The historic petition to the United Nations for relief from a crime of the United States government against the Negro people.* New York: International Publishers. (Original work published 1951)

Pattillo, M. E. (1998). Sweet mothers and gangbangers: Managing crime in a Black middle-class neighborhood. *Social Forces, 76,* 747–774.

Penn, E., Taylor Greene, H., & Gabbidon, S. L. (Eds.). (2005). *Race and juvenile justice.* Durham, NC: Carolina Academic Press.

Pennsylvania panel advises death penalty moratorium. (2003). Retrieved March 5, 2003, from www.CNN.com

Pennsylvania Supreme Court. (2003). *Pennsylvania Supreme Court Committee on Racial and Gender Bias in the Justice System* (Final report). Retrieved from www.courts.state.pa.us

Perloff, R. M. (2000). The press and lynchings of African Americans. *Journal of Black Studies, 30,* 315–330.

Perry, B. (2000). Perpetual outsiders: Criminal justice and the Asian American experience. In The Criminal Justice Collective of Northern Arizona University (Eds.), *Investigating difference: Human and cultural relations in criminal justice* (pp. 99–110). Boston: Allyn & Bacon.

Perry, B. (2002). Defending the color line: Racially and ethnically motivated hate crime. *American Behavioral Scientist, 46,* 72–92.

Perry, S. W. (2004). *Census of Tribal Justice Agencies in Indian Country, 2002.* Washington, DC: U.S. Department of Justice.

Petersilia, J. (1983). *Racial disparities in the criminal justice system.* Santa Monica, CA: Rand.

Petersilia, J. (2002). Community corrections. In J. Q. Wilson & J. Petersilia (Eds.), *Crime: Public policies for crime control* (pp. 483–508). Oakland, CA: Institute for Contemporary Studies.

Petersilia, J. (2003). *When prisoners come home: Parole and prisoner reentry.* New York: Oxford University Press.

Peterson, D., Miller, J., & Esbensen, F. (2001). The impact of sex composition on gangs and gang delinquency. *Criminology, 39*(2), 411–439.

Peterson, D., Taylor, T. J., & Esbensen, F. (2004). Gang membership and violent victimization. *Justice Quarterly, 21*(4), 794–815.

Peterson, R. D., & Krivo, L. J. (1993). Racial segregation and Black urban homicide. *Social Forces, 71,* 1001–1026.

Peterson, R. D., & Krivo, L. J. (2005). Macrostructural analyses of race, ethnicity, and violent crime: Recent lessons and new directions for research. *Annual Review of Sociology, 31,* 331–356.

Phillips, U. B. (1915). Slave crime in Virginia. *American Historical Review, 20,* 336–340.

Piquero, N. L., & Piquero, A. R. (2001). Problem-oriented policing. In R. G. Dunham & G. P. Alpert (Eds.), *Critical issues in policing* (4th ed., pp. 531–540). Prospect Heights, IL: Waveland Press.

Platt, A. (1969). *The child savers: The invention of delinquency.* Chicago: University of Chicago Press.

Plessy v. Ferguson, 163 U.S. 537 (1896).

Polk, W. R. (2006). The birth of America: From Columbus to the Revolution. New York: HarperCollins.

Pope, C. E., & Feyerherm, W. H. (1990, June/September). Minority status and juvenile justice processing: An assessment of the research literature, Parts I & II. *Criminal Justice Abstracts, 22,* 327–336 (Part I); *22,* 527–542 (Part II).

Pope, C. E., Lovell, R., & Hsia, H. M. (2002). *Disproportionate minority confinement: A review of the research literature from 1989 through 2001.* Washington, DC: Office of Juvenile Justice and Delinquency Prevention.

Potter, H. (Ed.). (2007). *Racing the storm: Racial implications and lessons learned from Hurricane Katrina.* Lanham, MD: Lexington Books.

Powell v. Alabama, 287 U.S. 45 (1932).

Pratt, T. C. (1998). Race and sentencing: A meta-analysis of conflicting empirical research results. *Journal of Criminal Justice, 26,* 513–523.

President's Commission on Law Enforcement and the Administration of Justice. (1967). *Task force report on the police.* Washington, DC: U.S. Government Printing Office.

Preuhs, R. R. (2001). State felon disenfranchisement policy. *Social Science Quarterly, 82,* 733–748.

Price, B. E. (2006). *Merchandizing prisoners: Who really pays for prison privatization?* Westport, CT: Praeger.

Pridemore, W. A., & Freilich, J. D. (2006). A test of recent subcultural explanations of white violence in the United States. *Journal of Criminal Justice, 34,* 1–16.

Public Policy Institute of California. (2008). *Poll data.* Retrieved March 24, 2008, from Polling the Nations database.

Public Report of the Forgotten Victims of Attica. (2003). Retrieved June 4, 2004, from http://newyorkv.homestead.com/files/fvoareportfinal.htm

Puzzanchera, C., Stahl, A., Finnegan, T. A., Tierney, N., & Snyder, H. (2003a). *Juvenile court statistics, 1998.* Washington, DC: Office of Juvenile Justice and Delinquency Prevention.

Puzzanchera, C., Stahl, A. Finnegan, T. A., Tierney, N., & Snyder, H. (2003b). *Juvenile court statistics, 1999.* Washington, DC: Office of Juvenile Justice and Delinquency Prevention.

Quetelet, A. Q. (1984). *Research on the propensity for crime at different ages.* Cincinnati, OH: Anderson. (Original work published 1833)

Quinney, R. (1970). *The social reality of crime.* Boston: Little, Brown.

Racial Profiling Data Collection Resource Center at Northeastern University. (2003). Retrieved from http://www.racialprofilinganalysis.neu.edu

Racial Profiling Data Collection Resource Center at Northeastern University. (2008). Retrieved February 23, 2008, from http://www.racialprofilinganalysis.neu.edu

Radelet, M. L. (1989). Executions of Whites for crimes against Blacks: Exceptions to the rule? *The Sociological Quarterly, 30,* 529–544.

Radelet, M. L., Bedau, H. A., & Putnam, C. E. (1992). *In spite of innocence: The ordeal of 400 Americans wrongly convicted of crimes punishable by death.* Boston: Northeastern University Press.

Radosh, P. F. (2008). War on drugs: Gender and race inequities in crime control strategies. *Criminal Justice Studies, 21,* 167–178.

Ramsey, R. J., & Frank, J. (2007). Wrongful convictions: Perceptions of criminal justice professionals regarding the frequency of wrongful convictions and system errors. *Crime & Delinquency, 53,* 436–470.

Rand, M., & Catalano, S. (2007). *Criminal victimizations, 2006.* Washington, DC: U.S. Department of Justice, Office of Justice Programs.

Rantala, R. R., & Edwards, T. J. (2000). *Effects of NIBRS on crime statistics.* Washington, DC: U.S. Department of Justice, Office of Justice Programs.

Raper, A. (1933). *The tragedy of lynching.* Chapel Hill: University of North Carolina Press.

Ray, M. C., & Smith, E. (1991). Black women and homicide: An analysis of the subculture of violence thesis. *The Western Journal of Black Studies, 15,* 144–153.

Reasons, C. E., Conley, D. C., & Debro, J. (Eds.). (2002). *Race, class, gender, and justice in the United States.* Boston: Allyn & Bacon.

Reaves, B. (2006). *Federal law enforcement officers, 2004.* Washington, DC: U.S. Department of Justice. Retrieved February 16, 2008, from www.ojp.usdoj.gov/bjs/welcome.html

Reaves, B. (2007). *Census of state and local law enforcement agencies, 2004.* Washington, DC: U.S. Department of Justice.

Reaves, B., & Bauer, L. (2003). *Federal law enforcement officers, 2002.* Washington, DC: U.S. Department of Justice.

Reaves, B., & Hickman, M. (2002a). *Census of state and local law enforcement agencies, 2000.* Washington, DC: U.S. Department of Justice.

Reaves, B., & Hickman, M. (2002b). *Police departments in large cities, 1990–2000.* Washington, DC: U.S. Department of Justice.

Reaves, B., & Hickman, M. (2003). *Local police departments, 2000.* Washington, DC: U.S. Department of Justice.

Reaves, B. A. (2001). *Felony defendants in large urban counties, 1998.* Washington, DC: Bureau of Justice Statistics.

Reaves, B. A., & Perez, J. (1994). *Pretrial release of felony defendants, 1992.* Washington, DC: Bureau of Justice Statistics.

Reid, I. De A. (1957). Race and crime. *Friends Journal, 3,* 772–774.

Reiman, J. (2004). *The rich get richer, and the poor get prison: Ideology, class and criminal justice* (7th ed.). Boston: Allyn & Bacon.

Reiman, J. (2007). *The rich get richer and the poor get prison: Ideology, class, and criminal justice* (8th ed.). Boston: Allyn & Bacon.

Reisig, M. D., & Parks, R. B. (2001). *Satisfaction with police—What matters?* Washington, DC: U.S. Department of Justice.

Rennison, C. M. (2001a). *Criminal victimization 2000 changes 1999–2000 with trends 1992–2000.* Washington, DC: U.S. Department of Justice, Office of Justice Programs.

Rennison, C. M. (2001b). *Violent victimization and race, 1992–1998.* Washington, DC: U.S. Department of Justice, Office of Justice Programs.

Rennison, C. M. (2002b). *Hispanic victims of violent crime, 1993–2000* (NCJ 191208). Washington, DC: U.S. Department of Justice, Office of Justice Programs.

Rennison, C. M., & Rand, M. R. (2003). *Criminal victimization, 2002.* Washington, DC: U.S. Department of Justice, Office of Justice Programs.

Renowned criminologist eschews alarmist theories. (1999, September 13). *The Patriot News,* p. B5.

Richardson, J. F. (1970). *The New York police.* New York: Oxford University Press.

Richie, B. (2002). The social impact of mass incarceration on women. In M. Mauer & M. Chesney-Lind (Eds.), *Invisible punishment: The collateral consequences of mass imprisonment* (pp. 136–149). New York: New Press.

Rippley, L. J. (1976). *The German-Americans.* Boston: Twayne Publishers.

Roane, K. R. (2001, April 30). Policing the police is a dicey business. *U.S. News & World Report,* p. 28.

Robinson, L. N. (1911). *History and organization of criminal statistics in the United States.* Boston: Houghton Mifflin.

Robinson, M. B. (2002). *Justice blind? Ideals and realities of American criminal justice.* Upper Saddle River, NJ: Prentice Hall.

Rocque, M. (2008). Strain, coping mechanisms, and slavery: A general strain theory application. *Crime, Law, and Social Change, 49,* 245–269.

Rodriguez, N. (2007). Juvenile court context and detention decisions: Reconsidering the role of race, ethnicity, and community characteristics in juvenile court processes. *Justice Quarterly, 24,* 629–656.

Roe v. Wade, 410 U.S. 113 (1973).

Roper v. Simmons, 543 U.S. 551 (2005).

Rose, D. R., & Clear, T. R. (1998). Incarceration, social capital, and crime: Implications for social disorganization theory. *Criminology, 36,* 441–479.

Ross, L. E. (1992). Blacks, self-esteem, and delinquency: It's time for a new approach. *Justice Quarterly, 9,* 609–624.

Ross, L. E. (2001). African-American interest in law enforcement: A consequence of petit apartheid? In D. Milovanovic & K. K. Russell (Eds.), *Petit apartheid in the U.S. criminal justice system: The dark figure of racism* (pp. 69–77). Durham, NC: Carolina Academic Press.

Rudwick, E. M. (1960). The Negro policeman in the South. *Journal of Criminal Law, Criminology, and Police Science, 51,* 273–276.

Rushton, J. P. (1995). Race and crime: International data for 1989–1990. *Psychological Reports, 76,* 307–312.

Rushton, J. P. (1999). *Race, evolution, and behavior* (special abridged ed.). New Brunswick, NJ: Transaction.

Rushton, J. P., & Whitney, G. (2002). Cross-national variation in violent crime rates: Race, *r-K* theory, and income. *Population and Environment, 23,* 501–511.

Russell, K. K. (1998). *The color of crime: Racial hoaxes, White fear, Black protectionism, police harassment, and other macroaggressions.* New York: New York University Press.

Russell, K. K. (1999). Critical race theory and social justice. In B. A. Arrigo (Ed.), *Social justice/ criminal justice: The maturation of critical theory in law, crime, and deviance* (pp. 178–188). Belmont, CA: West/Wadsworth.

Russell, K. K. (2002). "Driving while Black": Corollary phenomena and collateral consequences. In C. E. Reasons, D. J. Conley, & J. Debro (Eds.), *Race, class, gender, and justice in the United States* (pp. 191–200). Boston: Allyn & Bacon.

Russell-Brown, K. (2004). *Underground codes: Race, crime, and related fires.* New York: New York University Press.

Russell-Brown, K. (2006). While visions of deviance danced in their heads. In D. D. Trout (Ed.), *After the storm: Black intellectuals explore the meaning of Hurricane Katrina* (pp. 111–123). New York: New Press.

Sabol, W. J., & Minton, T. D. (2008). *Jail inmates at midyear 2007.* Washington, DC: Bureau of Justice Statistics.

Sabol, W. J., Minton, T. D., & Harrison, P. M. (2007). *Prison and jail inmates at midyear 2006.* Washington, DC: Bureau of Justice Statistics.

Sachar, H. M. (1993). *A history of Jews in America.* New York: Vintage.

Sale, K. (1990). *The conquest of paradise: Christopher Columbus and the Columbian legacy.* New York: Knopf.

Saleh-Hanna, V. (Ed.). (2008). *Colonial systems of control: Criminal justice in Nigeria.* Ottawa: University of Ottawa Press.

Sampson, R. J. (1985). Race and criminal violence: A demographically disaggregated analysis of urban homicide. *Crime and Delinquency, 31,* 47–82.

Sampson, R. J. (1987). Urban Black violence: The effects of male joblessness and family disruption. *American Journal of Sociology, 93,* 348–382.

Sampson, R. J., & Bean, L. (2006). Cultural mechanisms and killing fields: A revised theory of community-level racial inequality. In J. Hagan, R. Peterson, & L. Krivo (Eds.), *The many colors of crime: Inequalities of race, ethnicity, and crime in America* (pp. 8–36). New York: New York University Press.

Sampson, R. J., & Groves, W. B. (1989). Community structure and crime: Testing social-disorganization theory. *American Journal of Sociology, 94,* 774–802.

Sampson, R. J., Raudenbush, S. W., & Earls, F. (1997). Neighborhoods and violent crime: A multi-level study of collective efficacy. *Science, 277,* 918–924.

Sampson, R. J., & Wilson, W. J. (1995). Toward a theory of race, crime, and urban inequality. In J. Hagan & R. D. Peterson (Eds.), *Crime and inequality* (pp. 37–54). Stanford, CA: Stanford University Press.

Scalia, J. (1999). *Federal pretrial release and detention, 1996.* Washington, DC: Bureau of Justice Statistics.

Schafer, J. A., Carter, D. L., & Katz-Bannister, A. (2004). Studying traffic stop encounters. *Journal of Criminal Justice, 32,* 159–170.

Scheingold, S. A. (1984). *The politics of law and order: Street crime and public policy.* New York: Longman.

Schlesinger, T. (2005). Racial and ethnic disparity in pretrial case processing. *Justice Quarterly, 22,* 170–192.

Scott, G. (2001). Broken windows behind bars: Eradicating prison gangs through ecological hardening and symbol cleansing. *Corrections Management Quarterly, 5,* 23–36.

Sellin, T. (1928). The Negro criminal: A statistical note. *The American Academy of Political and Social Sciences, 130,* 52–64.

Sellin, T. (1935). Race prejudice in the administration of justice. *The American Journal of Sociology, XLI,* 212–217.

Sellin, T. (1938). *Culture conflict and crime.* New York: Social Science Research Council.

Sellin, T. J. (1976). *Slavery and the penal system.* New York: Elsevier.

Shane, S. (1999, April 4). Genetic research increasingly finds "race" a null concept. *The Baltimore Sun,* pp. 1A, 6A.

Shapiro, A. L. (1997). The disenfranchised. *American Prospect, 35,* 60–62.

Sharp, E. B. (2006). Policing urban America: A new look at the politics of agency size. *Social Science Quarterly, 87,* 291–307.

Shaw, C., & McKay, H. D. (1969). *Juvenile delinquency in urban areas.* Chicago: University of Chicago Press. (Original work published 1942)

Shelden, R. G. (2001). *Controlling the dangerous classes.* Needham Heights, MA: Allyn & Bacon.

Shelden, R. G., & Brown, W. (2003). *Criminal justice in America: A critical view.* Needham Heights, MA: Allyn & Bacon.

Shelden, R. G., & Osborne, L. T. (1989). "For their own good": Class interests and the child saving movement in Memphis, Tennessee, 1900–1917. *Criminology, 27,* 747–767.

Sherman, L. (1997). Introduction: The congressional mandate to evaluate. In L. W. Sherman, D. Gottfredson, D. MacKenzie, J. P. Eck, P. Reuter, & S. Bushway (Eds.), *Preventing crime: What works, what doesn't, what's promising.* Washington, DC: National Institute of Justice.

Sherman, L. W. (1974). *Police corruption: A sociological perspective.* Garden City, NY: Anchor Books.

Sherman, L. W. (1980). Execution without trial: Police homicide and the Constitution. *Vanderbilt Law Review, 33,* 71–100. (Reprinted in J. F. Fyfe, Ed., 1982, *Readings on police use of deadly force,* pp. 88–89.)

Sherman, L. W., Gottfredson, D., MacKenzie, D., Eck, J. P., Reuter, P., & Bushway, S. (Eds.). (1997). *Preventing crime: What works, what doesn't, what's promising.* Washington, DC: National Institute of Justice.

Short, J., & Sharp, C. (2005). *Disproportionate minority contact in the juvenile justice system.* Washington, DC: Child Welfare League of America.

Siegel, L. J. (2002). *Juvenile delinquency: The core.* Belmont, CA: Wadsworth/ Thomson Learning.

Sigelman, L., Welch, S., Bledsoe, T., & Combs, M. (1997). Police brutality and public perceptions of racial discrimination: A tale of two beatings. *Political Research Quarterly, 50,* 777–791.

Simmons v. Roper, S.C. 84454 (2003).

Simons, D. H. (2003, June 23). Genetic tests can reveal ancestry, giving police a new source of clues. *U.S. News & World Report, 134,* p. 50.

Simons, R. L., Chen, Y. F., Stewart, E. A., & Brody, G. H. (2003). Incidents of discrimination and risk for delinquency: A longitudinal test of strain theory with an African American sample. *Justice Quarterly, 20,* 827–854.

Simons, R. L., Gordon Simons, L., Harbin Burt, C., Brody, G. H., & Cutrona, C. (2005). Collective efficacy, authoritative parenting and delinquency: A longitudinal test of a model integrating community-and-family-level processes. *Criminology, 43,* 989–1029.

Sims, Y. (2009). Jena 6. In H. T. Greene & S. L. Gabbidon (Eds.), *Encyclopedia of race and crime.* Thousand Oaks, CA: Sage Publications.

Skogan, W. G., Steiner, L., DuBois, J., Gudell, J. E., & Fagan, A. (2002). *Community policing and "the new immigrants": Latinos in Chicago.* Chicago: Northwestern University Press.

Smith, J. (2009). Native American courts. In H. T. Greene & S. L. Gabbidon (Eds.), *Encyclopedia of race and crime.* Thousand Oaks, CA: Sage Publications.

Smith, S. K., Steadman, G. W., Minton, T. D., & Townsend, M. (1999). Criminal victimization and perceptions of community safety in 12 cities, 1998. Washington, DC: U.S. Department of Justice.

Snell, T. L. (2007). *Capital punishment, 2006.* Washington, D.C.: Bureau of Justice Statistics.

Snyder, H. N. (1997). *Juvenile arrests 1995.* Washington, DC: Office of Juvenile Justice and Delinquency Prevention.

Snyder, H. N. (2000). *Juvenile arrests 1999.* Washington, DC: Office of Juvenile Justice and Delinquency Prevention.

Snyder, H. N. (2003). *Juvenile arrests 2001.* Washington, DC: Office of Juvenile Justice and Delinquency Prevention.

Snyder, H. N., Finnegan, T., & Kang, W. (2007). *Easy access to the FBI's Supplementary Homicide Reports: 1980–2005.* Retrieved March 24, 2008, from http://ojjdp.ncjrs.gov/ojstatbb/esashr/

Snyder, M., & Sickmund, M. (1999). *Juvenile offenders and victims: 1999 national report.* Washington, DC: Office of Juvenile Justice and Delinquency Prevention.

Snyder, M., & Sickmund, M. (2008). *Juvenile offenders and victims: 2006 national report.* Washington, DC: Office of Juvenile Justice and Delinquency Prevention.

Sommers, S. R., & Norton, M. I. (2007). Race-based judgments, race-neutral justifications: Experimental examination of peremptory use and the *Batson* challenge procedure. *Law and Human Behavior, 31,* 261–273.

Sorenson, J., Hope, R., & Stemen, D. (2003). Racial disproportionality in state prison admissions: Can regional variation be explained by differential arrest rates? *Journal of Criminal Justice, 31,* 73–84.

Sowell, T. (1981). *Ethnic America: A history.* New York: Basic Books.

Spilde, K. (2001). *The economic development journey of Indian nations.* Retrieved August 6, 2004, from http://www.indiangaming.org/library/newsletters/index.html

Spohn, C. S. (2000). Thirty years of sentencing reform: The quest for a racially neutral sentencing process. In J. Horney (Ed.), *Policies, processes and decisions of the criminal justice system* (pp. 427–501). Washington, DC: National Institute of Justice.

Spohn, C. S. (2002). *How do judges decide? The search for fairness and justice in punishment.* Thousand Oaks, CA: Sage Publications.

Stahl, A. L. (2008a). *Delinquency cases in juvenile court, 2004* (OJDP Fact Sheet no. 01). Washington, DC: Office of Juvenile Justice and Delinquency Prevention.

Stahl, A. L. (2008b). *Petitioned status offense cases in juvenile court, 2004* (OJDP Fact Sheet no. 02). Washington, DC: Office of Juvenile Justice and Delinquency Prevention.

Stanford v. Kentucky, 492 U.S. 361 (1989).

Staples, R. (1975). White racism, Black crime, and American justice: An application of the colonial model to explain crime and race. *Phylon, 36,* 14–22.

Starbuck, D., Howell, J. C., & Lindquist, D. J. (2001). *Hybrid and other modern gangs* (Bulletin, Youth Gang Series). Washington, DC: U.S. Department of Justice, Office of Justice Programs, Office of Juvenile Justice and Delinquency Prevention.

State v. Soto, 324 N.J. Super 66, 734 A.2d 35 (1996).

Stauffer, A. R., Smith, M. D., Cochran, J. K., Fogel, S. J., & Bjerregaard, B. (2006). The interaction between victim race and gender on sentencing outcomes in capital murder trials: A further exploration. *Homicide Studies, 10,* 98–117.

Steffensmeier, D., & Britt, C. L. (2001). Judges' race and judicial decision making: Do Black judges sentence differently? *Social Science Quarterly, 82,* 749–764.

Steffensmeier, D., & Demuth, S. (2000). Ethnicity and sentencing outcomes in U.S. federal courts: Who is punished more harshly? *American Sociological Review, 65,* 705–729.

Steffensmeier, D., & Demuth, S. (2001). Ethnicity and judges' sentencing decisions: Hispanic-Black-White comparisons. *Criminology, 39,* 145–178.

Stewart, E. A. (2007). Either they don't know or they don't care: Black males and negative police experiences. *Criminology & Public Policy, 6,* 123–130.

Stewart, E., & Simons, R. L. (2006). Structure and culture in African-American adolescent violence: A partial test of the code of the street thesis. *Justice Quarterly, 23,* 1–33.

Stewart, E. A., Simons, R. L., & Conger, R. D. (2002). Assessing neighborhood and social psychological influences on childhood violence in an African American sample. *Criminology, 40,* 801–829.

Stewart, E. A., Schreck, C. J., & Brunson, R. K. (2008). Lessons of the street code: Policy implications for reducing violent victimization among disadvantaged citizens. *Journal of Contemporary Criminal Justice, 24,* 137–147.

Stewart, E. A., Schreck, C. J., & Simons, R. L. (2006). I ain't gonna let no one disrespect me: Does the code of the street reduce or increase violent victimization among African American adolescents? *The Journal of Research in Crime and Delinquency, 43,* 427–457.

Stowell, J. I. (2007). *Immigration and crime: Considering the direct and indirect effects of immigration on violent criminal behavior.* New York: LFB Scholarly Press.

Streib, V. (2004). *The juvenile death penalty today: Death sentences and executions for juvenile crimes.* Retrieved April 30, 2004, from http://www.law.onu.edu/faculty/streib/ JuvDeathApr302004.pdf

Suavecito's (2004). *Zoot suit riots.* Retrieved December 16, 2004, from http://www.suavecito.com/ history.htm

Subcommittee on Civil and Constitutional Rights. (1994). *Racial disparities in federal death penalty prosecutions: 1988–1994* (Staff report). Washington, DC: 103rd Congress, Second Session.

Sudbury, J. (Ed.). (2005). *Global lockdown: Race, gender and the prison industrial complex.* New York: Routledge.

Sutherland, E. H. (1947). *Principles of criminology* (4th ed.). Philadelphia: Lippincott.

Swain v. Alabama, 380 U.S. 202 (1965).

Swan, A. L. (1977). *Families of Black prisoners: Survival and progress.* Boston: G. K. Hall.

Takagi, P. (1974). A garrison state in a "democratic" society. *Crime and Social Justice: A Journal of Radical Criminology, 5,* 27–33. (Reprinted in J. F. Fyfe, Ed., 1982, *Readings on police use of deadly force,* pp. 195–213.)

Takagi, P. (1975). The Walnut Street jail: A penal reform to centralize the powers of the state. *Federal Probation, 39,* 18–26.

Takaki, R. (1989). *Strangers from a different shore: A history of Asian Americans.* Boston: Little, Brown.

Tarlow, M., & Nelson, M. (2007). The time is now: Immediate work for people coming home from prison as a strategy to reduce their reincarceration and restore their place in the community. *Federal Sentencing Reporter, 20,* 138–140.

Tarver, M., Walker, S., & Wallace, H. (2002). *Multicultural issues in the criminal justice system.* Boston: Allyn & Bacon.

Tatum, B. L. (1994). The colonial model as a theoretical explanation of crime and delinquency. In A. Sulton (Ed.), *African-American perspectives on: Crime causation, criminal justice administration, and crime prevention* (pp. 33–52). Englewood, CO: Sulton Books.

Tatum, B. L. (2000). Deconstructing the association of race and crime: The salience of skin color. In M. W. Markowitz & D. Jones-Brown (Eds.), *The system in Black and White: Exploring the connections between race, crime, and justice* (pp. 31–46). Westport, CT: Praeger.

Taylor, B. M. (1989). *New directions for the National Crime Survey.* Washington, DC: U.S. Department of Justice, Office of Justice Programs.

Taylor, D. L., Biafora, F. A., Warheit, G., & Gil, A. (1997). Family factors, theft, vandalism, and major deviance among a multiracial/multiethnic sample of adolescent girls. *Journal of Social Distress and the Homeless, 6,* 71–87.

Taylor, P. (1931). The problem of the Mexican. In *Crime and the foreign born* (National Commission on Law Observance and Enforcement Report No. 10, pp. 199–243). Washington, DC: U.S. Government Printing Office.

Taylor Greene, H. (2004). Do African American police make a difference? In M. D. Free, Jr. (Ed.), *Racial issues in criminal justice: The case of African Americans* (pp. 207–220). Monsey, NY: Criminal Justice Press.

Taylor Greene, H., & Gabbidon, S. (2000). *African American criminological thought.* Albany, NY: State University of New York Press.

Taylor Greene, H., Gabbidon, S. L., & Ebersole, M. (2001). A multi-faceted analysis of the African American presence in juvenile delinquency textbooks published between 1997 and 2000. *Journal of Crime and Justice, 24,* 87–101.

Taylor Greene, H., & Penn, E. (2005). Reducing juvenile delinquency: Lessons learned. In E. Penn, H. Taylor Greene, & S. L. Gabbidon (Eds.), *Race and juvenile justice* (pp. 223–241). Durham, NC: Carolina Academic Press.

Tennessee v. Garner, 471 U.S. 1 (1985).

Texeira, E. (2005, August 28). Should the term "minority" be dropped? Word confuses, insults people some critics say. *Patriot-News,* p. A15.

Thompson v. Oklahoma, 487 U.S. 815 (1988).

Thornberry, T. P. (1998). Membership in youth gangs and involvement in serious and violent offending. In R. Loeber & D. P. Farrington (Eds.), *Serious and violent juvenile offenders: Risk factors and successful interventions* (pp. 147–166). Thousand Oaks, CA: Sage Publications.

Thornberry, T. P., Huizinga, D., & Loeber, R. (2004). The causes and correlates studies: Findings and policy implications. *Juvenile Justice, 10*(1), 3–19.

Thornberry, T. P., Krohn, M. D., Lizotte, A. J., Smith, C. A., & Tobin, K. (2003). *Gangs and delinquency in developmental perspective.* New York: Cambridge University Press.

Thurman, Q., Zhao, J., & Giacomazzi, A. (2001). *Community policing in a community era.* Los Angeles: Roxbury.

Tita, G., & Abrahamse, A. (2004, February). *Gang homicide in LA, 1981–2001.* Sacramento, CA: California Attorney General's Office.

Tolnay, S., & Beck, E. (1995). *A festival of violence: An analysis of southern lynchings, 1882–1930.* Urbana: University of Illinois Press.

Tonry, M. (1995). *Malign neglect: Race, crime, and punishment in America.* New York: Oxford University Press.

Travis, J. (2005). *But they all come back: Facing the challenges of prisoner reentry.* Washington, DC: Urban Institute Press.

Travis, J. (2007). Reflection on the reentry movement. *Federal Sentencing Reporter, 20,* 84–87.

Trujillo, L. (1995). La evolucion del "bandido" al "pachuco": A critical examination and evaluation of criminological literature on Chicanos. In A. S. Lopes (Ed.), *Criminal justice and Latino communities* (pp. 21–45). New York: Garland. (Reprinted from *Issues in Criminology, 74*[9], 44–67, 1974.)

Trulson, C. R., Marquart, J. W., & Kawuncha, S. K. (2006). Gang suppression and institutional control. *Corrections Today, 68,* 26–31.

Tso, T. (1996). The process of decision making in tribal courts. In M. O. Nielson & R. A. Silverman (Eds.), *Native Americans, crime, and justice* (pp. 170–180). Boulder, CO: Westview Press.

Tulchin, S. H. (1939). *Intelligence and crime.* Chicago: University of Chicago Press.

Turk, A. T. (1969). *Criminality and legal order.* Chicago: Rand McNally & Company.

Turner, K. B., & Johnson, J. B. (2005). A comparison of bail amounts for Hispanics, Whites, and African Americans: A single county analysis. *American Journal of Criminal Justice, 30,* 35–53.

Tyler, S. L. (1973). *A history of Indian policy.* Washington, DC: U.S. Department of Interior.

Uchida, C. (1997). The development of American police. In R. G. Dunham & G. P. Alpert (Eds.), *Critical issues in policing* (pp. 18–35). Prospect Heights, IL: Waveland Press.

Uggen, C., & Manza, J. (2002). Democratic contraction? Political consequences of felon disenfranchisement in the United States. *American Sociological Review, 67,* 777–803.

Uggen, C., Manza, J., & Behrens, A. (2003). Felon voting rights and the disenfranchisement of African Americans. *Souls: A Critical Journal of Black Politics, Culture & Society, 5,* 47–55.

Ulmer, J. T., & Johnson, B. (2004). Sentencing in context: A multilevel analysis. *Criminology, 42,* 137–177.

United States v. Barry, 938 F.2d, 1327, 1329 (D.C. Cir. 1991).

United States v. Booker, 543 U.S. 220 (2005).

Unnever, J. D. (2008). Two worlds far apart: Black-White differences in beliefs about why African-American men are disproportionately imprisoned. *Criminology, 46,* 511–538.

Unnever, J. D., & Cullen, F. T. (2007). Reassessing the racial divide in support for capital punishment: The continuing significance of race. *Journal of Research in Crime and Delinquency, 44,* 124–158.

Urban, L. S., St. Cyr, J. L., & Decker, S. H. (2003). Goal conflict in the juvenile court: The evolution of sentencing practices in the United States. *Journal of Contemporary Criminal Justice, 19,* 454–479.

U.S. Bureau of the Census. (1995). *Top 25 American Indian tribes for the United States: 1990 and 1980.* Retrieved December 12, 2003, from www.census.gov/population/socdemo/race/indian/ailant1.txt

U.S. Bureau of the Census. (2000). Retrieved April 15, 2003, from http://www.census.gov/main/www/cen2000.html

U.S. Bureau of the Census. (2001). *Overview of race and Hispanic origin. Census 2000 brief.* Retrieved April 15, 2003, from http://www.census.gov/prod/ 2001pubs/c2hbr01-1.pdf

U.S. Bureau of the Census. (2003). *Statistical abstracts of the United States.* Washington, DC: Author.

U.S. Bureau of the Census. (2004). *U.S. interim projections by age, race, and Hispanic origin.* Retrieved January 18, 2008, from www.census.gov.

U.S. Bureau of the Census. (2007). *Minority population tops 100 million.* Retrieved January 17, 2008, from www.census.gov.

U.S. Bureau of the Census. (2008). *Statistical abstract of the U.S.* Retrieved from www.census.gov/compendia/statab/

U.S. Bureau of Justice Statistics. (2006). *Compendium of federal justice statistics, 2004.* Washington, DC: U.S. Department of Justice.

U.S. Department of Justice. (2000). *The federal death penalty system: A statistical survey (1988–2000)*. Washington, DC: Author.

U.S. General Accounting Office. (1997). *Drug courts: Overview of growth, characteristics, and results*. Washington, DC: Author.

U.S. Senate. (2002). *Senate hearing before the subcommittee on crime and drugs of the committee on the judiciary, United States (107–911)*. Washington, DC: U.S. Government Printing Office.

U.S. Sentencing Commission. (2002). *Cocaine and federal sentencing policy*. Retrieved from www.ussc.gov/r_congress/02crach/2002crackrpt.pdf

U.S. Sentencing Commission. (2007). *Cocaine and federal sentencing policy*. Retrieved March 19, 2008, from http://www.ussc.gov/r_congress/cocaine2007.pdf

van den Haag, E. (1975). *Punishing criminals: Concerning a very old and painful question*. New York: Basic Books.

Vandiver, M. (2006). *Lethal punishment: Lynchings and legal executions in the South*. New Brunswick, NJ: Rutgers University Press.

Van Stelle, K. R., Allen, G. A., & Moberg, D. P. (1998). Alcohol and drug prevention among American Indian families: The family circles program. In J. Valentine, J. A. De Jong, & N. J. Kennedy (Eds.), *Substance abuse prevention in multicultural communities* (pp. 53–60). Binghamton, NY: Haworth Press.

Vega, W. A., & Gil, A. G. (1998). Different worlds: Drug use and ethnicity in early adolescence. In W. A. Vega & A. G. Gil (Eds.), *Drug use and ethnicity in early adolescence* (pp. 1–12). New York: Plenum Press.

Velez, M. B. (2006). Toward an understanding of the lower rates of homicide in Latino versus Black neighborhoods: A look at Chicago. In J. Hagan, R. Peterson, & L. Krivo (Eds.), *The many colors of crime: Inequalities of race, ethnicity, and crime in America* (pp. 91–107). New York: New York University Press.

Venkatesh, S. A. (2006). *Off the books: The underground economy of the urban poor*. Cambridge, MA: Harvard University Press.

Visher, C. A. (2007). Returning home: Emerging findings and policy lessons about prisoner reentry. *Federal Sentencing Reporter, 20,* 93–102.

Vold, G. B., Bernard, T. J., & Snipes, J. B. (1998). *Theoretical criminology* (4th ed.). Oxford, UK: Oxford University Press.

Wacquant, L. (2002). Scrutinizing the street: Poverty, morality, and the pitfalls of urban ethnography. *American Journal of Sociology, 107,* 1468–1532.

Wakeling, S., Jorgensen, M., Michaelson, S., Begay, M., Hartmann, F., & Wiener, M. (2001). *Policing on Indian reservations*. Washington, DC: U.S. Department of Justice.

Walker, S. (1989). *Sense and nonsense about crime: A policy guide* (2nd ed.). Pacific Grove, CA: Brooks/Cole.

Walker, S. (2001). *Police accountability: The role of citizen oversight*. Belmont, CA: Wadsworth/ Thomson Learning.

Walker, S., & Katz, S. M. (2002). *Police in America*. New York: McGraw-Hill.

Walker, S., Spohn, C., & DeLone, M. (2004). *The color of justice: Race, ethnicity, and crime in America* (3rd ed.). Belmont, CA: Thomson Learning.

Walker, S., Spohn, C., & DeLone, M. (2007). *The color of justice: Race, ethnicity, and crime in America* (4th ed.). Belmont, CA: Thomson Learning.

Walker-Barnes, C. J., Arrue, R. M., & Mason, C. A. (1998). *Girls and gangs: Identifying risk factors for female gang involvement*. Retrieved from http://www.unc.edu/~cwalkerb/present1/pdf

Walsh, A. (2004). *Race and crime: A biosocial analysis*. New York: Nova Science.

Walsh, A., & Ellis, L. (Eds.). (2003). *Biosocial criminology: Challenging environmentalism's supremacy.* New York: Nova Science.

Ward, G. (2001). *Color lines of social control: Juvenile justice administration in a racialized social system, 1825–2000.* Unpublished doctoral dissertation, University of Michigan, Ann Arbor.

Warner, S. B. (1929). *Survey of criminal statistics in the United States for National Commission on Law Observance and Enforcement.* Washington, DC: Government Printing Office.

Warner, S. B. (1931). Crimes known to the police: An index of crime? *Harvard Law Review, 45,* 307.

Warnshuis, P. L. (1931). Crime and criminal justice among the Mexicans of Illinois. In *Crime and the foreign born* (National Commission on Law Observance and Enforcement, Report No. 10, pp. 265–329). Washington, DC: Government Printing Office.

Warren, P. Y., Tomaskovic-Devey, D., Smith, W. R., Zingraff, M., & Mason, M. (2006). Driving while Black: Bias processes and racial disparity in stops. *Criminology, 44,* 709–736

Washington, L. (1994). *Black judges on justice: Perspectives from the bench.* New York: New Press.

Websdale, N. (2001). *Policing the poor: From slave plantation to public housing.* Boston: Northeastern University Press.

Weisburd, D., Greenspan, R., Hamilton, E., Williams, H., & Bryand, K. (2000). *Police attitudes toward abuse of authority: Findings from a national study.* Washington, DC: U.S. Department of Justice.

Weisheit, R. A., & Wells, L. E. (2004). Youth gangs in rural America. *NIJ Journal, 251,* 2–6.

Weitzer, R. (1999). Citizens' perceptions of police misconduct: Race and neighborhood context. *Justice Quarterly, 16,* 819–846.

Weitzer, R., & Tuch, S. (2002). Perceptions of racial profiling: Race, class, and personal experience. *Criminology, 40,* 435–456.

Welch, K. (2007). Black criminal stereotypes and racial profiling. *Journal of Contemporary Criminal Justice, 23,* 276–288.

Wenzel, S., Longshore, D., Turner, S., & Ridgely, M. S. (2001). Drug courts: A bridge between criminal justice and health services. *Journal of Criminal Justice, 29,* 241–253.

Western, B. (2006). *Punishment and inequality in America.* New York: Russell Sage Foundation Publications.

Whitaker, C. (1990). *Black victims.* Washington, DC: U.S. Department of Justice.

Whren v. United States, 517 U.S. 806 (1996).

Wickham, D. (2007, April 24). Madman, not Koreans, to blame for shootings. *USA TODAY,* p. 11A.

Wilbanks, W. (1987). *The myth of a racist criminal justice system.* Pacific Grove, CA: Brooks/Cole.

Wilder, K. (2003). *Assessing resident's satisfaction with community policing: A look into quality of life and interaction with police.* Unpublished master's thesis, Old Dominion University, Norfolk, VA.

Wilkins v. Missouri (1989). See *Stanford v. Kentucky.*

Williams, E. (1944). *Capitalism and slavery.* London: Andre Deutsch.

Williams, F. P., & McShane, M. D. (2004). *Criminological theory* (4th ed.). Upper Saddle Rivser, NJ: Prentice Hall.

Williams, H., & Murphy, P. (1990). *The evolving strategy of police: A minority view.* Washington, DC: National Institute of Justice.

Williams, M. R., & Holcomb, J. E. (2004). The interactive effects of victim race and gender on death sentence disparity findings. *Homicide Studies, 8,* 350–376.

Williams, S. (2007, October 8). Police urged not to check legal status: Activists want immigration standing off-limits in stops; some chiefs agree. *The Milwaukee Journal Sentinel.* Retrieved March 7, 2008, from http://find.galegroup.com/ips/retrieve.do

Willmott, D. (2000). It's time to bring our political prisoners home. In J. James (Ed.), *States of confinement: Policing, detention, and prisons* (pp. 312–321). New York: St. Martin's Press.

Wilson, J. Q., & Herrnstein, R. (1985). *Crime and human nature.* New York: Simon & Schuster.

Wilson, W. J. (1987). *The truly disadvantaged.* Chicago: University of Chicago Press.

Wilson, W. J. (1996). When work disappears: The world of the new urban poor. New York: A. A. Knopf.

Withrow, B. L. (2006). *Racial profiling: From rhetoric to reason.* Upper Saddle River, NJ: Prentice Hall.

Wolfgang, M. (1963). Uniform crime reports: A critical appraisal. *University of Pennsylvania Law Review, 111,* 708–738.

Wolfgang, M. E. (1958). *Patterns in criminal homicide.* Philadelphia: University of Pennsylvania Press.

Wolfgang, M. E., & Ferracuti, F. (1967). *The subculture of violence: Towards an integrated theory in criminology.* London: Tavistoc.

Wolfgang, M. E., & Riedel, M. (1973). Race, judicial discretion, and the death penalty. *The Annals of the American Academy of Political and Social Sciences, 407,* 119–133.

Wolf Harlow, C. (2000). *Defense counsel in criminal cases.* Washington, DC: Bureau of Justice Statistics.

Woodward, C. (1971). *The origins of the New South, 1877–1913.* Baton Rouge: Louisiana University Press.

Wooldredge, H., Hartman, J., Latessa, E., & Holmes, S. (1994). Effectiveness of culturally specific community treatment for African American juvenile felons. *Crime & Delinquency, 40,* 589–598.

Wordes, M., Bynum, T. S., & Corley, C. J. (1994). Locking up youth: The impact of race on detention decisions. *Journal of Research in Crime and Delinquency, 31,* 149–165.

Work, M. (1900). Crime among the Negroes of Chicago. *American Journal of Sociology, 6,* 204–223.

Work, M. (1913). Negro criminality in the South. *Annals of the American Academy of Political and Social Sciences, 49,* 74–80.

Work, M. (1939). Negro criminality in the South. *Annals of the American Academy of Political and Social Sciences, 49,* 74–80.

Wright, B. (1987). *Black robes, White justice.* New York: Carol Publishing.

Wright, J. P. (2008). Inconvenient truths: Science, race, and crime. In A. Walsh & K. M. Beaver (Eds.), *Contemporary biosocial criminology* (pp. 137–153). New York: Routledge.

Wright, R. R. (1969). *The Negro in Pennsylvania: A study in economic history.* New York: Arno Press and *The New York Times.* (Original work published 1912)

Wu, F. H. (2002). *Yellow: Race in America beyond Black and White.* New York: Basic Books.

Wyrick, P. A., & Howell, J. C. (2004). Strategic risk-based response to youth gangs. *Juvenile Justice, 10*(1), 20–29.

Xu, Y., Fiedler, M. L., & Flaming, K. H. (2005). Discovering the impact of community policing: The broken windows thesis, collective efficacy, and citizens' judgement. *The Journal of Research in Crime and Delinquency, 42,* 147–186.

Young v. Robinson, 29 Cr. L. 2587 (1981).

Young, V. (1986). Gender expectations and their impact on Black female offenders and victims. *Justice Quarterly, 3,* 305–327.

Young, V. (1993). Punishment and social conditions: The control of Black juveniles in the 1800s in Maryland. In A. G. Hess & P. F. Clement (Eds.), *History of juvenile delinquency: A collection of essays on crime committed by young offenders, in history and in selected countries* (pp. 557–575). Aalen, Germany: Scientia Verlag.

Young, V. (1994a). Race and gender in the establishment of juvenile institutions: The case of the South. *Prison Journal, 74,* 244–265.

Young, V. (1994b). The politics of disproportionality. In A. T. Sulton (Ed.), *African-American perspectives on: Crime causation, criminal justice administration and crime prevention* (pp. 69–81). Boston: Butterworth-Heinemann.

Young, V. (2007, October 25). Immigration crackdown feasts on motorists. *St. Louis Post-Dispatch.* Retrieved March 7, 2008, from General Reference Center Gold, Gale, Texas Southern University.

Young, V., & Reviere, R. (2006). *Women behind bars: Gender and race in U.S. prisons.* Boulder, CO: Lynne Rienner Publishers.

Zalman, M. (2006). Criminal justice system reform and wrongful convictions: A research agenda. *Criminal Justice Policy Review, 17,* 468–492.

Zane, N., Aoki, B., Ho, T., Huang, L., & Jang, M. (1998). Dosage-related changes in a culturally-responsive prevention program for Asian American youth. In J. Valentine, J. A. De Jong, & N. J. Kennedy (Eds.), *Substance abuse prevention in multicultural communities* (pp. 105–125). Binghamton, NY: Haworth Press.

Zangrando, R. L. (1980). *The NAACP crusade against lynching, 1900–1950.* Philadelphia: Temple University Press.

Zatz, M. S. (1987). The changing form of racial/ethnic biases in sentencing. *Journal of Research in Crime and Delinquency, 24,* 69–92.

Zimmerman, C. S. (2002). From the jailhouse to the courthouse: The role of informants in wrongful convictions. In S. Westervelt & J. A. Humphrey (Eds.), *Wrongly convicted: Perspectives on failed justice* (pp. 55–76). New Brunswick, NJ: Rutgers University Press.

Zimring, F. (2003). *The contradictions of American capital punishment.* New York: Oxford University Press.

Index

ABC News Poll, 233
Abolition of slavery, 103, 139
Abortion-crime connection, 95–96
Abril, J. C., 78
Abu-Jamal, M., 261
Adams, A. T., 84, 88
Adkins, S., 197
African Americans, xii, 4, 5, 6 (table), 9, 35 (table)
　arrest rates, 47–48, 49–51, 51–52 (tables), 58,
　　300–306 (tables)
　Black codes and, 14, 89, 107
　Black compulsive masculinity alternative, 94–95
　Black flight, 18–19, 73
　bombings of, 17
　civil rights movement and, 18, 108, 240
　code of the street and, 85–88, 85–87 (box)
　collective efficacy and, 76
　colonial model and, 92–94
　colonial society and, 10, 11
　convict-lease system and, 14–15, 89
　court system/United States and, 138, 139–141,
　　140 (box)
　crime and, 11, 12 (table), 17, 18, 19
　death penalty and, 205, 205–207 (box)
　differential justice and, 11, 12–13 (box), 12 (table)
　domestic labor of, 17
　Emancipation Proclamation/Thirteenth
　　Amendment and, 14, 15
　family structure and, 74–75 (box)
　genocide against, 18
　Great Migration and, 15, 17, 70, 240
　hate/bias crimes against, 59
　indentured servitude and, 10
　middle class groups of, 76
　Plessy v. Ferguson/separate-but-equal decision
　　and, 15
　policing practices and, 107–108
　rape accusations and, 11, 12 (table), 17
　religious practices of, 80
　research on, xi, xii, 18
　slavery history of, 10–14
　social disorganization and, 71, 72–73
　Southern penal system and, 14–15
　strain/anomie theory and, 78–79
　strain theory and, 80
　strikebreaker role and, 17
　structural-cultural theory and, 94–95
　subculture of violence theory and, 81–88, 83
　　(figures), 85–87 (box)
　truly disadvantaged populations and, 18–19, 73
　unemployment and, 72
　victimization rates, 39, 52–53, 54–55 (tables)
　White supremacy/hate groups and, 15, 63
　See also Crime statistics; Lynching; Race; Race
　　riots; Racial minorities; Racial profiling;
　　Sentencing; Slavery
African holocaust, 10
Agency, 81
Aggravated assault, 49–50, 52
Aggression, 67
　inner city populations and, 86
　slaves, sexual aggression against, 17, 67
　White colonizers, 67
Agnew, R., 80
Agozino, B., 93, 94
Akers, R. L., 78
Alaska Natives, 4, 6 (table)
　arrest rates, 302–303 (table), 305–306 (table)
　See also Crime statistics; Native Americans
Alcohol-related offenses, 58
　juvenile alcohol abuse, 278, 280
　Mexican Americans and, 109
　Native Americans and, 77, 107
Alienation, 85, 93, 283
American Civil Liberties Union (ACLU), 121, 156
American Dream, 23, 78, 79
American Indian Movement (AIM), 261
American Indians. *See* Native Americans
Ancestral roots, 3–4, 47
Anderson, C., 11
Anderson, E., 292
Anderson, E. A., 85–87

Anomie, 78
    See also Strain/anomie theory
Anti-Drug Abuse Act of 1988, 208
Apartheid, 140
Aptheker, B., 260
Argersinger v. Hamlin, 147
Argothy, V., 221, 222
Arrests, xi
    arrest/victimization data, 46, 47 (table)
    DNA evidence and, 3–4 (box)
    juvenile arrests, 271–272, 272(table), 275,
        276–277 (table), 286
    overrepresentation of minorities in, xii, xiii, 19,
        38–39, 58
    trends in arrests, 47–48, 49–51, 51–52 (tables),
        300–306 (tables)
    See also Corrections; Incarceration; Policing
Arrue, R. M., 287
Arthur, J. A., 218
Aryan Brotherhood, 243, 243–244 (box)
Ashley-Montague, M. F., 64
Asian Americans, xii, 4, 5, 6 (table), 29, 30 (table)
    anti-Chinese sentiment, 30, 31
    arrest rates, 47–48, 49–51, 51–52 (tables),
        300–306 (tables)
    bachelor society and, 30
    Chinese Americans, 29–31
    court system/United States and, 142
    Filipino Americans, 29, 32
    immigrant groups of, 29–34, 30 (table), 33–34 (box)
    Japanese Americans, 29, 31–32
    Korean Americans, 29, 32, 33–34 (box)
    model minority label and, 34
    organized crime and, 30–31
    Oriental crime and, 72
    outsider status of, 34
    policing practices and, 108–109
    victimization data, 49
    Vietnamese Americans, 29
    See also Crime statistics; Race; Racial minorities
Asian Indians:
    arrest rates, 300–301 (table), 304–305 (table)
    See also Asian Americans
Assault:
    aggravated assault, 49–50, 52
    Black vs. White commission of, 12 (table)
    data on, 39, 45
    hate/bias crime trends, 55, 57, 58 (table), 59
    prehomicidal assaults, 83
    prison assault, 244–245
    sexual assault, 17, 23, 46, 67
    See also Rape; Sexual assault; Violent crime
Atkins v. Virginia, 288, 289 (box)
Attica prison, 241–242
Autonomy, 81

Bachman, J. G., 279
Bachman, R., 73
Bail Reform Act of 1984, 143, 144, 145
Baker, D. V., 201, 202, 205
Baker, S. R., 292
Baldus, D. C., 204
Baldus Study, 204
Ball, R. A., 88
Banner, S., 205, 207
Barkan, S. F., 219
Barlow, D. E., 99, 122
Barlow, M. H., 99, 122
Barry, M., 155–156
Batson v. Kentucky, 153
Battle v. Anderson, 241
Batzer, M., 3
Bean, L., 73
Beaumont, G., 236
Becker, S., 121
Becton, J. B., 118
Bedau, H. A., 224
Behrens, A., 258, 259
Bell, D., 96
Bell, S., 117
Beniean, F., 1
Bennett, R. L., 104
Berkowitz, D., 111
Bernard, T. J., 266
Biafora, F. A., 287
Bias:
    IQ tests and, 66
    jury selection, 150–152
    policing, selective enforcement and, 48, 59, 64,
        116, 118
    See also Bias crimes; Culture conflict theory;
        Discrimination; Prejudice; Racial profiling
Bias crimes, xii, 45
    hate/bias crime trends, 55, 57, 58 (table)
    single-bias incidents, 55
Big Brothers/Big Sisters of America, 291
Biological factors in crime, 63–64, 65, 70, 89
    human nature factors, 64–65
    intelligence-based theories of crime, 65–66
    r/K life history theory of crime, 66–69
    skin color, 67, 68–69
    See also Population genetics; Theoretical
        perspectives
Bjerregaard, B., 214
Black child savers, 263
Black codes, 14, 89, 107, 295
Black Guerilla Family, 243
Black Liberation Movement, 261
Black Muslim movement, 240–241
Black Panther movement, 240, 241–242, 243, 261
Black power movement, 92, 240–242, 261

Blacks. *See* African Americans
Blakely v. Washington, 185
Blalock, H. M., 90
Blauner, R., 90, 92, 93
Bloods, 244
Bloom, B., 287
Blowers, A. N., 182
Blueprint Model Programs, 291
Blumenbach, J. F., 1, 2
Blumstein, A., 95, 250, 252
Bobo, L. D., 219
Bohm, R. M., 62, 79, 92, 202, 217
Bombings, 17, 111
Bonczar, T. P., 254
Bonger, W. A., 64
Booher Feagin, C., 7–8, 8, 24, 35
Bracero Program, 26
Bradley, M., 67
Brewer, T. W., 222
Brody, G. H., 80
Brooks, C., 260
Brown, H. R., 240
Brown, W., 183
Brunson, R. K., 83, 87
Brushway, S., 290
Bryand, K., 118, 128
Bundy, T., 111
Bureau of Indian Affairs (BIA), 8, 105, 107, 122, 138
Bureau of Justice Statistics (BJS), 37, 38 (table)
    hate crimes, 57
    homicide patterns/trends, 53
    juvenile crime, 271
    See also Crime statistics
Burgess, E. W., 71
Burrus, S., 160
Bush, President G. H., 176, 258
Bush, President G. H. W., 44
Butler, P., 155, 156

Cahalan, M., 172, 173
California Department of Corrections, 253
California Youth Authority, 88
Calnon, J. M., 126
Cao, L., 84, 88
Capital crime. *See* Death penalty
Capitalism, 89, 235
Capital Jury Project (CJP), xiii, 214, 221–222
Caribbean peoples, 5
Carmichael, S., 92, 240
Carroll, C., 63
Carter, President J., 190
Castro, F., 28
Catalano, S. M., 54
Catterall, H. T., 166, 167, 168
Census information. *See* U. S. Census Bureau

Census of Juveniles in Residential Placement, 281
Center for Employment Opportunities (CEO), 258
Centers for Disease Control and Prevention
        (CDC), 275
Center for the Study and Prevention of Violence, 291
Central Park Jogger case, 224–225
Cernkovich, S. A., 79
Chabrán, R., 109
Chabrán, R, 109
Chambliss, W. J., 89
Chamlin, M. B., 221
Chapin, B., 166, 235
Charles, M. T., 118
Chen, Y. F., 80
Chesney-Lind, M., 286, 287
Chicago Alternative Policing Strategy (CAPS), 128
Chicago School, 70–72, 76
Child abuse/neglect, 80, 257, 278
Child Welfare League of America, 289
Childhood:
    child savers and, 263, 268–269, 270
    incarcerated parents and, 257
    inner city childhood, criminalization of, 85
    See also Child abuse/neglect; Families
Children's Defense Fund, 289
Chinatowns, 30
Chinese Americans, 29–31, 30 (table), 35 (table)
    arrest rates, 300–301 (table), 304–305 (table)
    court system/United States and, 142
    Oriental crime and, 72
    policing practices and, 108–109
Chinese Exclusion Act, 295
Chircos, T. G., 179, 180
Christianson, S., 237
Circular migration, 27
Civil law, 87
Civil liberties, 100
Civil rights, 115–116
Civil rights movement, 18, 108, 240
Civil War (United States), 104, 205, 267
Clarke, A., 220
Class structure:
    crime and, 89
    justice outcomes and, 131, 132–133 (box)
    policing activities and, 111, 115
    power threat concept and, 90
    See also Lower class; Middle class; Upper class
Clear, T. R., 75, 232, 236
Clement, P. T., 267
Clinton, President W. J., 176, 184, 190
Cloward, R. A., 81
Cochran, J. K., 214, 221
Code of the street, 85–88, 85–87 (box)
    decent families and, 85
    prison experience and, 88

rap music and, 87
street families and, 87
victimization and, 87
Cohen, A. K., 81
Cohen, F., 104
Cohen, S., 225
Cohn, S. F., 219
COINTELPRO, 261
Coker v. Georgia, 202
Cole, D., 153
Cole, G. F., 232, 236
Collective efficacy concept, 75–76
Collins, C. F., 235
Colonialism, 92
    aggression and, 67
    colonial-era policing, 102–103
    cultural conflict and, 77
    internal colonialism, 93
    United States settlement, 19, 21
    See also Historic context
Colonial model, 92–94, 283
Color-blind jurors, 157, 222
Communities:
    Black flight and, 18–19, 73
    citizen satisfaction with policing, 112–115,
        113–114 (table), 128
    collective efficacy and, 75–76
    kin/neighbors, social control and, 76
    social buffers in, 73
    social disorganization in, xii, 70–75
    social isolation and, 73
    truly disadvantaged populations and, 18–19, 73
    See also Policing; Urban communities
Community policing (COP), 76, 103, 112, 127
Compromise of 1850, 13–14
Computer-aided mapping, 99
Conflict gangs, 81
Conflict theory, 88–89
    Black codes and, 89
    convict-lease system and, 89
    development of, 89–90
    discrimination-disparity continuum and, 91–92,
        91 (box), 92 (figure)
    dominant power structure and, 89
    no discrimination thesis and, 90, 91
    power threat approach and, 90
    race/crime and, 89–92
    race discrimination, punishment and, 90
    race privilege, Whites in power and, 90
    weaknesses of, 92
    See also Culture conflict theory; Strain/anomie
        theory; Theoretical perspectives
Conley, D. C., 240
Constitutional Union Guards, 15
Convict-lease system, 14–15, 89, 139, 237–238, 253

Conyers, Congressman J., 204
Coontz, S., 74
Corporate crime, 90
Corrections, xiii, 231–232
    American corrections, overview of, 232–238
    Attica prison riot, 241–242
    Black power movements and, 240–242
    chivalry factor and, 236
    contemporary issues, race/corrections and, 250–253
    contemporary state of, 245–247, 246–247
        (tables)
    convict-lease system and, 237–238, 253
    disparities in corrections and, 234, 240, 249, 250–253
    early national prison statistics, 238–242, 239
        (table)
    Eastern Penitentiary/Western Penitentiary
        and, 236
    employment in, 232, 252–253
    faith-based initiatives and, 258
    felon disenfranchisement and, 258–260
    gender and, 236, 238
    historical overview, race/corrections and, 234–238
    homosexual involvement and, 238
    jail populations and, 247–248, 248–249 (tables)
    lawsuits, management changes and, 241–242, 245
    mass incarceration, 233–234, 253–254
    Negro courts and, 235
    political prisoners and, 260–261
    prevention programming and, 298
    prisoner reentry concerns, 253–258, 254 (table),
        255–256 (box)
    prison gangs and, 242–245, 243–244 (box)
    prison-industrial complex and, 253
    probation/parole and, 249, 254, 254 (table)
    public opinion and, 233–234
    rage riots and, 242
    reentry court model, 257–258
    rehabilitation approach and, 233, 238
    solitary confinement/hard labor and, 235
    Walnut Street Jail and, 235–236
    workhouses and, 235
    See also Courts; Incarceration
Corrections Corporation of America, 253
Cose, E., 75
Council of Safety, 15
Court of Indian Offenses, 138
Courts, 131
    African American history and, 138, 139–141,
        140 (box)
    American slavery jurisprudence, 11, 12–13 (box)
    Asian American history and, 142
    bail/pretrial process, 141, 143–147
    bench trials, 125
    class/social status and, 131, 132–133 (box)
    color-blind jurors, 157, 222

contemporary issues in, 143–157
discrimination thesis/no discrimination thesis
    and, 145–146
drug courts, 157–160
dual-court system, 134–135
gender, pretrial release and, 146–147
Hispanic history and, 141–142
historic context, racial minorities and, 138–142
indigent defendants, legal counsel/defense
    counsel and, 147–148
jury nullification and, 155–157
jury selection process, 141, 150–152, 150–151 (box)
juvenile courts, 269–270
minority judges and, 189–195, 191–194 (table),
    196–197 (x)
Native American court system, 136–138,
    136–137 (tables)
Native American history and, 138–139
Negro courts, 235
Peacemaker Courts/mediation, 137
Pennsylvania public defense system and, 148–149
plea-bargaining process, 141, 149–150
public defense, 141, 147–149
race, pretrial release and, 144–146
racism in, 140–141
release on own recognizance and, 143
Sleepy Lagoon Case and, 141–142
social control function and, 138
United States courts, actors/processes in, 134–138
voir dire process and, 152–155
See also Corrections; Government; Sentencing;
    Supreme Court; Wrongful convictions
Courtwright, D. T., 109
Covington, J., 84
Craig, R., 119
Crank, J. P., 127
Crawford, C., 179, 180
Crime, xi–xiv
differential justice and, 11, 12–13 (box),
    12 (table)
disadvantaged populations and, 18–19
future research on, 296–298
human nature and, 64–65
lynching as punishment for, 15
organized crime, 22, 25
positivist explanations of, xi–xii
race and, xi–xii, 18
racialization of, 39, 58
separate-but-equal decision and, 15
slave trade, White crime of, 13
Southern penal system and, 14–15
vice crimes, 23, 30–31
See also Crime statistics; Female delinquency;
    Historic context; Immigrant populations;
    Theoretical perspectives; Violent crime

Crime Control Act of 1994, 127
Crime index offenses, 43–44, 45
Crime statistics, xiii, 37–39
arrest trends by race, 49–51, 51–52 (tables),
    57–58, 300–303 (table)
arrest/victimization data, 46, 47 (table)
automated data collection and, 49
citizen satisfaction with policing,
    112, 113–114 (table)
crime index offenses, 43–44, 45
estimations, utilization of, 49
eugenics movement and, 40
federal/state law/participation and, 41, 44
foreign vs. native-born convicts, 37, 40, 47
General Social Survey, 84
hate/bias crime trends, 55, 57, 58 (table)
Hate Crime Statistics, 44
history of data collection, 40–46, 42–43 (table)
homicide victimization trends by race, 53,
    56–57 (tables), 84
limitations of, 46–48, 47 (table), 48 (box), 49,
    58–59
lynching, 41, 42–43 (table)
murders/arrests by race, 304–306 (table)
National Crime Victimization Survey, 45–46
National Incident-Based Reporting System and, 44
quality/timeliness of, 44
race-crime relationship, 18, 58
racial categories, definitions of, 46–48, 48 (box)
racial/ethnic minorities and, 37, 38–39, 58
racialized crime and, 39
reporting/recording, variations in, 48
self-reports and, 45
sources of, 37, 38 (table), 39, 46, 57
Supplemental Homicide Report and, 43
trend analysis data, 46
Uniform Crime Reports, 37, 38 (table), 39, 41,
    43–45
victimization data, 39, 45–46
violent crime data, 49–51, 51–52 (table),
    304–306 (table)
violent victimization trends by race, 52–53,
    54–55 (tables)
weapons information, 44
See also Criminal investigators; Criminal justice
    system; Policing; Theoretical perspectives
Crime Survey Redesign Consortium, 45
Crime in the United States, 38 (table), 39
Criminal investigators, 4
illegal immigration and, 27
See also Crime statistics; Policing
Criminal justice system, xi
African American males and, 85–86 (box)
Black codes and, 14, 89
colonial model and, 92–94

crime control/due process models of, 165
differential treatment and, 11, 12–13 (box), 12
    (table), 85, 90
discrimination-disparity continuum and, 91–92,
    91 (box), 92 (figure)
DNA evidence, 2, 3–4 (box)
gringo justice, 92
minorities, overrepresentation of, xii, xiii, 19
no discrimination thesis, 90, 91
parallel systems, unfair treatment and, 85, 86, 90–91
prejudice/discrimination and, xi, xii, 5–6, 90
slave codes, plantation justice and, 10–11
social control and, 89–90
Southern penal systems, 14–15
utilitarian philosophy of, 223
White privilege and, 90
Wickersham Commission Report and, 26
wrongful convictions, xii, 96, 222–228,
    226–227 (box)
See also Corrections; Courts; Crime statistics;
    Death penalty; DNA evidence; Juvenile
    justice system; Policing; Sentencing
Criminal profiling, 111, 118–119, 121
Criminals:
    biological factors and, 63–64
    convict-lease system and, 14–15
    immigrant groups and, 25
    Irish immigrants, 22
    Jewish immigrants, 25
    Mexican immigrants and, 26
    Southern penal system, 14–15
    truly disadvantaged populations and, 18–19
Criminals, See also Crime; Criminal justice system
Criminal Victimization, 39
Criminology:
    African American scholars in, xii
    colonial model and, 92–94
    criminal anthropology and, 40
    critical criminology, 89, 96
    New Criminology, 89
    race-crime relationship and, xi–xii, 18
    radical criminology, 96
    See also Crime statistics; Theoretical perspectives
Crips, 244
Critical legal studies movement, 96
Critical race theory (CRT), 96
Critical resistance movement, 253
Cubans, 28
Cullen, F. T., 88, 219, 220
Culture conflict theory, 76
    conduct norms and, 77
    cultural norms/rules and, 76–77
    Native Americans, example of, 77
    primary conflicts, cultural origins and, 77
    secondary conflicts, social differentiation and, 77

sources of conflict, 77
See also Conflict theory; Sociological
    explanations; Strain theory; Subcultural
    theory; Subculture of violence theory;
    Theoretical perspectives
Curran, D. J., 62
Curtis Act of 1898, 104

Daniels, R., 29, 30
Data. See Crime statistics
Davis, A. Y., 17, 90, 240, 261
Dawes Act of 1887, 8, 104
Death in Custody Reporting Act of 2000, 116–117
Death penalty, xii, xiii, 201
    Black on White rape accusation and, 17
    Capital Jury Project and, 214, 221–222
    college student views on, 220–221
    color-blind jurors and, 222
    contemporary issues, race/death
        penalty and, 220–229
    cruel/unusual punishment and, 202, 203, 288
    current statistics on, 208, 211–212 (tables),
        213–214
    due process and, 202
    ethnic deceit concept and, 222
    ethnic threat concept and, 222
    federal death penalty statistics, 208, 213–214
    guided discretion statutes and, 202
    historical overview, race/death penalty and,
        204–208, 205–207 (box), 209–210 (tables)
    Innocence Project and, 224, 226
    juveniles and, xiii, 288–289, 289 (box)
    kidnapping and, 202–203
    life in prison and, 218
    Marshall Hypotheses and, 220, 221
    Native Americans, executions of, 205
    protocol for, 213
    public opinion on, 217–220
    race and, 146, 203–204
    race-of-the-victim effect and, 204, 214
    scholarship on race/death penalty,
        214–217, 216 (figure)
    significant cases, 201–204
    slave-era capital crimes, 205, 205–207 (box), 206
        (table), 207 (figure)
    state death penalty statistics, 208,
        211–212 (tables)
    state-sanctioned executions/illegal lynchings,
        207–208, 209–210 (tables)
    substitution thesis, Zimring analysis and,
        215–217, 216 (figure)
    support of, race and, 218–220
    wrongful convictions and, xii, xiii, 96, 222–228,
        226–227 (box)
    See also Courts; Sentencing

Death penalty moratorium movement, xiii, 228–229
Debro, J., 240
DeFrances, C. J., 147
De Las Casas, B., 9, 10
Delgado, R., 96
Delinquency:
    discrimination and, 80
    female delinquency, 286–288
    opportunity structure and, 81
    subcultural theory and, 81
    See also Crime; Delinquency prevention; Gangs;
        Juvenile delinquency
Delinquency prevention, xiii, 283, 290–292, 290 (box)
    Black child savers and, 263
    comprehensive strategy framework for, 291
    culturally specific programs, 291–292
    gang prevention/suppression, 274
    rites of passage programs, 291
    violence-victimization relationship and, 292
DeLone, M., 91, 92, 155
Demuth, S., 146
Denton, N. A., 17
Department of Homeland Security (DHS), 101
Department of Justice. *See* U. S. Department of
        Justice (DOJ)
Deschenes, E. P., 287
Desmond v. Blackwell, 241
Deviance:
    deviant values, 88
    police deviance, accountability and, 115–127
    See also Crime; Delinquency
Dewees, M. A., 224
Diallo, A., 117
Disadvantage. *See* Poverty; Socioeconomic status;
        Truly disadvantaged populations
Discrimination, 6, 67, 89
    contextual discrimination, 91 (box)
    delinquency and, 80
    discrimination thesis, 145, 146
    discrimination-disparity continuum, 91–92, 91
        (box), 92 (figure)
    disproportionate minority confinement, xii, xiii,
        263, 281–286
    institutional discrimination, 91 (box)
    Irish immigrants and, 22–23
    Jewish immigrants and, 24–25
    no discrimination thesis, 90, 91
    policing practices and, 115–117
    race discrimination, punishment and, 90
    religious practice and, 22, 24
    separate-but-equal decision and, 15
    See also Conflict theory; Immigrant populations;
        Racial profiling
Discrimination thesis (DT), 145, 146
Disenfranchisement. *See* Felon disenfranchisement

Disproportionate minority confinement (DMC),
        263, 281–286
DMC Index, 283, 285
DNA evidence, 2, 3–4 (box)
    biogeographical ancestry admixture, 3, 4
    DNA photofitting technology, 3
    Duke University lacrosse players case, 226
    innocence protection policy and, 228
    National DNA Index System, 99
    privacy issues and, 4
    See also Population genetics
DNAPrint Genomics, 3, 4
Domestic terrorists, 111
Domestic violence, 46, 84
Donohue, J. J., 95
Donzinger, S. R., 150
Douglas, J. E., 118
Draft Riot, 23
Drapella, L. A., 159
Dred Scott decision, 14
Drug abuse:
    arrests for, 58
    Asian gangs and, 109
    drug courier profile, 121–122
    drug kingpin law, 213
    juvenile drug use, 278–280, 279 (table), 280 (table)
    opium establishments, 30, 108–109
    organized crime and, 30
    police drug dealing, 115
    retreatist gangs and, 81
    war on drugs, 176, 182, 183–189
    See also Anti-Drug Abuse Act of 1988; Drug courts
Drug courts, 157
    drug court teams, 157–158
    effectiveness of, 158–160
    family treatment drug courts, 160
    juvenile drug courts, 266
    non-violent felony offenders, 159, 176
    philosophy of, 158
    progress monitoring and, 158
    recidivism and, 159–160
    structure of, 157–158
    treatment/rehabilitation services and, 158
    See also Courts
Drug Enforcement Administration, 101
Dubner, S. J., 95
Du Bois, W. E. B., 11, 13, 15, 70, 89, 163
Due process, 165, 202, 297
Dugdale, R. L., 65
Duke University lacrosse players case, 225–227,
        226–227 (box)
Dulaney, W. M., 107
Durkheim, E., 78
Durose, M. R., 123
Dyer, J., 253

Earls, F., 75
Eastern Penitentiary, 236
Eberheart v Georgia, 202–203
Eck, J. P., 290
Eddings v. Oklahoma, 288, 289 (box)
Education:
    model minority and, 34
    school failure, 66
Edwards, T. J., 48
Eigenberg, H., 154
Eighth Amendment, 143, 202, 288
Einstein, A., 25
Eitle, D., 80
Emancipation Proclamation of 1863, 14, 15, 107,
        139, 237
Employment:
    African American males and, 17, 86
    Bracero Program and, 26
    Chinese Americans and, 29–30
    circular migration and, 27
    correctional officers, 232
    domestic labor, 17, 23
    Great Migration, urban influx and, 17
    indentured servitude, 10, 19, 22
    Japanese Americans and, 31
    Jewish Americans and, 25
    Latinos and, 26–27, 28
    low-skill/low-wage jobs, 17, 28
    Mexican immigrants and, 26–27
    migrant workers, 109
    police agencies, 101
    prisoner reentry concerns and, 253–258, 254
        (table), 255–256 (box)
    Puerto Ricans and, 27
    racialized minority groups, dominant group
        needs and, 96
    reentry court model and, 257–258
    strikebreaker role and, 17
    underground economy and, 18–19
    workhouses and, 235
    See also Unemployment
Emsley, C., 103
Engel, R. S., 126
Environmental crime, 90
Epps, E. G., 78
Esbensen, F. A., 287
Escobedo v. Illinois, 103
Ethnicity, 4–5
    ethnic groups, suspicions among, 28–29 (box)
    ethnic threat/ethnic deceit, 222
    organized crime and, 22, 25
    See also Immigrant populations; Racial minorities
Eugenics movement, 37, 40, 63
Excobar, E. J., 109
Executive Order 9066, 31

Facial invariance thesis, 73
Faith-based initiatives, 258
Families:
    close integration of, 72
    decent families, 85
    disruption of, 72, 257
    female-headed households, 72, 73, 74–75 (box)
    street families, 87
Family treatment drug courts (FTDCs), 160
Fanon, F., 92
Fate, 81
Faust, A. B., 20
Feagin, R. F., 7, 8, 24, 35
Fearlessness, 81
Federal Bureau of Investigation (FBI), 37, 38
        (table), 39, 101, 125
    counterintelligence, 261
    National Hate Crime Data Collection
        Program, 44
    National Incident-Based Reporting System and, 44
    Uniform Crime Reports, 37, 38 (table), 39, 41,
        43–45, 173, 271
Federal Law Enforcement Officers, 101
Federation of Colored Women's Clubs, 263, 269
Feld, B., 269, 283, 292
Felon disenfranchisement, xii, 258–260
Female delinquency, xiii, 286–288
Ferracuti, F., 81, 82, 84
Feyerherm, W. H., 263, 284, 286
Fielding, J., 159
Filipino Americans, 26, 29, 30 (table), 32
    Oriental crime and, 72
    See also Asian Americans
Finger, B., 9, 10
Fingerprint cards, 47
Finighan, M. W., 160
Finkelstein, N. H., 25
Finnegan, T., 275
First Amendment, 241
Flanagan, T. J., 233
Flowers, R. B., 2
Fluery-Steiner, B., 221, 222
Focal concerns theory, 81
Fogel, S. J., 214
Forst, B., 99
Fourteen Amendment, 202
Fox, C., 157, 158
Fox, J. A., 56, 57
Frazier, E. F., 270
Fredrickson, D. D., 120
Free, M. D., 145, 146
Friedman, L. M., 138, 141, 142
Frudakis, T., 3, 4
Fugitive Slave Law, 14
Fulwood v. Clemmer, 241

Furman v. Georgia, 201–202, 204, 220
Furrer, C., 160

Gabbidon, S. L., 61, 119–120, 122, 154, 296
Gacy, J. W., 111
Gallagher, C. A., 2
Galloway, A. L., 159
Gallup Poll, 112, 218, 223, 289
Gallup Poll Social Audit, 112
Gambling:
    Chinese Americans and, 30, 108
    Filipinos and, 26
    Native American facilities, 9
Gang Resistance Education And Training
        (G.R.E.A.T.), 274
Gangs:
    Asian gangs, 109
    conflict gangs, 81
    gang membership, risks for, 274
    gang migration, 273
    gender-mixed gangs, 273, 287
    Latinos and, 28, 39
    opportunity structure and, 81
    prevention programs, 274
    prison gangs, 242–245, 243–244 (box)
    retreatist gangs, 81
    serious/violent crimes and, 273–274
    subcultural theory and, 81
    threat assessment, 39
    White mobs, 107
    youth gangs, 272, 272–274 (box)
    See also Organized crime
Gans, H. J., 1
Garland, T. S., 154
Geis, G., 37, 46
Gender:
    bail/pretrial release and, 146–147
    biology of crime and, 63
    corrections and, 236, 238
    homicide commission and, 84
    homicide victimization trends and, 53
    jury selection and, 152
    unemployment/crime and, 72
    violent victimization trends and, 52–53, 54 (table)
General Social Survey (GSS), 84, 218, 219
Genetics. *See* DNA evidence; Population genetics
Genocide, 7, 8, 18, 205
Georges-Abeyie, D., 48, 68
German Americans, 19, 20–21, 110
Get-tough approach, 173, 233, 290
Gideon v. Wainwright, 147
Gil, A., 287
Giordano, P. C., 79
Glaze, L. E., 254
Glendening, Governor P., 228

Goddard, H. H., 65
Goldkamp, J. S., 116, 231
Gore, A., 259
Gossett, T., 1
Gottfredson, D., 290
Gould, L. A., 34
Government:
    American slavery jurisprudence, 11, 12–13 (box)
    anti-narcotics legislation, 30
    Bracero Program, 26–27
    Emancipation Proclamation, 14
    genocide accusations and, 18
    homeland security issues, 99–100
    law and order campaigns, 37
    Native American history and, 7–9
    religious discrimination, ban on, 24
    Thirteenth Amendment, 14
    wrongful conviction, public policy and, 227–228
    See also Courts; Crime statistics; Policing;
            Political crime; Slavery; Supreme Court
Great Migration from the South, 15, 17, 70, 240
Green, B. L., 160
Greenberg, D., 169, 170, 171
Greenleaf, R. G., 121
Greenspan, R., 118, 128
Greenwald, F., 266
Gregg v Georgia, 202, 204, 208
Griffin, M. L., 244
Groves, W. B., 72
Growette Bostaph, L. M., 121
Guevara, L., 287
Guilt, 82
Gun trade, 87
Gypsy Jokers gang, 242

Hamilton, C., 92
Hamilton, E., 118, 128
Hampton Conference, 263, 269
Hardy, K. A., 251
Harer, M. D., 84
Harlow, W., 148
Harrell, E., 55
Harris, D. A., 121
Harris Poll, 218
Hart, F. M., 104
Hart, H. L. A., 164
Harvey, J. T., III, 156
Hate crimes, xii, 15, 45
    hate/bias crime trends, 55, 57, 58 (table)
    single-bias incidents, 55
Hate Crime Statistics, 38 (table), 44, 45
Hate Crime Statistics Act of 1990, 44
Hate groups:
    White supremacy groups, 11, 12 (box), 15, 63
    See also Gangs

Hawkins, D. F., 82, 88, 90, 251
Hayner, N., 71, 72
Healey, J. F., 7, 8
Henderson, C. R., 64
Hepburn, J. R., 244
Heritage Foundation, 196–197
Hernandez v. New York, 153
Herrnstein, R., 64, 65, 66, 68
Herz, D., 287
Higginbotham, A. L., 11, 12–13, 138, 140, 195
Higgins, G. E., 119, 121
Hilton, P., 132–133 (box)
Hindelang, M., 66
Hirschi, T., 66
Hispanic Americans, xii, 4, 5, 6 (tables)
    arrest rates, 47–48, 49–51, 51–52 (tables)
    classification of, 46
    court system/United States and, 141–142
    Cubans, 28
    hate/bias crimes against, 55, 58 (table), 59
    immigrant groups of, 25–28
    Jewish immigrants and, 24
    lynching of, 15
    Mexican immigrants, 26–27
    numbers of, 5
    pachuco killers and, 141–142
    policing practices and, 109–110
    Puerto Ricans, 27–28
    social disorganization and, 73–74
    strain/anomie theory and, 79
    subculture of violence theory and, 81–88, 83
        (figures), 85–87 (box)
    violent victimization trends and, 52–53,
        54–55 (tables)
    See also Crime statistics; Mexican Americans;
        Race; Racial minorities; Racial profiling;
        Sentencing
Historic context, xii–xiii, 1, 295–296
    African American history, 9–19, 12–13 (box),
        12 (table), 16 (figure)
    Asian Americans, 29–34, 30 (table), 33–34 (box)
    census statistics on race, 2–3, 4, 5, 6 (tables)
    Chinese Americans, 29–31
    civil rights movement, 18
    Civil War (United States), 14–15
    court systems, racial minorities and, 138–142
    crime-race relationship hypothesis, xi–xii, 18
    crime/victimization statistics collection, 40–46,
        42–43 (table)
    differential justice, 11, 12–13 (box), 12 (table)
    DNA evidence and, 2, 3–4 (box)
    ethnicity/ethnic groups, 4–5
    Filipino Americans, 26, 32
    future research and, 296–298
    German Americans, 19, 20–21

hate groups and, 15
invention of race, 1–2
Irish Americans, 22–23, 23 (box)
Italian Americans, 19, 21–22
Japanese Americans, 31–32
Jewish Americans, 24–25
Korean Americans, 32, 33–34 (box)
Latino American groups, 25–28
migration patterns, 2, 2 (figure), 15, 17
Native American history, 7–9
Plessy v. Ferguson decision, 15
policing history, race and, 102–111, 106 (table)
Puerto Ricans, 27–28
races, categorization of, 2, 4, 35, 35 (table)
sentencing practices, race and, 165–176
Southern penal system, 14–15
White ethnics and, 19–25
See also Slavery; Theoretical perspectives
Historically Black colleges and universities
    (HBCUs), 263
Holcomb, J. E., 214
Homeland security, 100, 297–298
Homeland Security Act of 2002, 101
Homicide, xi
    arrests by race and, 49–51, 51–52 (tables),
        304–306 (table)
    Black vs. White commission of, 12 (table)
    disaggregated data on, 84
    gender, crime commission and, 84
    homicide victimization trends by race,
        53, 56–57 (tables)
    juvenile victims of, 275, 275 (box)
    Native American homicides, 73
    pachuco killers and, 141–142
    police homicides, 115, 116, 117
    police officer deaths, 117, 203
    prehomicidal assaults and, 83
    types of, 53, 57 (table)
    See also Crime statistics; Mass killings; Serial
        killers; Subculture of violence theory;
        Violent crime
Hooton, E. A., 64, 65
Hoover, President H., 26, 115
Hopelessness, 84, 85
House Judiciary Subcommittee on Civil and
    Constitutional Rights, 208
Houses of refuge, 267
Howell, J. C., 270, 290, 291
Hsia, H. M., 284
Huddleston, W., 157, 158
Hudson v McMillian, 195
Huff, C. R., 224, 227
Huff-Corzine, L., 73
Huling, T., 252
Human nature, 64–65

Human trafficking, 10
Hurricane Katrina, 19

Illegal immigrants, 27, 109, 110, 122–123,
    124–125 (box)
Illegal Immigration Reform and Immigrant
    Responsibility Act of 1996, 27
Illegitimacy, 66, 74
Imam, I., 159
Immigrant populations:
    anti-immigration movement, 26
    Asian Americans, 29–34, 30 (table)
    Chinese Americans, 29–31, 108–109
    circular migration and, 27
    colonization of United States, 19, 21
    competition for work and, 17
    Eastern European immigrants, 24
    Filipino Americans, 26, 32
    German Americans, 19, 20–21
    hate/bias crimes and, 55
    illegal immigrants, 27, 109, 110, 122–123,
        124–125 (box)
    Irish Americans, 22–23, 23 (box)
    Italian Americans, 19, 21–22
    Japanese Americans, 31–32
    Jewish Americans, 24–25
    Korean Americans, 32, 33–34 (box)
    Latino groups, 25–28
    Mexicans, 26–27
    new immigrants, 110
    promised land and, 27
    Puerto Ricans, 27–28
    quotas on, 21–22, 23, 25, 26, 27, 32
    social problems, social disorganization and, 70–72
    White ethnics, 19–25, 35
    See also Historic context; Migration patterns
Immigration Act of 1924, 21
Immigration and Naturalization Service (INS), 101
Immigration and Reform and Control Act of 1986, 27
Impulse control, 66, 67
Incarceration, xi
    code of the street and, 88
    convict-lease system and, 14–15, 89
    crime statistics collection an, 40
    disciplinary process and, 84
    disproportionate minority confinement, xii, xiii, 19
    felon disenfranchisement, xii, 258–260
    institutional violence and, 84
    mass incarceration, 253–254
    prison writings, 90
    sentencing approaches and, 176
    social disorganization and, 75
    war on drugs and, 176, 184
    See also Corrections; Courts; Criminal justice
        system; Political prisoners; Sentencing

Indentured servitude, 10, 19, 22
Indian Boarding Schools, 8
Indian Citizenship Act of 1924, 8
Indian Removal Act of 1830, 8, 295
Indian Reorganization Act of 1934, 8, 138
Indian Self-Determination and Education
    Assistance Act of 1975, 9
Industrial Workers of the World (Wobblies), 261
Innocence Project, 224, 226
Innocence Protection Act of 2003, 228
Intelligence factor:
    criminal behavior and, 65–66, 67
    skin color and, 68, 68 (table)
    street smarts, 81
International Association of Chiefs of Police
    (IACP), 41, 118
International Covenant on Civil and Political rights, 288
Internet:
    computer-aided mapping, 99
    police department sites, 99
Intervention. See Courts; Criminal justice system;
        Delinquency prevention; Incarceration;
        Juvenile justice system
Invention of race, xiii, 1–2
Ioimo, R., 118
IQ tests, 65, 66
Irish Americans, 22–23, 23 (box), 35 (table),
    110, 111, 139
Italian Americans, 19, 21–22

Jackson, G., 90, 261
James, J., 261
Jang, S. J., 80
Japanese Americans, 21, 29, 30 (table), 31–32, 35 (table)
    arrest rates, 300–301 (table), 304–305 (table)
    attitudes about crime, 72
    immigration quotas and, 32
    Oriental crime and, 72
    relocation centers and, 31
Jena 6 case, 227
Jensen, V. J., 84, 88
Jewish Americans, 22, 24–25, 35 (table)
Jim Crow, 46
Johnson, B. R., 80
Johnson, D., 219
Johnson, G. B., 82
Johnson, J. B., 146
Johnson v. California, 245
Johnston, L. D., 279
Jones, J., 218
Jones v. Willingham, 241
Jones-Brown, D., 83
Jordan, K. L., 154
Jukes study, 65
Justice is blind, xi

Juvenile delinquency, xii, 266
    Asian Americans and, 72
    Chicago immigrant populations and, 71
    child savers and, 263, 268–269, 270
    delinquency prevention, 290–292, 290 (box)
    female delinquents, xiii, 286–288
    hate crimes and, 57
    houses of refuge and, 267–268
    inner city childhood and, 85–86 (box)
    intelligence and, 66
    Jewish immigrants and, 25
    risk-taking activities, 81
    social context of, 266
    subcultural theory and, 81
    See also Delinquency; Delinquency prevention;
        Gangs; Juvenile justice system
Juvenile Justice and Delinquency Prevention (JJDP)
    Act of 1974, 283, 285
Juvenile justice system, xii, xiv, 263–264
    behavioral problems and, 266
    Central Park Jogger case and, 224–225
    child savers and, 263, 268–269, 270
    colonial model and, 283
    death penalty and, xiii, 288–289, 289 (box)
    delinquency prevention and, 290–292, 290 (box)
    disproportionate minority confinement and,
        263, 281–286, 292
    female delinquency and, 286–288
    historical context, race/juvenile justice and, 267–270
    historically Black colleges/universities and, 263
    houses of refuge and, 267–268
    jury trial and, 266
    juvenile courts and, 269–270
    juvenile crime/victimization and, 270–280, 272
        (table), 275 (box), 276–277 (table), 280 (table)
    juvenile drug use, 278, 279 (table), 280, 280 (table)
    overview of, 264–266, 265 (figure)
    punitive approach to, 266, 270, 283, 290, 292
    reformatories and, 268
    rehabilitation approach and, 266, 270, 292
    serious crime, adult court and, 266
    superpredators and, 270
    youth gangs, 272, 272–274 (box)
    youth in the system, data on, 280–281, 282 (figure)
    See also Corrections; Crime statistics; Criminal
        justice system

Kang, W., 275
Katz, S. M., 102
Kelling, L., 103
Kennedy, R., 116, 139, 183
Kidnapping, 202–203
Kimbrough v. United States, 188
King, R., 115, 117
Knepper, P. E., 40, 46, 48, 61
Knights of White Camellia, 15

Korean Americans, 29, 30 (table), 32, 33–34 (box)
Kowal, L., 154
Krauss, E., 156, 157
Krisberg, B., 89, 90
Krivo, L. J., 73
Ku Klux Klan (KKK), 15
Kusow, A., 128

Labeling issues, xi, 34
LaFree, G., xii
Lambert, E., 220
La Neustra Familia, 243
Langan, P. A., 123
Lanier, C., 73
Latinos. *See* Hispanic Americans
Law-abiding citizens, xi, 102
Law enforcement. *See* Courts; Criminal
    investigators; Criminal justice system;
    Juvenile justice system; Policing
Law Enforcement Assistance Administration, 45
Law Enforcement Management and
    Administrative Statistics (LEMAS),
    38 (table)
Laws:
    abortion, legalization of, 95–96
    anti-narcotics legislation, 30
    anti-opium smoking legislation, 109
    Black codes, 14, 89
    civil law, street justice and, 87
    critical legal studies movement and, 96
    deadly force in policing and, 116–117
    felony disenfranchisement laws, 258–260
    formal laws, crime incidence and, 77
    illegal activities, guidelines for, 297
    race neutrality and, 295–296
    religious discrimination ban, 24
    See also Courts; Criminal justice system;
        Government; Policing; Supreme Court
Lee, C., 214, 215
Lee, M. T., 73
Lee, N., 77
Lehrer, E., 298
Leiber, M. J., 182
Levitt, S. D., 95
Life history theory. *See* r/K life history theory
Life Skills Training, 291
Lilly, R. J., 88
Linnaeus, C., 1
Litras, M. F. X., 147
Lombroso, C., 63, 64, 65
Long, A., 159
Longmire, D. R., 219
Lurigio, A. J., 121
Lopez, J., 261
Louima, A., 117
Lovell, R., 284

Lower class:
    Black compulsive masculinity alternative
        and, 94
    fate vs. autonomy and, 81
    street smarts and, 81
    values of, 81
    See also Class structure; Middle class;
        Upper class
Lublin, J., 255
Lurigio, A. J., 121
Lynching, xii, 15, 16 (figure), 17, 59, 89
    crime statistics on, 41, 42–43 (table)
    Mexican American victims of, 109
    policing practices and, 107
    state-sanctioned executions/illegal lynchings,
        207–208, 209–210 (tables)
    substitution thesis, Zimring analysis and,
        215–217, 216 (figure)
Lynn, R., 68

MacKenzie, D., 290
Mafia, 22
Manatu-Rupert, N., 146, 147
Mandak, J., 151
Manhattan Bail Project, 143
Mann, C. R., 30, 61, 84, 91, 143
Manza, J., 258, 259, 260
Map v. Ohio, 103
Marriage and Incarceration Act, 255
Mars, K., 88
Marshall Hypotheses, 220, 221
Marshall, Justice T., 220
Martin, D. E., 128
Martinez, R., 73
Martinson, R., 165
Marzette, L. N., 121, 122
Mason, C. A., 287
Mason, M., 121
Mass incarceration, 253–254
Mass killings:
    Native American populations, 7, 8, 9–10
    Virginia Tech massacre, 32, 33–34 (box)
    workplace murders, 53, 57 (table)
    See also Homicide; Violent crime
Massey, D. S., 17
McCarthy-era witch hunts, 261
McCleskey v. Kemp, 203–204
McCluskey, C. P., 79
McGee, Z. T., 292
McGuffee, K., 154, 155
McIntyre, C. C. L., 10, 235, 236
McMorris, B. J., 182
McShane, M. D., 76
McVeigh, T., 111
Meadows, L. A., 118
Meagher, T. J., 22

Media, xi
    Black women, stereotypes of, 146–147
    lynchings, reporting on, 41
Meier, A., 15
Mentoring programs, 291
Merton, R. K., 64, 78
Mexican Americans, 26–27, 35 (table)
    arrest rates, 300 (table)
    colonial model and, 92–94
    court system/United States, 141–142
    gringo justice, 92
    policing practices and, 109–110
Mexican Mafia (La Eme), 243
Middle class:
    African American middle class, 76, 115
    anomie theories, middle class bias of, 79
    criminality in, 88
    death penalty, support of, 218
    middle class values, subcultural theory and, 81
    policing practices and, 111, 115
    See also Class structure; Lower class; Upper class
Middle Easterners, 100
Miethe, T. D., 49, 61
Migration patterns, 2, 2 (figure), 15, 17
    circular migration, 27
    mass South to North migration, 15, 17, 70, 240
    r/K life history theory of crime and, 66–69
    See also Immigrant populations
Miller, E., 257
Miller, J., 88
Miller, J. G., 251
Miller, W., 81
Minnesota Multiphasic Personality Inventory
        (MMPI), 65
Minnesota Sentencing Guidelines, 174, 175 (figure)
Minorities. *See* Immigrant populations; Middle
        Easterners; Race; Racial minorities
Miranda v. Arizona, 103
Mitchell, A. D., 221
Mitchell, O., 181
Model minority label, 34
Monitoring the Future survey, 271, 278
Montgomery boycott, 18
Moore, M. H., 103
Morin, J. L., 153
Morris, R., 189
Mosher, C. J., 49, 61
Moynihan Report, 74–75 (box)
Moynihan, Senator Daniel Patrick, 74
Mukasey, Attorney General M., 189
Multiracial population, 5
Multisystemic Therapy, 291
Munoz, E. A., 182
Murder Incorporated (New York City), 25
Murphy, P., 103
Murray, C., 66, 68

Muslims, 100
Myers, M. A., 238
Myrdal, G., 108

National Adolescent Survey, 271
National Association for the Advancement of
        Colored People (NAACP), 41, 44, 204, 269
National Association of Drug Court Professionals, 158
National Center for Injury Prevention
        and Control, 275
National Center on Institutions and Alternatives
        (NCIA), 251–252
National Center for Juvenile Justice, 264
National Coalition to Abolish the Death Penalty, 289
National Commission on Law Observance and
        Enforcement, 41, 115, 238
National Crime Survey, 45
National Crime Victimization Survey (NCVS),
        38 (table), 39, 45–46
    estimations, utilization of, 49
    hate crimes and, 59
    homicide victimization trends, 53
    juvenile victimization, 271, 278
    recording/reporting discrepancies and, 48
    self-report studies, 271
    traffic stops, 121–122, 125
    violent victimization trends by race, 52–53,
        54–55 (tables), 58
    See also Crime statistics; Racial profiling
National DNA Index System, 99
National Drug Control Strategy, 160, 176
National Hate Crime Data Collection Program
        (NHCDCP), 39, 44–45
National Household Survey on Drug Abuse, 271
National Incident-Based Reporting System
        (NIBRS), 38 (table), 44
    hate crimes, 57
    juvenile crime, 271
    recording/reporting discrepancies and, 48, 59
    See also Crime statistics
National Institute of Justice's Arrestee Drug Abuse
        Monitoring Program, 280
National Minority Advisory Council on Criminal
        Justice, 140, 142
National Opinion Research Center (NORC), 45, 68, 112
National Opinion Survey on Crime and Justice
        (NOSCJ), 219, 233
National Organization of Black Law Enforcement
        Executives (NOBLE), xi, 116, 120, 127
National Police Association, 41
National Survey of Black Americans, 80
National Survey on Drug Use and Health, 271, 278
National Youth Gang Survey, 272–274 (box)
Nation of Islam, 240
Nation of Islam Prison Reform Ministry, 241

Native Americans, xiii, 4, 5, 6 (table)
    arrest rates, 47–48, 49–51, 51–52 (tables),
        302–303 (table), 305–306 (table)
    boarding schools and, 8, 72
    Bureau of Indian Affairs and, 8
    civil/cultural rights for, 8
    collective efficacy and, 76
    colonial model and, 92–94
    court system of, 136–138, 136–137 (tables)
    court system/United States and, 138–139
    cultural conflict theory and, 77
    Dawes Act of 1887, 8, 104
    diverse/advanced societies of, 7
    execution of, 205
    gambling facilities and, 9
    historic context for, 7–9
    land rights of, 7, 8, 9
    lynching of, 15
    massacre/enslavement of, 7, 8, 9–10, 138
    policing practices and, 104–107, 106 (table)
    removal of, 8, 9, 107
    reservations for, 8, 9, 105, 107
    restorative justice and, 137–138
    self-determination legislation, 9
    social disorganization and, 72, 73
    termination policy and, 9
    treaties with, 7–8, 9
    tribal governments and, 8, 9
    tribal law enforcement agencies,
        105, 106 (table)
    victimization data, 39, 48, 49, 52–53,
        54–55 (tables), 58
    White aggression and, 7–8, 19
    See also Crime statistics; Race; Racial minorities
Native Hawaiians, 4, 6 (table)
Natural disasters, 19
Negro courts, 235
Neufeld, P., 224
New Jerseyans for a Death Penalty Moratorium,
        228, 229
News. See Media
Newton, H., 240
Nichols, T., 111
No discrimination thesis (NDT), 90, 145, 146
Norms:
    anomie and, 78
    code of the street and, 85–88, 85–87 (box)
    conduct norms, 77
    cultural norms, 76–77
    suicide, normlessness and, 78
Northern v. Nelson, 241
Norton, M. I., 154

Office of Civil Rights (OCR), 116
Office of Federal Statistical Policy and Standards, 46

Office of Juvenile Justice and Delinquency
        Prevention (OJJDP), 37, 263, 271, 275,
        280–281, 283–284, 285
Ogawa, P., 159
Ohlin, L. E., 81
Okafo, N., 119
Oklahoma City bombing, 111
Oliver, W., 94
Olweus Bullying Prevention Program, 291
O'Malley, P. M., 279
Organized crime:
    Asian groups, 30–31
    Italian groups, 22
    Jewish groups, 25
    Latino gangs and, 28
Osborne, L. T., 268
Oshinsky, D. M., 237
Ossorio, P., 4
Owen, B., 287

Pachuco killers, 141–142
Pacific Islanders, 4, 6 (table)
    arrest rates, 302–303 (table), 305–306 (table)
    Filipino workers and, 32
    Japanese workers and, 31
    See also Asian Americans; Crime statistics
Packer, H., 165, 297
Pager, D. I., 256
Pale Faces, 15
Palmer raids, 261
Parchman Farm, 14–15
Parenti, C., 253
Park, R., 71
Parker, K. F., 224
Parsons, L., 172, 173
Patterrollers, 107
Pattillo, M. E., 76
Peltier, L., 261
Penn, E., 290, 291
Perez, J., 144
Perry, B., 30, 34, 55
Petersilia, J., 231, 254, 257
Peterson, R. D., 73
Peterson, S. A., 119, 122
Phillips, D. M., 49, 61
Phillips, U. B., 205, 206, 207
Phrenology, 63
Plantation justice, 10–11, 139, 140 (box)
Platt, A., 268
Plessy v. Ferguson, 15
Police brutality, 115–116, 117, 118, 126, 240
Policing, xii, 99–100
    accountability in, 126–127
    African American history and, 107–108
    Asian American history and, 108–109

citizen satisfaction with, 112–115,
        113–114 (table), 128
    civil liberties and, 100
    civil rights movement and, 18, 108
    colonial-era policing, 102–103
    colonial model and, 93
    community policing, 76, 103, 112, 127
    computerized statistics, 118, 128
    contemporary innovations in, 127–128
    contemporary issues, race/policing and, 111–115
    crime statistics and, 41, 46, 118, 128
    criminal profiling, 111, 118–119
    deadly force and, 115, 116–117, 121
    discriminatory practices and, 115–117, 118
    DNA evidence and, 2, 3–4 (box)
    dominant social values and, 86
    fingerprint cards, 47
    Hispanic history and, 109–110
    historic context, racial minorities and, 102–111,
        106 (table)
    illegal activities, police profit from, 30, 115
    illegal immigrants, policing of, 122–123,
        124–125 (box)
    Irish immigrants and, 22–23, 110, 111
    Ku Klux Klan activities and, 15
    lynchings and, 107
    minority employment in, 101, 102 (figure), 103,
        107, 108, 116
    Native American history and, 104–107, 106 (table)
    police brutality, 115–116, 117, 118, 126, 240
    police deviance, accountability and, 115–127
    police officer deaths and, 117
    political control over, 103, 111
    power threat approach and, 90
    prehomicidal assaults and, 83
    private citizens, posses and, 14
    problem-oriented policing, 127, 128
    professionalization of, 103
    race riots and, 107–108
    racial profiling and, xii, 96, 112, 119–126,
        119–120 (box), 123 (table)
    scanning/analysis/response/assessment process, 128
    selective enforcement/bias, 48, 59, 64, 83, 85, 86
    slave patrols, 10, 103, 107
    state-sanctioned segregation and, 15
    terrorist-era policing, 99–100
    third degree tactic, 115
    tribal law enforcement agencies, 105, 106 (table)
    United States policing practices, 100–102,
        102 (figure)
    use-of-force databases, 118
    White immigrant history and, 110–111
    zero-tolerance policing, 127, 128
    See also Arrests; Corrections; Crime statistics;
        Criminal investigators; Incarceration

Polish Americans, 22
Political crime, 66, 79
Political prisoners, xii-xiii, 260–261
Poorhouses, 235
Pope, C. E., 263, 284, 286
Popular culture:
    Black women, stereotypes of, 146–147
    code of the street and, 85–88, 85–87 (box)
    rap music, 87
Population genetics, 4
    genetic inferiority, crime and, 63
    intelligence, race/crime and, 65–66
    r/K life history theory, 66–69
    skin color, criminal inclination and, 67, 68–69
Positivism, xi-xii
Potter, D. M., 40
Poverty, 66, 71
    code of the street and, 85–88, 85–87 (box)
    powerlessness, climate of, 83–84
    racialized poverty, 85, 86, 89
    social disorganization/violent
        crime and, 72–73
Powell v. Alabama, 147
Power differential. See Conflict theory
Power threat approach, 90
Powerlessness, 83–84
Pratt, G., 261
Pratt, T. C., 180
Prejudice, 5–6
    Mexican immigrants and, 26
    See also Bias; Discrimination
President's Commission on Law Enforcement and
        the Administration of Justice, 45, 112
Prevention. See Delinquency prevention; Education
Prison-industrial complex, 253
Privacy rights, 3, 4, 100, 298
Privilege, 90
Problem-oriented policing (POP), 127, 128
Profiling. See Criminal profiling; Racial profiling
Progressive Era reforms, 111, 269
Project Towards No Drug Abuse, 291
Promising Programs, 291
Property victimization, 39, 53
Prostitution, 23, 30, 108, 109
Public defenders, 141, 147–149
Public Law 280, 105
Public opinion, xiii
    citizen satisfaction with policing,
        112, 113–114 (table)
    correctional system operations, 233–234
    crime statistics, publishing of, 41
    death penalty and, 217–220, 228–229
    juvenile death penalty and, 289
    racial profiling polls, 119
    sentencing approaches and, 173
    slave system and, 10

Puerto Rican nationalists, 261
Puerto Ricans, 27–28, 35 (table)
Pulaski, C. A., 204
Punishment:
    physical punishment, 80, 235
    race discrimination and, 90
    See also Corrections; Courts; Incarceration;
        Sentencing
Purple Gang (Detroit), 25

Quality-of-life policing, 127
Quetelet, A. Q., 69
Quinney, R., 89

Race, xi-xiv
    categorization of, 2, 4, 35, 35 (table), 46–48
    census statistics on, 2–3, 4, 5, 6 (tables)
    crime and, xi-xii, 18
    differential justice and, 11, 12–13 (box), 12 (table)
    DNA evidence of, 2, 3–4 (box)
    ethnicity/ethnic groups and, 4–5
    future research on, 296–298
    invention of race, xiii, 1–2
    measurement of, 4
    migration patterns and, 2, 2 (figure)
    multiracial category, 5
    prejudice/discrimination and, 5–6
    purity of race, 12 (box), 67
    racialization of crime, 39, 58
    r/K life history theory and, 66–69
    social construct of, 2
    See also Crime statistics; Historic context;
        Theoretical perspectives
Race privilege, 90
Race riots, xii, 15, 17, 74, 103, 107
Racial Justice Act of 1988, 204
Racial minorities, xi-xiv, 1
    arrests, overrepresentation of
        minorities, xii, xiii, 19, 38–39, 58
    biological factors in crime and, 63–64, 65
    colonial model and, 92–94
    critical race theory and, 96
    historic record of, xii
    human nature and crime, 64–65
    model minority label, 34
    oppression of, 22–23, 24–25
    people-of-color term, 5
    physical characteristics of, xi-xii
    police accountability and, 126–127
    police brutality and, 115–116, 117, 118
    policing history and, 107–110
    power threat concept and, 90
    racial hierarchy and, 1–2
    racialization of, dominant group needs and, 96
    scholarly research on, xi, xii
    skin color and, 67, 68–69

social control, criminal justice practices and, 89–90
terrorist-era policing and, 99–100
victimization estimates and, 49
voice-of-color perspective and, 96
See also African Americans; Asian Americans;
    Crime statistics; Immigrant populations;
    Race; Racial profiling
Racial profiling, xii, 96, 112, 119–126,
    119–120 (box), 123(table)
definition of, 119, 120
driving while Black/Brown, 119, 121
drug courier profile, 121
illegal immigrants and, 122–123, 124–125 (box)
research on, methodological problems with, 125–126
traffic stops and, 121–123, 123 (table)
walking/shopping while Black, 119, 119–120 (box)
See also Policing
Racial Profiling Data Collection Resource Center at
    Northeastern University (RPDCRC), 121, 122, 125
Racialized crime, 39, 58
Racialized groups, 96, 283
Racialized poverty, 85
Racist beliefs/ideologies, xi, 22
American society/legal structures and, 96
color-blind jurors and, 157
interest convergence, racism and, 96
racial classification systems and, 46–48
r/K life history theory of crime, 66–69
violence, racial differences in, 84
wrongful convictions and, 224
Radelet, M. L., 224
Radical political movements, 260–261
Rage riots, 242
Ramirez, D., 120
Rantala, R. R., 48
Rap music, 87
Rape, xi, 46
Black vs. White commission of, 11, 12
    (table), 17, 58
domestic laborers and, 23
See also Victimization; Violent crime
Raper, A., 43
Rattner, A., 224
Raudenbush, S. W., 75
Ray, M. C., 84
Re Davis, 167
Re Negro Emmanuel, 167, 168
Re Negro John Punch, 167–168
Re Sweat, 167
Reagan, President R., 173, 176
Reasons, C. E., 240
Reaves, B. A., 102, 144
Reconstruction Era, 15, 17, 107, 190, 240
Reentry court model, 257–258
Rehabilitation approach, 165, 173, 233
Reid, I. De A., 63

Reiman, J., 90
Relative Rate Index (RRI), 285
Religion-based discrimination, 22
Religious practice, 80
Relocation centers, 31
Relocation Program (Bureau of Indian Affairs), 107
Reno, Attorney General J., 213
Renzetti, C. M., 62
Report on Lawlessness in Law Enforcement, 115
Research, xi-xii, 296–298
    See also Crime statistics; Theoretical perspectives
Respect:
    African Americans, disadvantage of, 140
    street culture and, 85, 86–87
Retreatist gangs, 81
Reuter, P., 290
Revolutionary War (United States), 139, 172
Riedel, M., 214
Riots, xii, 15, 17, 23, 74, 103, 107, 110, 140, 242
Risk-taking activities, 81
r/K life history theory, 66–69
Roberts, J. L., 154
Robinson, L. N., 40
Robinson, M., 149
Rocque, M., 80
Roe v. Wade, 95
Role models, 73
Roosevelt, President Franklin D., 31
Roper v. Simmons, 288, 289 (box)
Rose, D. R., 75
Rosenbaum, J., 287
Rothstein, A., 25
Rudolph, J. L., 79
Rudwick, E., 15
Rushton, J. P., 66, 67, 68
Russell, K. K., xii, 10, 12, 96, 146, 183, 184
Russell-Brown, K., 99, 117, 238, 251
Ryan, Governor G., 228

Sacks, B., 25
Sagarin, E., 224
Sailor riots, 110
Sale, K., 7
Saleh-Hanna, V., 94
Sampson, R. J., 72, 73, 75, 84, 283
San Quentin, 240–241
Scalia, J., 145
Scanlan, K., 131
Scheck, B., 224
Schlesinger, T., 146
Schreck, C. J., 87
Schulenberg, J. E., 279
Schulman, M., 156, 157
Schultz, C., 226–227
Scottsboro Boys, 17, 26, 147, 150
Second Chances Act of 2007, 258

Segregation, xii, 15, 89
  civil rights movement and, 18
  Irish immigrants and, 22–23
  post-World War II era, 18
  race riots and, 17
  See also Urban communities
Self-reporting, 45
Sellin, T., 76, 77, 78, 237
Sengeral Social Survey (GSS), 84
Sentencing, 135, 163–164
  colonial New York, crime/justice in, 169–171,
    169–171 (tables)
  contemporary approaches/1980s-2000s,
    173–176, 175 (figure)
  contemporary issues, race/sentencing and,
    177–197, 178 (table)
  crime control/due process models of criminal
    justice and, 165
  cumulative disadvantage and, 179
  disparities in, xii, 183–189, 186 (figures),
    187 (table), 188–189 (box)
  early colonial cases of, 166–168, 168 (figure)
  early national sentencing statistics,
    172, 172–173 (tables)
  felony convictions, 177, 178 (table)
  get-tough approach, 173, 233
  hierarchy of social precedence and,
    168, 168 (figure)
  historical overview, race/sentencing and, 165–176
  incarceration rates and, 176
  life in prison, 218
  mandatory minimum sentences, 174
  minority judges and, 189–195, 191–194 (table),
    196–197 (box)
  misdemeanor sentencing, race and, 182
  philosophies of, 164–165
  political influence on, 164–165
  presumptive sentencing guidelines and,
    174, 175 (figu)
  process-related factors and, 181
  punishment, colonial America and, 165–166
  race/sentencing, research on, 177, 179–182
  rehabilitation approach and, 165, 173, 233
  retributive sentencing, 220
  three strikes sentencing, 96, 173, 174
  truth-in-sentencing laws, 173, 174
  war on drugs and, 176, 182, 183–189,
    186 (figures), 187 (table), 188–189 (box)
  See also Corrections; Courts; Death penalty;
    Wrongful convictions
Sentencing Project, 251
Serial killers, 111
Serious and Violent Offender Reentry Initiative of
    2003, 255
'76 Association, 15

Sexual assault:
  domestic laborers and, 17, 23
  slave victims of, 17, 67
  violent crime, 46
  See also Rape; Victimization
Shane, S., 2
Shelden, R. G., 183, 235, 268, 286, 287
Sherman, L. W., 290
Shriver, M., 4
Sickmund, M., 265, 278, 287
Siljander, R. P., 120
Simmel, G., 88
Simons, D. H., 4
Simons, R. L., 80, 87
Simpson, O. J., 156
Skin color, 67, 68
  intelligence and, 68, 68 (table)
  social distance and, 68–69
Skogan, W. G., 121
Slavery:
  anti-slavery movement, 11, 13
  Black codes and, 14, 107
  chattel slavery, 11
  colonial society and, 10, 11
  Compromise of 1850 and, 13–14
  control strategies in, 11, 139, 140 (box)
  criminal activity, punishment of, 11
  death penalty and, 205, 205–207 (box),
    206 (table), 207 (figure)
  differential justice and, 11, 12–13 (
    box), 12 (table)
  Emancipation Proclamation and, 14, 15, 107
  free Blacks, 13 (box), 235, 237
  Fugitive Slave Law, 14
  incarceration and, 235
  international movement against, 11
  international slave trade, 11, 13
  Negro courts and, 235
  plantation justice and, 10–11, 139, 140 (box)
  racial classification and, 46–47
  slave codes, 10–11, 139, 140 (box), 295
  slave-free states, 13–14
  slave patrols, 10, 103, 107
  slave trade, 10, 11, 13, 19
  Spanish origins of, 9–10
  Thirteenth Amendment and, 14, 15
  White sexual aggression and, 17, 67
  See also African Americans
Sleepy Lagoon Case and, 141–142
Smith, E., 84
Smith, E. L., 123
Smith, J., 136, 138
Smith, M. D., 214
Smith, W. R., 121
Snyder, H. N., 275

Snyder, M., 265, 278, 287
Social buffers, 73
Social class. *See* Class structure
Social constructionism:
   race categories and, 2, 35, 35 (table), 96
   See also Class structure
Social control:
   criminal justice practices and, 89–90
   kin/neighbors and, 76
   power threat approach and, 90
   social buffers and, 73
   strain theory and, 79
   See also Courts; Criminal justice system;
      Juvenile justice system; Policing
Social disorganization, xii, 70–72
   African Americans and, 72–73
   collective efficacy and, 75–76
   contemporary theory of, 72–75
   female-headed families and, 72, 73, 74–75 (box)
   Hispanic communities and, 73–74
   mass incarceration and, 75
   Native Americans and, 73
   racial invariance thesis and, 73
   social buffers/role models and, 73
   social isolation and, 73
   See also Code of the street; Sociological
      explanations; Theoretical perspectives
Social distance, 68–69
Social isolation, 73
Social structure. *See* Class structure;
   Strain/anomie theory
Society for the Prevention of Pauperism, 267
Society for the Reformation of Juvenile
   Delinquents, 267
Socioeconomic status, 67, 89, 296
   powerlessness, violent crime and, 83–84
   See also Class structure; Employment; Poverty
Sociological explanations, 67, 69–70
   collective efficacy concept, 75–76
   female-headed families, 72, 73, 74–75 (box)
   racial invariance thesis and, 73
   Seventh Ward study (Philadelphia), 70
   social disorganization theory, 70–75
   truly disadvantaged populations and, 18–19, 73
   See also Theoretical perspectives
Sommers, S. R., 154
Sourcebook of Criminal Justice Statistics, 38 (table)
Sowell, T., 20, 21, 22, 26, 31, 35
Spohn, C. S., 91, 92, 155, 164, 174, 180, 181, 195, 287
Stanford v. Kentucky, 288, 289 (box)
Staples, R., 92
State v. Soto, 121
Statistics. *See* Crime statistics
Stauffer, A. R., 214
Steffensmeier, D. J., 84, 146

Stereotypes:
   Black women, pretrial process and, 146–147
   ethnic groups, suspicion among, 28–29 (box)
   Italian organized crime and, 22
   Jewish immigrants and, 25
   Mexican immigrants and, 26
   policing practices and, 115–117
   Puerto Ricans/Latinos and, 27–28
   races, categorization of, 35, 35 (table)
   racial profiling incidents and, 122
   racial stereotyping scale, 219
   religious practice and, 22
   truly disadvantaged populations and, 19
   wrongful convictions and, 224
Stewart, E. A., 80, 83, 87
Strain/anomie theory, 78
   African Americans and, 78–79
   American Dream and, 78, 79
   culturally specific models and, 79
   dominant group, values of, 78
   Hispanics and, 79
   limitations of, 79–80
   middle class bias of, 79
   occupational/educational aspirations and, 78–79
   overprediction and, 79–80
   See also Conflict theory; Culture conflict theory;
      Strain theory; Theoretical perspectives
Strain theory, 80
   positive/negative stimuli and, 80
   religiosity, coping mechanism of, 80
   See also Strain/anomie theory; Theoretical
      perspectives
Street crime, 90
Street smarts, 81
   See also Code of the street
Structural-cultural theory, 94–95
Subcultural theory, 81
   code of the street and, 85–88, 85–87 (box)
   focal concerns theory and, 81
   lower class values and, 81
   middle class values and, 81
   middle/upper class criminality and, 88
   opportunity structure and, 81
   weaknesses of, 88
   See also Subculture of violence theory;
      Theoretical perspectives
Subculture of violence theory, 81–82
   disaggregated crime data and, 84
   economic deprivation, powerlessness and, 83–84
   gender and, 84
   guilt, feelings of, 82
   hierarchy of homicide and, 82, 83 (figures)
   hopelessness and, 84
   justice administration, inequities in, 83
   prison misconduct/institutional violence and, 84

propositions of, 82
values favorable to violence, 84
weaknesses of, 82, 84, 88
See also Subcultural theory; Theoretical
    perspectives
Substitution thesis, 215–217, 216 (figure)
Suicide, 78
Superpredators, 270, 295
Supplemental Homicide Reports (SHR), 43, 53, 275
Supreme Court, xiii, 134, 135, 172
    Argersinger v. Hamlin, 147
    Atkins v. Virginia, 288, 289 (box)
    Batson v. Kentucky, 153
    Battle v. Anderson, 241
    Blakely v. Washington, 185
    Coker v. Georgia, 202
    death penalty cases and, 201–204, 288–289, 289 (box)
    Desmond v. Blackwell, 241
    Eberheart v. Georgia, 202–203
    Eddings v. Oklahoma, 288, 289 (box)
    Escobedo v. Illinois, 103
    Fulwood v. Clemmer, 241
    Furman v. Georgia, 201–202, 204, 220
    Gideon v. Wainwright, 147
    Gregg v. Georgia, 202, 204, 208
    Hernandez v. New York, 153
    Hudson v. McMillian, 195
    indigent defendants, legal counsel for, 147
    Japanese relocation centers and, 31
    Johnson v. California, 245
    Jones v. Willingham, 241
    Kimbrough v. United States, 188
    Mapp v Ohio, 103
    McCleskey v. Kemp, 203–204
    Miranda v. Arizona, 103
    Native American rights and, 8
    Navajo Nation Supreme court, 137
    Northern v. Nelson, 241
    Plessy v. Ferguson/separate but equal decision, 15
    Powell v. Alabama, 147
    Roe v. Wade, 95
    Roper v. Simmons, 288, 289 (box)
    Stanford v. Kentucky, 288, 289 (box)
    Swain v. Alabama, 152–153
    Tennessee v. Garner, 116
    Thompson v. Oklahoma, 288, 289 (box)
    United States v. Barry, 155–156
    United States v. Booker, 185
    Whren v. United States, 121
    Wilkins v. Missouri, 288, 289 (box)
    Young v. Robinson, 241
    See also Courts
Surveillance, 297–298
Sutherland, E. H., 18, 76
Swain v. Alabama, 152–153

Takagi, P., 261
Tarver, M., 27, 136, 137
Tatum, B. L., 68, 93
Taylor, D. L., 287
Taylor, P., 109
Taylor Greene, H., 290, 291
Tears, R. S., 118
Technology:
    DNA evidence and, 2, 3–4 (box)
    DNA photofitting technology, 3
Tennessee v. Garner, 116
Ten Precepts of American Slavery Jurisprudence,
    11, 12–13 (box)
Terrorism, 99–100
    domestic terrorists, 111
    public surveillance and, 297–298
    United States policing and, 101
    zero-tolerance policing and, 128, 297
Texas Syndicate, 243
Theft victimization rates (TVRs), 52, 58
Theoretical perspectives, 61–62, 96–97, 97 (table)
    abortion-crime connection, 95–96
    biological factors, race/crime and, 63–64, 65
    bridging theories, 62
    code of the street and, 85–88, 85–87 (box)
    collective efficacy concept, 75–76
    colonial model, 92–94
    conflict theory, 88–92, 91 (box), 92 (figure)
    critical race theory, 96
    culture conflict theory, 76–78
    future research and, 296–298
    human nature, crime and, 64–65
    intelligence-based theories, 65–66, 68
    macrotheories/microtheories, 62
    racial invariance thesis, 73
    rational choice theory, 62
    r/K life history theory, 66–69
    scientific theory, 62
    social disorganization theory, 70–76
    social distance concept, 68–69
    sociological explanations, 67, 69–76
    strain/anomie theory, 78–80
    strain theory, 80
    structural-cultural theory, 94–95
    subcultural theory, 81, 85–88
    subculture of violence theory, 81–88, 83
        (figures), 85–87 (box)
    theory, definitions of, 62–63
    theory, utility of, 62
    See also Crime statistics; Criminology
Thirteenth Amendment, 14, 237, 1589
Thomas, Justice C., 195
Thompson v. Oklahoma, 288, 289 (box)
Three strikes sentencing, 96, 173, 174, 233
Thrill seeking, 81, 94

Till, Emmett, 18
Tocqueville, A., 236
Toleration Act of 1649, 21
Tomsakovic-Devey, D., 121
Tongs, 30
Tonry, M., 183
Trail of Tears, 8
Transportation Security Agency, 101
Trujillo, L., 109
Truly disadvantaged populations, 18–19, 73
  powerlessness, violent crime and, 83–84
  See also Poverty; Socioeconomic status
Truth in Sentencing Act of 1997, 173, 174
Tulchin, S. H., 66
Turk, A. T., 89
Turner, K. B., 146
Turner, R. J., 80
Tuskegee Institute, 41
2005 National Gang Threat Assessment, 39
Tye, G., 159

Uggen, C., 258, 259, 260
Unemployment, 66
  African Americans and, 72
  social disorganization and, 72, 73
  See also Employment; Poverty
Uniform Crime Reports (UCR), 37, 38 (table),
    39, 41, 43–45, 173
  arrest trends data, 49–51, 51–52 (tables)
  estimations, utilization of, 49
  juvenile crime, 271, 272 (table), 280, 280 (table)
  reporting/recording, variations in, 48
  Supplementary Homicide Reports and, 43, 53
  See also Crime statistics; Federal Bureau of
    Investigation (FBI)
United National Convention on the Rights of
    Children, 288
United Nations, 18
United States v. Barry, 155–156
United States v. Booker, 185
Unnever, J. D., 219, 220, 234
Upper class:
  criminality in, 88
  death penalty, support of, 218
  policing practices and, 111
  See also Class structure; Lower class; Middle class
Urban communities:
  arrest rates in, 50, 51 (table)
  Black flight and, 18
  code of the street and, 85–88, 85–87 (box)
  collective efficacy concept and, 75–76
  concentrated urban poverty, 85
  death penalty, support for, 219
  ecological environment view of, 71
  emigration from, 18, 25

Irish immigrants and, 22
Jewish immigrants and, 25
organized crime and, 22, 25
social disorganization and, 70–75
truly disadvantaged and, 18–19
Uniform Crime Reports and, 37
See also Communities; Great Migration
    from the South
USA PATRIOT Act of 2001, 100
U. S. Census Bureau:
  crime statistics, 39, 40
  multiracial category, 5
  race, categorization of, 2–3, 4, 5, 6 (tables)
  racial/ethnic groups, 4
  victimization survey and, 45–46
U. S. Customs and Border Protection (CBP), 101
U. S. Department of Justice (DOJ), 37, 39, 116, 118,
    126, 158, 213
U. S. General Accounting Office (GAO), 158–159
U. S. Immigration and Customs Enforcement (ICE), 101
U. S. Marshals Services, 101

Values:
  deviant values, 88
  lower class values, 81
  middle class values, 81
  values favorable to violence, 84
Vandalism, 55
van den Haag, E., 164
Velez, M. B., 73
Vera Institute, 143
Vice crimes, 23, 30–31
Victimization, xii, 39, 80
  code of the street and, 87
  death penalty, support of, 219–220
  estimates of victimization, 49
  homicide victimization trends, 53, 56–57 (tables)
  minorities, police brutality and, 115–116
  Native American victimizations, 48
  property victimization, 39, 53
  self-reports of, 45
  theft victimization rates, 52, 58
  trend analysis data, 46
  victimization surveys, 45–46, 46
  violent victimization trends by race,
    52–53, 54–55 (tables)
  See also Crime; Crime statistics; Juvenile justice
    system
Vietnamese Americans, 29, 30 (table), 34
Vigilantes, 109
Vincenzi, N., 154
Violent crime, 39
  aggravated assault, 49–50
  arrest trends data, 49–51, 51–52 (tables),
    304–306 (tables)

code of the street and, 85–88, 85–87 (box)
domestic violence, 46, 84
homicide victimization trends by race, 53,
    56–57 (tables)
lynching, xi, 15, 16 (figure), 17, 41, 42–43 (table)
subculture of violence theory and, 81–88, 83
    (figure), 85–87 (box)
violent victimization trends by race, 52–53,
    54–55 (tables)
See also Crime statistics; Homicide; Mass
    killings; Rape; Sentencing
Violent Crime Control and Law Enforcement Act of
    1994, 157, 176
Violent Crime Index (VCI), 271
Violent victimization rates (VVRs), 52–53, 54–55
    (tables)
Virginia State Federation of Colored Women's
    Clubs, 263, 269
Vito, G. F., 121
Voice-of-color perspective, 96
Voir dire process, 152–155
Voting rights, 258–260

Wacquant, L., 88
Walker, S., 27, 91, 92, 102, 126, 136, 137, 155
Walker-Barnes, C. J., 287
Wallace, H., 27, 136, 137
Walnut Street Jail, 235–236
Walsh, W. F., 121
War on drugs, 176, 182, 183–189, 186 (figures),
    187 (table), 188–189 (box)
War resistance, 261
Ward, G., 268, 269, 284
Warheit, G., 287
Warner, S. B., 41
Warnshuis, P. L., 141
Warren, P. Y., 121
Watts riots, 74
Weaver, R., 111
Weber, M., 88
Websdale, N., 128
Weisburd, D., 118, 128
Weitzer, R., 115
Welfare dependency, 66, 71
Western Penitentiary, 236
W. Haywood Burns Institute, 285
White Brotherhood, 15
White House Office of National Drug Control
    Policy, 176
White privilege, 90
White supremacy, 11, 12 (box), 15, 63, 96, 111, 243
White-collar crime, 66, 79, 90
Whites, xi, 2, 4, 5, 6 (tables), 19
    anti-Chinese sentiment and, 30, 31
    anti-slavery movement and, 11

arrest rates, 47–48, 49–51, 51–52 (tables), 58,
    300–306 (tables)
Black domestic laborers and, 17
convict-lease system and, 14
hate/bias crimes and, 55
immigration, New World and, 19–25
indentured servitude and, 10
interest convergence, racism and, 96
lynchings, 15, 16 (figure), 41, 42–43 (table), 59
race, differential justice and,
    11, 12–13 (box), 12 (table)
separate-but-equal decision and, 15
slave codes and, 11
slave trade, White involvement in, 13, 19
violent victimization trends, 52–53, 54–55 (tables)
See also Crime statistics
Whren v. United States, 121
Wickersham Commission Report, 26, 41, 141, 238
Wickersham, G., 26
Wilbanks, W., 90, 91, 146
Wilkins v. Missouri, 288, 289 (box)
Williams, F. P., 76
Williams, H., 103, 118, 128
Williams, M. R., 214
Wilson, E. O., 66
Wilson, J. Q., 64, 65
Wilson, L. C., 128
Wilson, W. J., 18, 72, 73, 256, 283
Wines, F. H., 40
Wirth, L., 76
Wolfgang, M. E., 41, 48, 81, 82, 84, 214
Woodworth, G., 204
Worcel, S., 160
Workhouses, 235
Wright, B., 190
Wrongful convictions, xii, xiii, 96, 222–228
    Central Park Jogger case, 224–225
    Duke University lacrosse players case, 225–227,
        226–227 (box)
    Innocence Project and, 224, 226
    Jena 6 case, 227
    public policy and, 227–228
    rush to judgment and, 227
    See also Courts; Death Penalty; Sentencing

Young, V., 267
Young v. Robinson, 241
Youth Risk Behavior Surveillance System, 271, 278

Zatz, M. S., 177, 179
Zawitz, M. W., 56, 57
Zero-tolerance policing (ZTP), 127, 128
Zimring, F., 215, 216, 217
Zingraff, M., 121
Zoot suit riots, 110, 141

About the Authors

Shaun L. Gabbidon is Professor of Criminal Justice in the School of Public Affairs at Penn State Harrisburg. In addition to having authored numerous peer-reviewed articles, he is the author or editor of ten books. Professor Gabbidon's most recent books are *Criminological Perspectives on Race and Crime* (2007; Routledge) and *W. E. B. Du Bois on Crime and Justice: Laying the Foundations of Sociological Criminology* (2007; Ashgate). His book, *Race, Ethnicity, Crime, and Justice: An International Dilemma*, is slated for publication by Sage Publications in spring 2009. In 2007, Dr. Gabbidon was the recipient of Penn State Harrisburg's Faculty Award for Excellence in Research and Scholarly Activity. In 2008, he received the Distinguished Scholar Alumni Award from Indiana University of Pennsylvania's Doctoral Program in Criminology. Dr. Gabbidon can be contacted at slg13@psu.edu.

Helen Taylor Greene is a Professor in the Department of Administration of Justice in the Barbara Jordan–Mickey Leland School of Public Affairs at Texas Southern University. She is an author, coauthor, and coeditor of numerous articles, book chapters, and books. Most recently, she coedited the *Encyclopedia of Race and Crime* with Dr. Gabbidon that is forthcoming in 2009 from Sage Publications.

Photo Credits
and Permissions

For permission to reprint from the following, grateful acknowledgment is made to the publishers and copyright holders:

Figure 1.1: "Genetic Research Increasingly Finds 'Race' A Null Concept," by S. Shane from *The Baltimore Sun,* April 4, 1999, p. 6A. Copyright © 1999 by Baltimore Sun Company. Reproduced with permission of Baltimore Sun Company via Copyright Clearance Center.

Highlight Box 1.1: "Genetic Tests Can Reveal Ancestry, Giving Police a New Source of Clues," by D. H. Simmons from *U.S. News & World Report,* 134 (22), June 23, 2003, p. 50. Copyright © 2003 U.S. News & World Report, L.P. Reprinted with permission.

Highlight Box 1.2: From *Shades of Freedom: The Racial Politics and Presumptions of the American Legal Process* by A. Leon Higginbotham, Jr. Copyright © 1996 by A. Leon Higginbotham, Jr. Used by permission of Oxford University Press, Inc. pp. 195–196.

Figure 1.2: From "Lynching and the Status Quo," by Oliver C. Cox in *The Journal of Negro Education, 14*(4), pp. 577–578. Copyright © 1945 by Howard University. All rights reserved. Reprinted with permission of *The Journal of Negro Education.*

Highlight Box 1.3 Photo credit: ©Landov.

Highlight Box 1.4: "Minorities Suspicious of Ethnic Groups: Poll Finds Stereotypes, Resentments Are Mutual," by Lesli Clark from the *Miami Herald,* December 13, 2007. Copyright © 2007 by McClatchy Interactive West. Reproduced with permission of McClatchy Interactive West via Copyright Clearance Center.

Highlight Box 1.5: "Madman, not Koreans, To Blame for Shootings," by DeWayne Wickman from *USA Today,* April 24, 2007, p. 11A. Copyright © 2007. *USA Today* is a division of Gannett Co., Inc. Reprinted with permission.

Photo 1.1: © Corbis.

Photo 2.1: Courtesy of the Library of Congress.

Photo 2.2: © Associated Press.